A History of Humanity

Humanity today functions as a gigantic, world-encompassing system. Renowned world historian Patrick Manning traces how this Human System evolved from *Homo sapiens'* beginnings over 200,000 years ago right up to the present day. He focuses on three great shifts in the scale of social organization – the rise of syntactical language, of agricultural society, and today's newly global social discourse – and links processes of social evolution to the dynamics of biological and cultural evolution. Throughout each of these shifts, migration and social diversity have been central, and social institutions have existed in a delicate balance, serving not just their own members but undergoing regulation from society. Integrating approaches from world history, environmental studies, biological and cultural evolution, social anthropology, sociology, and evolutionary linguistics, Patrick Manning offers an unprecedented account of the evolution of humans and our complex social system and explores the crises facing that Human System today.

Patrick Manning is Andrew Mellon Professor of World History, Emeritus at the University of Pittsburgh, Founding Director of the World History Center, and former president of the American Historical Association. He is a multidisciplinary scholar in African and world history and in historical datasets.

A History of Humanity

The Evolution of the Human System

Patrick Manning

University of Pittsburgh

CAMBRIDGE
UNIVERSITY PRESS

CAMBRIDGE
UNIVERSITY PRESS

University Printing House, Cambridge CB2 8BS, United Kingdom

One Liberty Plaza, 20th Floor, New York, NY 10006, USA

477 Williamstown Road, Port Melbourne, VIC 3207, Australia

314–321, 3rd Floor, Plot 3, Splendor Forum, Jasola District Centre, New Delhi – 110025, India

79 Anson Road, #06–04/06, Singapore 079906

Cambridge University Press is part of the University of Cambridge.

It furthers the University's mission by disseminating knowledge in the pursuit of education, learning, and research at the highest international levels of excellence.

www.cambridge.org
Information on this title: www.cambridge.org/9781108478199
DOI: 10.1017/9781108784528

First published 2020

A catalogue record for this publication is available from the British Library.

Library of Congress Cataloging-in-Publication Data
Names: Manning, Patrick, 1941– author.
Title: A history of humanity : the evolution of the human system / Patrick Manning, University of Pittsburgh.
Description: Cambridge, United Kingdom ; New York, NY : Cambridge University Press, 2020. | Includes bibliographical references and index.
Identifiers: LCCN 2019033819 (print) | LCCN 2019033820 (ebook) | ISBN 9781108478199 (hardback) | ISBN 9781108747097 (paperback) | ISBN 9781108784528 (epub)
Subjects: LCSH: Human ecology – History. | Nature – Effect of human beings on – History. | Technological innovations – Social aspects – History.
Classification: LCC GF13 .M34 2020 (print) | LCC GF13 (ebook) | DDC 304.209–dc23
LC record available at https://lccn.loc.gov/2019033819
LC ebook record available at https://lccn.loc.gov/2019033820

ISBN 978-1-108-47819-9 Hardback
ISBN 978-1-108-74709-7 Paperback

Additional resources for this publication at www.cambridge.org/Humanity

Contents

Maps, Figures, and Tables

Maps

Figures

Tables

Preface

During a long career as a historian, I have watched the dramatic changes in what is known as "world history." At first, I read writings on empires, wars, great men, and European expansion. In time these topics were set into the background by studies of globalization, gender, migration, environmental change, decolonization, and genomes. My own studies – focusing on Africa in world history and on economy, migration, and culture through multiple disciplines – changed along with those of other writers. Throughout, I planned to prepare a new review of world history, surveying the full length of the human experience. In completing this volume during the past three years, I have been repeatedly surprised by the new problems to study and new knowledge about them. The result, by no means definitive, is an update for 2020, addressing much of what is new in world history.

I began work on this volume by identifying a "Human System," to emphasize the numerous elements of human life, past and present, that interact with one another as parts of an elaborate, evolving organism. I chose to set the Human System in coevolution with "Gaia" – the complex natural environment with which humans interact as individuals and in groups. I also assumed that biological evolution, in giving rise to humankind, somehow linked up with a process of social evolution that gradually created the social groups of life today – cities, schools, nations, companies, teams. I assumed that the rise of speech – syntactic language – was central to the transition to social evolution.

One surprise was to discover that a theory arose in the 1980s on cultural evolution, in between biological and social processes. This theory successfully shows how individual learning was able to build human cooperation to a higher level. But the same theory assumes that no further steps were necessary to achieve large-scale human societies. In contrast, a still-newer theory of social evolution shows how agreements to join groups, especially via spoken language, enabled people to form social institutions with inherent dynamics. While the two new theories have not yet been fit together, I believe their combination can give a satisfactory account of the rise of the Human System. Sorting out these and other theories requires debate on big questions. Can one treat humanity as a growing and functioning social system? Does there exist a process of social evolution that can explain main trends in history? Can we link social evolution and biological evolution? Is "cultural evolution" part of overall human evolution? How widespread and divergent are the meanings of "culture"? Was spoken language the first social institution? How do we understand our behavior in social groups – does the behavior of human groups bring improved decisions over individual behavior? Is migration helpful or harmful? Is popular culture as important today as capitalism?

My interpretation is the individual-level project of a researcher with a small-scale research design to explore a very big topic. But how can a single scholar assemble knowledge from so many disciplines? Needless to say, my deployment of the various disciplines will be uneven. Nevertheless, in what I expect is not a truly unusual experience in recent years, I have undergone training in several of these areas; I am an autodidact in others. I had undergraduate training in chemistry and biology, followed by graduate training in African history and economic history with a world-historical approach, including some anthropology. Later I conducted a formal study of demography and migration; I have also conducted reading and research on aspects of historical linguistics. I have done reading in genetics and climatology with guidance from specialists; I am self-trained in my recent research in history of science. Further, thanks to work by dedicated scholars, the basic principles and advances of many fields of study have been very well written up in the articles and books I have cited here. A final argument for taking on so many disciplines at once is systemic: the Human System functions through interrelated processes, and studying them at once should lead the researcher to encounter resonances that might not be visible with a piecemeal exploration. In addition, the assertions that I advance here, while I do stand by them, are to be understood as hypotheses rather than confirmed research results.

Research on human social and biological change, in my opinion, is moving in the direction that I have staked out, and I offer this book to point to additional possibilities, hoping to speed the process of research. My central hypothesis is that humans developed speech in Northeast Africa some 70,000 years ago. That assertion raised big questions about the time before and the time after that transformation – questions that are explored here. My second hypothesis, a Human System based on social evolution, gained reinforcement from James Miller's *Living Systems*, a framework at once biological, social, and systemic. Then began my program of reading and writing on how to link social, biological, and cultural evolution.

At every stage in research, I encountered an incredible wealth of knowledge and debate unfolding in multiple disciplines. My theory and narrative underwent endless twists and shifts as I learned new research results. I have encountered the boldness of physicists, the skill of chemists in locating multiple scales of existence, the tension in biology between theory and empiricism, the struggle of anthropology and sociology to make use of ideology, the strategic but difficult standpoint of psychology, the caution of historians, the extremes of systems and game theory, and the perplexity of "groups" in the minds of all. At one stage I hoped to include, in this book, a review of this two-century intellectual evolution, as an accompaniment to my narrative of human evolution. Instead I have written up that process separately, in *Methods for Human History*. Here, I reaffirm that the developments and exchanges across disciplines have been a fundamental and exciting aspect of learning about the many dimensions of human evolution.

World history is not simply a story. It is a set of puzzles, for which the solutions require the very best in our theoretical analysis and empirical data collection, in all the disciplines of knowledge that address aspects of human experience. In this volume I offer my understanding of major problems, data, and answers in world history, thinking of readers with several types of interests. Readers with an interest in world history will find interpretations of social change tied to innovation and migration. Readers concerned with social dynamics will find analyses of the origins of current crises in society and the environment. Readers attentive to research in disciplines of physical, biological, and social sciences will see the theories applied to human history. Further, students at graduate, undergraduate, and secondary levels will find here an argument that a single author or reader, taking the time to read and discuss the wonderful knowledge resources now available, can explore and reach conclusions on the interplay of big issues in human society. There remain many more big issues to explore.

Acknowledgments

I begin by expressing appreciation to my late father, John V. Manning, a trade unionist intellectual, and to my brother, Curtis V. Manning, an astronomer, for a lifetime of discussions on systems thinking in the human context. During undergraduate days at Caltech, majoring in chemistry (and informally in history), I had exposure to such scientific luminaries as Linus Pauling, Max Delbrück, Roger Sperry, James Bonner, and George Beadle. At the University of Wisconsin–Madison I had the great fortune to do graduate study in African and world history with Philip D. Curtin, Jan Vansina, John Smail, Stuart Schaar, and, in economic history, Jeffrey Williamson.

I benefited from close contact with three great linguists: Christopher Ehret of my own generation and Joseph Greenberg and Noam Chomsky of the previous generation. I had a year of formal study of demography under a Guggenheim Fellowship at the University of Pennsylvania, where I learned from Samuel Preston, Jane Mencken, Susan Watkins, and Tukufu Zuberi. At the Northeastern University World History Center, I worked closely with Alfred W. Crosby and Andre Gunder Frank. Study of Africa and the African diaspora, a world-historical realm in itself, was reinforced by 22 years in the Department of African-American Studies at Northeastern University, working especially with Ronald Bailey and Robert L. Hall and also with Michael Gomez, Kim Butler, Maghan Keita, Chouki El Hamel, and Jean-Jacques Sène. The International Institute of Social History led me to Leo Lucassen, Jan Lucassen, Marcel van der Linden, Karin Hofmeester, Filipa Ribeiro da Silva, Pepijn Brandon, and others. The Global Economic History Network, under a grant from the Leverhulme Foundation led by Patrick O'Brien, put me back in touch with economic historians as their global work began to flourish.

At the University of Pittsburgh, the World History Center has supported scholars worldwide, advancing research on this and other projects: it is worthy of continued and expanded support. The Andrew Mellon Foundation sponsored a three-volume publication project in world history of science at the Center and the University of Pittsburgh Press, for which my coeditors were Daniel Rood, Mat Savelli, and Abigail Owen. Center Director Ruth Mostern leads the World-Historical Gazetteer project, a resource central to spatial analysis. The Center brought me into contact with Daniel Bain and Aubrey Hillman for consultation in geology; with Monica Green on disease history; with Jason Carson on genomics; and with R. Charles Weller, who encouraged me to complete a chapter-length overview of the Human System in 2016. Pitt grad students James Hommes and Lars Peterson critiqued my world history manuscript; Chris Eirkson, Jack Bouchard, and especially Madalina Veres helped me map language distributions; Ahmet Izmirlioglu assisted me on language and network analysis; Matt Drwenski brought key insights in economic history. Jeremy Black and Felipe Fernandez-Armesto each kindly provided me with copies of their book manuscripts in process. Boris Michev and Daniel Andrus of Pitt's University Library System skillfully prepared the maps for this volume.

My daughter, Pam Manning, showed me the benefits of wide reading on humans and animals. Liu Jiafeng and the Shandong University Institute for Global and Transnational History generously hosted me for seven lectures and support for writing in 2017. Readers of drafts of this study have provided valuable commentary: Eugene Anderson, Iris Borowy, Lincoln Paine, Christopher Ehret, Felipe Fernández-Armesto, Matthias van Rossum, Pim de Zwart, Ahmet Izmirlioglu, Bennett Sherry, Molly Warsh, and two anonymous referees for Cambridge University Press. Michael Watson of Cambridge University Press has skillfully shepherded this volume through the past two years of its revisions.

I dedicate this work with deep love to Susan Manning, my wife, with whom I have shared many happy years. She has been warmly and imaginatively supportive of the preparation of this book.

Introduction

1 The Human System

How did the human condition of today come to be? We live in a world created by human energies and activities, in which "nature" is receding steadily. The cities in which most of us live are the results of human construction – out of concrete, asphalt, iron, glass, and bricks. Even the wooden elements of cities are cut and reshaped by human energy. Water is piped in or transported in bottles; we bring gasoline in tanks and natural gas by pipeline. We communicate by electronic telephones, televisions, and computers that are built in factories. Even the countryside depends heavily on human construction and creativity – while the wonders of nature are a pleasure to see, the rural world is charted and exploited by humanity. The crops on farmlands have been bred and protected by chemical and biological engineering. Our cattle, sheep, pigs, and chickens live and die under human control: these domesticated species are the majority of all the large and medium-sized animals. Even the insects and the bacteria fall increasingly under human control. Fishing has transformed the populations of oceans, while plastic waste materials mark the oceanic currents and shores. Of course, Earth remains in its orbit so that the sun appears to rise and fall each day, yet even the seasons are changing.

The achievements of human energies have created a Human System, a complex set of social interactions and structures from local to global levels. This system reproduces and transforms itself on every continent, creating social institutions, material goods, and new knowledge, including science and culture. The system generates achievements and distributes benefits – yet the course of human expansion has brought oppression and destruction, leading in some cases to destruction of whole societies. The natural world, though increasingly marginalized, has not yet been tamed. New strains of bacteria and viruses overwhelm pharmaceuticals and vaccines, spreading disease where medical science thought it had achieved conquest. Cancers arise in response to the new creations of chemistry and petroleum. The burning of petrochemicals and the methane created by domestic animals are raising Earth's temperature at a rate that is bringing short-term climate fluctuations and long-term disaster for human life.

In this book I trace four agencies of change in human history. The first three are the processes of biological, cultural, and social evolution, which add up to the Human System. In a grand process of coevolution, these three types of changes in humans have consistently interacted with each other and with the natural environment. Biological evolution, as theorized by Darwin, brought the rise of bipedal species with large brains. Cultural evolution, as theorized since the 1980s, explores the learning processes of big-brained human individuals, especially through cooperation. Social evolution, for which I present an expanded theory in this book, results from conscious group behavior in which humans create a succession of social institutions. Institutions include language, communities, migration, agriculture, writing, and economic systems – each growing in its own way. The fourth agent of historical change is Gaia – the overall system of life on Earth. Gaia, as theorized in the 1970s by James Lovelock, is the part of the natural environment that generates oxygen through plant forms in an interaction of biosphere, atmosphere, and lithosphere, producing fluctuations in climate but also long-term stability in temperature.[1] The Human System began as a tiny portion of Gaia but has grown and created new tensions. Humanity is now the greatest influence on Gaia yet Gaia still has great influence over humanity. Human history is then a story of exploitation – of the natural resources of Gaia, of the energies and imagination of individual humans, and of humans in groups as they found new ways to combine.[2]

In the gradual evolution and coevolution of the human order, I argue that many past choices have accumulated sufficiently to expand the dangers we face today. At the same time, I argue that current conditions provide important new patterns of group behavior with the potential to address and even resolve the great crises of today. I treat human groups as the governing factor in coevolution: group behavior has created immense advances in which humans have formed huge groups sharing common interests – languages spoken by millions of people; nations of millions sharing common citizenship; corporations of thousands of workers producing goods and services under a small but expert and privileged management. Yet the creation of larger groups has permitted the rise of larger-scale hatreds, such as between pro-communists and pro-capitalists or Christians and Muslims. Group behavior also involves debate among alternative views of social priorities, so that ideology and ideological dispute have been with us from the beginning.

Humanity and its system now appear to be in a period of crisis. How and when did we reach this point, with its strange mix of triumphal advance for some, relentless discrimination against others, and threatened collapse or war for all? Are these dilemmas the consequence of capitalism and the last few generations of rapid change? Could it be that the founding days of great empires and great religions, 2000 years ago, started us on a path to this difficult situation? Or could it be that the early days of human emergence and expansion instilled habits in our ancestors that prevented them – and us – from guarding against the difficulties that now threaten us so deeply?[3] In any case, I argue that humanity faces two great crises. In a crisis of environmental degradation, we face climate fluctuation, predictions from the scientific community of acceleration in climate disasters, extinctions of species at all levels, destruction of water supplies, and more. The crisis of social inequality is bringing divergence and deprivation in social welfare, economic life, employment, and participation in governance. Health conditions, life course, and education levels have narrowed in their discrepancies, but the recognition of human equality has only diverged.

Further, I argue, we face a third dimension in our current crisis – the inability of our system of knowledge to respond to the crises in environment and social inequality. To a large degree, it appears that humans prefer to ignore and deny the challenges facing Earth and humanity. Why is it that leaders of corporations, governments, and communities deny the existence of climate change? Why is it that they put so little energy into policies to limit climate change or adjust to it? Why do these same institutions deny the negative effects of social inequality and deprivation? And why is it that the knowledge specialists, those who have generated so much learning, seem unable to diagnose the crises in environment and society and unable to investigate the reasons for denial?

In response to these crises, I offer a concluding emphasis on further transformations in the Human System – the contemporary development of larger-scale and even planetary human groups. That is, we have now experienced the rise of popular culture (drawing inspiration from multiple communities and not just from one's parents) and dramatic expansions in knowledge (at both general and specialized levels). Out of these changes is arising a global popular discourse – agreements and disagreements on social priorities that have the potential to generate global consensus on the great choices we face. I acknowledge the disappointments brought by the negative sides of national discrimination and exclusion, the turn of formally constitutional government to dictatorship, the recurring violation of elegant proclamations of human rights, and the narrowness of corporate profiteering. We proclaim the equal rights of women and children, yet the stories from every part of the world tell of the exploitation of these and others. Yet I emphasize that global standards, articulated by steadily broader groupings of humans, have been met in some cases and are gradually gaining in prestige even though they are still violated with impunity. Despite the denial of crisis that remains widespread, humanity is experiencing a steady rise in mutual understanding that is increasingly critical of discrimination and exploitation by gender, age, or racial categorization.

One of the choices before the human community is whether to focus attention on learning more about the Human System – its evolution, its flaws, and the questions we face in seeking to guide it. As humans formed groups to achieve a common purpose, did they seek to assess how the common welfare would be affected? What is it that enables leading figures in institutions to profit for themselves and ignore the general welfare? To address this question, throughout this book, I trace the expansion of networks connecting humans for common purposes, and the parallel expansion of hierarchies – vertical networks that might also yield a division of labor that benefits human welfare. As I argue, human nature, though deeply embedded, has changed with time. A global consensus on social welfare, should it arise, might modify human nature once again and, with it, the direction of human history.

The adventures of *Homo sapiens* began more than 200,000 years ago and reach up to our day. The story traces drama in family life, in learning to speak, building social institutions, migrating to new lands, creating a Human System through maintaining connections, and encountering global crises in today's urban society. More than a story, the book presents puzzles to solve. I hope readers will join in working to reveal how our ancestors took each turn in their complex trajectory. We will explore data, logic, and terminology across many disciplines. My narrative centers on processes of evolution and migration in human history. I divide the narrative into four main periods, beginning with hominin biological and cultural evolution from the early Pleistocene epoch up to 70,000 years ago. I then trace three epochs dominated by social evolution – the late Pleistocene, from 70,000 years ago; the Holocene, from 12,000 years ago to 1800 CE; and the Anthropocene with its new levels of social evolution. I treat the issue of migration as subordinate to the overall question of social evolution, yet I see migration as essential to the human trajectory: it maintains diversity and originality in every region. More generally, my approach to human evolution highlights diversity in populations, to guard against the reasoning that there might have existed selected groups that carried forth the essence of human excellence. I emphasize genetic diversity, in reasoning analogous to that of Theodosius Dobzhansky who, writing in the 1930s to advance the neo-Darwinian hypothesis, emphasized the broad diversity of human populations in contrast to the arguments of eugenicists that excellence could be purified in a few key genes.[4] Further, I emphasize diversity in the learning processes of cultural evolution, in the institutions of social evolution, and in the expressions of emotions at individual and group levels. This is one reason for my emphasis on migration and comprehensive regional coverage in this volume: it is an effort to point to the full range of human diversity over time.

Evolution of the Human System: Key Questions, Key Assumptions

Between the crowded world of today and the days of our ancestors on the East African savanna lies a missing link – a disconnect in knowledge linking the biological emergence of humankind to the social complexity of humanity today. The missing link is temporal, disciplinary, and theoretical. Temporal, because some 70,000 years separate humans today from our ancestors as they began speaking and migrating to new lands. Disciplinary, since the training and reading of biologists, social scientists, and historians address significantly different materials. The gap is also theoretical. That is, biologists are guided by Darwinian theory, by the recent advances in genetics and epigenetics, and by connection to disciplines including physics and biochemistry; historians and social scientists have worked less with theories than with narrative, though their facility with theories is advancing. Between the natural sciences and the social sciences lies the gap on which I seek to focus – although I emphasize the thread of erudite discussion among scholars linking those disciplines, whose contributions may be resolving aspects of the big questions about human growth and change.[5]

How can one hope to bridge the gap between biological study of humanity (for early times) and historical study of humanity (for early and especially recent times)? The first step in bridging is to identify major questions that point us toward exploring the gap. To begin this study of the Human System, I pose four questions that will be pursued throughout the book:

- System. How does humanity function as a system?
- Evolution. What are the processes of human evolution?
- Natural World. How are humanity and the natural world linked?
- Transformation. What major transformations has humanity faced in the past and the present?

These four questions, in varying ways, focus on two big issues: system behavior and human evolution. Such questions lead us deeply into exploring processes and events in human history. Study of growth and change in the Human System involves identifying the system's elements, tracing their interplay, and analyzing their transformations up to today – especially with attention to the roles of individual and collective consciousness – our "behavior" and "human nature." By "human evolution" I mean several overlapping processes: not only biological evolution but also the processes of cultural and social change, right up to the present. Is the Human System now prepared with adaptive responses that will respond automatically to current crises? Is the system capable of changing its direction in time to limit the damage and threat that it faces from within and without? Or, to anticipate a question that will be central in this book: can human nature change, either at the level of individuals or in group behavior?

The four questions just posed will not answer themselves: they must be addressed by analysts who frame their inquiry with well-chosen *assumptions.* In the selection of assumptions, I have been inspired (as have been many others) by Charles Darwin's assumptions and analyses in the study of biological change. Darwin specified with remarkable clarity the key assumptions in his *On the Origin of Species by Means of Natural Selection* (1859):

> Owing to this struggle for life, any variation, however slight and from whatever cause proceeding, if it be in any degree profitable to an individual of any species, in its infinitely complex relations to other organic beings and to external nature, will tend to the preservation of that individual, and will generally be inherited by its offspring . . . . I have called this principle, by which each slight variation, if useful, is preserved, by the term of Natural Selection.[6]

Darwin thus emphasized his assumptions on the struggle for existence, variation, and inheritance, summarizing them as natural selection. The principal effect of natural selection was *divergence* among the species.[7]

In contrast to Darwin's specificity, Herbert Spencer, the wide-ranging sociologist, published a vaguer and more general overview of "progress" in 1857. Spencer ranged across the natural and social sciences, highlighting "evolution" and "progress" as characteristic of every field of study.

> The advance from the simple to the complex, through a process of successive differentiations . . . is seen in the geologic and climatic evolution of the earth, and of every single organism on its surface; it is seen in the evolution of humanity, whether contemplated in the civilized individual or in the aggregation of races; it is seen in the evolution of society in respect alike of its political, its religious, and its economical organization; and it is seen in the evolution of all those endless concrete and abstract products of human activity which constitute the environment of our daily life.[8]

Spencer's logic was to lump together, under the term "evolution," every sort of transformation, assuming that "progress" was inherent at each level. While he would soon accept the mechanism that Darwin proposed for biological change, labeling it "evolution," Spencer did not propose equivalent mechanisms for other sorts of change. He described what he saw as a universal result, then vaguely referred to a universal cause but without offering specifics. Spencer was the theorist of progress in every domain; Darwin was the theorist of divergence in the biological domain. Darwin's theory, since it included a specific mechanism to back up its overall hypothesis, was testable and ultimately verifiable, while Spencer provided no mechanism for "the transformation of the homogeneous into the heterogeneous" and simply repeated his overall hypothesis. Darwin launched a concrete research project on biological evolution, while Spencer's speculations fueled debate yet did not launch organized study of human social change. The problem of ensuring that analytical assumptions are sufficiently specific will occupy us significantly in this book.

Darwin's analysis provoked a related question: how does the history of human social and cultural change fit into biological evolution? Edward B. Tylor, a founding figure in anthropology, began his *Primitive Culture* (1871) with a definition that has since remained famous: "Culture or Civilization, taken in its wide ethnographic sense, is that complex whole which includes knowledge, belief, art, morals, law, custom, and any other capabilities and habits acquired by man as a member of society."[9] This definition of culture left unspecified whether its locus is the human individual or the social group, though Tylor's focus on law and custom implies that he saw culture as existing at the level of groups.[10] Tylor proceeded to discuss racial differences in humans, concluding that distinct races had arisen from a uniform original human species, and he emphasized the special roles of language and consciousness in human history. Tylor supported scientific study of human change but expressed his methods in terms of narrative and description rather than proposing mechanisms of change. Thus, Tylor saw anthropology, the study of humanity, as scientific study that was separate from the study of biological evolution.[11]

The specificity of Darwin's theorization meant that, with time, his errors have been identified and addressed. Darwin assumed a set of workings of natural selection, relying on individual level motivation and action. In the time since these initial statements, biological analysis has developed with great complexity, mostly reinforcing rather than negating Darwin's early insights. In contrast, the imprecision of nineteenth-century social-science analysis hid both the strengths and the weaknesses of social science. Spencer assumed an inevitable result of progress from unity to diversity, but without assuming a process that would yield the result. Many analysts of change in human society, in addition to Spencer, tended to assume a general and unspecified impulse to progress rather than any more specific mechanism.[12] Tylor assumed the existence of culture that relied in unspecified ways on human groups, but he described culture rather than analyzing it. Social scientists in general, lacking specific theoretical mechanisms for social change, failed also to establish a basis for their studies that could match that of biology.[13] This failure resulted not from weakness of the scholars but, at least in part, from the complexity of consciousness and the sensitivity of social science to ideological claims.

While today's analysis of human biological evolution is coherent though imperfect, we do not yet have a coherent analysis of human social evolution. My understanding of social evolution relies significantly on the insights of psychologist Donald T. Campbell, who asserted the centrality of social evolution in human society and modeled it in Darwinian terms.[14] Yet I revise Campbell's approach, based on advances since his time in fields that are discussed throughout this book. The analysis of social evolution, in my opinion, should center on group behavior and social institutions. Still-unresolved issues include: How do we explain the functioning of human groups? What is the role of spoken language in human evolution? What is the place of migration in human history and evolution? At a larger scale, we do not yet have a coherent analysis linking the roles of individual and group behavior in human change, the functioning of humanity as a system, the place of cultural production in that system, the interplay of humanity and the environment of the natural world, and the major transformations in the human experience – especially the expansion of human knowledge about our past. To that end, this book is to propose a framework for analysis of overall human evolution, focusing especially on the logic of social evolution. In an Appendix, I provide a compact summary of methods within that framework.[15]

A World-Historical Approach

The discipline of world history provides an appropriate institutional framework for exploring this set of big questions. With this framework, I seek to provide a new analysis and a new narrative of world history. The world-historical framework includes working with such valuable conceptual resources as multiple perspectives (disciplinary and ideological), multiple scales of existence (in humanity and in the natural world), and systemic interactions among the elements of society – those thought to be central but also other elements. World history, while a relatively new field of study and not yet a large one, presents advantages for this task because of its focus on coordinating study across multiple disciplines. Here I offer a statement of the broad scope of the field. At the birth of world-historical studies, the founding analysts participated actively and successfully in advancing environmental history, including climate history and history of biological change, but also including the human dimensions of those histories. Within this framework, world historians have also analyzed the global economy, empires and nations, migratory paths across the world, patterns in gender relations, issues of race and slavery, and global connections in science.[16] I argue in favor of seeking now to identify the missing link in the human past – the analytical separation of human origins from our life today. The incentive is growing for us to learn about human social processes in greater historical and analytical depth, since we face expanding social and environmental crises. In one sense, preparing this study has been an individual voyage for me as author, as I have encountered surprising new types of evidence and skillful formulations of analysis by a growing number of analysts. In another sense, I find that this volume reveals the fundamentally collaborative nature of interdisciplinary research, in which debate and assembly of knowledge have combined to identify and ultimately resolve the big questions. Nevertheless, a satisfactory picture of human evolution and the Human System is not necessarily near to clarification.

In this book I trace the rise and transformation of the Human System from our early hominin ancestors to today, with a focus on processes of evolution, migration of populations, the creation of social institutions, and the interplay of humanity with Gaia. While I offer arguments based on the most recent evidence and on distinctive theories, I owe a great deal to the authors who have previously composed books at roughly this scale, surveying the full scope of the human experience. Major such works are Jared Diamond, *Guns, Germs, and Steel* (1997); John R. McNeill and William H. McNeill, *The Human Web* (2003); David Christian, *Maps of Time* (2004); Fred Spier, *Big History and the Future of Humanity*, 2nd ed. (2010); the edited collection of Andrew Shryock and Daniel Lord Smail, *Deep History* (2011); Christopher Chase-Dunn and Bruce Lerro, *Social Change* (2014); and Yuval Harari, *Sapiens* (2014).[17] Taken together, these works have been highly effective in laying out a long-term narrative of the transformations leading from bipedal hominin species to the cosmopolitan world of today. In particular, they have shown the interplay of geological and climatological change with biological evolution and have summarized the paleontological work documenting the expansion of hominin brains. The McNeills developed the metaphor of the web or network to trace the steady tightening in human connections; Christian distinguished between the "flat" networks of the Paleolithic and hierarchical networks of the agricultural age.[18] Christian and Spier highlighted the context of the physical universe and traced the steady expansion in energy consumed by humanity. In these broad and smooth narratives, certain transitions stand out: bipedalism, large brains, agriculture, and complex civilizations. Chase-Dunn and Lerro, a historical sociologist and a psychologist, present a comparative world-systems approach to human history. Harari gives more emphasis than most to the rise of spoken language, and traces shifting philosophical dilemmas in succeeding stages of human society. Shryock, Smail, and their colleagues present a wide range of methodological explorations: they emphasize methodology more than narrative and identify numerous and valuable tools for exploring the deep past. Overall, these works lay out a historical and multidisciplinary framework for study of the long-term human past. It remains to implement the framework in various ways.

In addition to the historic distance between biology and social sciences, there is a potential area of significant disagreement: Is human behavior to be explained entirely in terms of individual experience and choices? Or are groups of conscious humans significant in joining to make collective choices at the group scale? Is culture best understood as individual choices or as choices by groups in debate? As I argue, the creation of language is necessarily a group phenomenon, as the acts of creating vocabulary and syntax rely necessarily on interaction and compromise among members of a sustained group. It would be difficult indeed for any evolutionary theory restricted to the analysis of individuals to explain plausibly the rise or especially the maintenance of spoken language. This logic leads me to seek out an appropriate model of group behavior in human populations.

Yet many biologists are skeptical that groups can be a basis of evolutionary change: they have argued and theorized that efforts to build biological or cultural evolution around social groups collapse back to the level of individual evolution.[19] Meanwhile, theorists of cultural evolution argue that processes of individual social learning, in coevolution with biological evolution, can facilitate cooperation at a steadily higher level, perhaps sufficient to explain the large-scale social and cultural institutions of humanity today.[20] Analysts of large-scale social institutions and of cultural collaboration, however, are unlikely to believe that such a mechanism can explain the richness of today's cultural production. The discussions will surely be interesting. To highlight these distinctive scales and contending perspectives, I have chosen to divide the general notion of "culture" into two subcategories, distinguishing "individual-level culture" from "group-level culture." Put simply, "individual-level culture" is analyzed through the theory of cultural evolution and a focus on learning, while "group-level culture" is analyzed in the theory of social evolution and a focus on representation – but is also explored in the disciplines of the arts and humanities.[21] In fact, I think, the two categories overlap significantly.

To illustrate my claim that human behavior is a mix of individual motivation and group solidarity, I offer two brief historical examples, each of which links individual identities to national groups. In 1789, as the French Revolution unfolded, the National Assembly adopted the Declaration of the Rights of Man and the Citizen, a memorable statement of individual rights in contrast to monarchical domination. Yet three years later, as France entered war with invading powers, the focus shifted from the individual to the nation and to the commonality of its military forces as expressed in the group-focused popular anthem, "La Marseillaise." A parallel drama of the balance between individual and group identity unfolded after the World War II. After the defeat of Nazi and Imperial Japanese campaigns of racial hierarchy, the newly formed United Nations adopted in 1948 the Universal Declaration of Human Rights: this was another memorable statement of individualistic philosophy. Yet as the Declaration was adopted, a great era of decolonization took place. More than 100 new national collectivities arose, in which national identities expanded as never before. In both cases, one sees a formal statement of the primacy of the individual and individual rights, linked to the practical development of powerful communities in which participants (or most of them) agree to devote themselves to the common welfare. In sum, theorization and analysis in terms of human groups are as important as the analysis of individual human behavior.

Organization and Hypotheses of the Book

Following this introduction, the book is organized into three sections to ask: what governs human behavior and how do we know? This opening chapter has introduced the Human System we live in – the multiple levels and accelerating change of its functioning, but also its troubles in environmental degradation and social conflict. It proposes the task of explaining the evolution of human society and the Human System.[22] The book's remaining three sections are a narrative that documents three geological ages of transformations in human society: the Pleistocene epoch from nearly 3 million years ago to 12,000 years ago (Chapters 2 through 5); the Holocene epoch from 12,000 years ago to 1800 CE (Chapters 6 through 8); and the Anthropocene epoch of the past two hundred years (Chapters 9 and 10). In the latter, humanity became the dominant force for change on Earth.

Chapter 2 begins my narrative with the biological and cultural evolution of hominin species in the Pleistocene epoch, up to 70,000 years ago, in the context of ecological change, documenting these hypotheses:

1. *Hominin brain size.* Hominin brain size grew rapidly, beginning 2 million years ago, through coevolution of genetic, epigenetic, cultural, and environmental processes.
2. *Expanding hominin capacities.* Expanding brains enabled individual-level advances in hominin capabilities, through varying mechanisms: new capabilities advanced reasoning, learning, communication, imagination, and emotional expression.
3. *Individual-level human nature.* The sum of these hominin capabilities may be defined as "individual-level human nature," including personality; emotions; family-level behavior; and capacities for reasoning, learning, nonverbal communication, and cooperation.

Chapter 3, the most fully theorized, presents three hypotheses arguing for the sudden rise of spoken, syntactic language in Northeast Africa and the formation of the Human System, roughly 70,000 years ago:

4. *Syntactic speech.* Syntactic speech emerged through a complex process to become the first community-scale institution, as it relied on explicit group behavior. Spoken language enabled the creation of other institutions and it accelerated the benefits of individual-level human capabilities.
5. *Group behavior and group-level human nature.* Group behavior, reinforced by spoken language, arose and facilitated construction of social institutions to perform needed tasks. The community of speaking humans, substantially larger than preceding family groups, emerged to sustain language. The sum of these emerging capabilities may be defined as "group-level human nature," including language, discourse, collective intentionality, emotional expressions, ideology, cultural production, and networking.
6. *Social evolution and emergence of the Human System.* The Human System emerged, linking all human activities and communities through coevolution among biological, cultural, and social evolutionary processes, in interplay with the environment. It expanded geographically and in its activities.

The narrative continues with the expansion of the Human System to the end of the Pleistocene in Chapter 4 (from 65,000 to 25,000 years ago) and Chapter 5 (from 25,000 to 12,000 years ago). The hypotheses for these two chapters are:

7. *Representation.* Early speaking communities emphasized representation of their world and group-based development of their group-level culture.
8. *Cross-community migration.* Human migration into multiple habitats brought networks of connection among communities.

9. *Incorporation of nonspeaking humans.* Nonspeaking humans (including *Homo sapiens*, Neanderthals, and Denisovans) were incorporated in significant numbers into communities of speaking humans.
10. *Production and confederation.* During the severe cooling and rapid warming of the Last Glacial Maximum, communities sought to protect themselves by combining to expand their numbers and adding productive activities to their foraging lifestyle. Confederation of communities built larger-scale social groupings to coordinate production.

For the early and mid-Holocene epochs, from 12,000 years ago to 1000 years ago, Chapter 6 traces the innovative expansion of social institutions, advancing these hypotheses:

11. *Expanded production.* Institutions of artisanal, agricultural, and pastoral production arose in the Holocene epoch, relying on new technology and elaborated social organization.
12. *Societies.* Societies arose out of communities and confederations in the Holocene epoch, coordinating production with substantially larger populations than the preceding social groupings.
13. *Network and hierarchy.* Networks and hierarchies grew in tension with each other during Holocene-era production and institutional change. As institutions of urbanism, commerce, religion, and empire arose, expanding hierarchy modified group-level human nature, encouraging a mix of coordination, conflict, oppression, and revolt.

The late Holocene epoch, 1000–1800 CE, brought contractions that ended human growth for a time, followed by growth. Chapter 7, on the years from 1000 to 1600 CE, traces collisions of populations and institutions, within humanity and Gaia. Chapter 8, on the years from 1600 to 1800, analyzes expansion and transformation in commerce and group-level culture. The hypotheses advanced for the late Holocene are

14. *Collisions and contractions.* Gaia brought warming then cooling while hierarchy in the Human System brought warfare and social oppression. The two met through migration of living things by terrestrial and oceanic routes, reducing human population while expanding environmental diversity.
15. *Global economic network.* Commercial links nevertheless tightened in regional economies. A global economic network arose in the Old World from 1200, circulating textiles, slaves, and silver. This network spread by 1500 to Africa and the Americas. Knowledge and cultural practice spread by the same network.
16. *Capitalism.* Capitalism – a regional mix of socioeconomic institutions and hierarchies – arose in Western Europe, especially in the

seventeenth century. It expanded in wealth and power during the eighteenth century, relying on ties throughout the world.

The Anthropocene epoch, from roughly 1800, brought unparalleled success and excess. Chapter 9 traces growth and crisis in the Human System. Chapter 10 identifies two social processes that may enable limitation of the crises.

17. *Accelerating growth.* Accelerating growth – unevenly distributed – expanded population, capitalist production, knowledge, the scale of warfare, and environmental degradation. Hierarchies and networks both expanded.
18. *Institutional and environmental crises.* Success and crisis, based on narrow self-interest, reached new extremes. Ideological disputes reached a global level. Crises broke out, highlighting environmental degradation and social inequality.
19. *Popular culture network worldwide.* Networks of global popular culture created new alliances, perhaps enabling reorganization of the Human System.
20. *Knowledge network worldwide*; *democratic discourse worldwide.* Exchange of knowledge brought a trend toward the sharing of knowledge through expanding worldwide literacy and the efforts of some scientific specialists. Democratic discourse may enable widespread consensus on social objectives. In it, a modified group-level human nature might rely on a larger scale of community to regulate institutions intended to fit the general welfare.

In the conclusion to this work, I suggest a certain parallel between humanity's current crises and the moment, long ago, when humans created a speaking community. Present-day creation of human networks – unified for the common good by popular culture and sharing available knowledge – would be quite a surprise. But it would be a surprise not much bigger than the rapid and imaginative creation of a community of speaking humans, 70,000 years ago, in which individuals and groups brought change to themselves and the world.

Pleistocene Evolution

2 Biological and Cultural Evolution

Hominin species emerged some 4 million years ago: in this chapter I trace their evolution up to 70,000 years ago. These were recent results of the extraordinary process of biological evolution. The beginnings of eukaryotic cells unfolded some 2 billion years ago and, since then, DNA has been the central constituent of multicellular life. DNA has served as the archive containing the core biological information for reproducing multicellular organisms and their behavior. The remarkable ability of DNA to reproduce itself with accuracy guarantees continuity in species. DNA and its evolution supported a recurring divergence of species across Earth's middle and recent geological epochs, continuing to the present. It is the sequelae of DNA that are of greatest interest: that is, the individual organisms and the groups of organisms that arose, lived, and died.

Our distant ancestors, the first primates, arose some 60 million years ago as inhabitants of tropical trees, developing along with the other mammalian species that replaced dinosaurs at the top of Earth's food chain. Primate lineages developed in many directions, including eventually the great apes some 12 million years ago. For the great apes in Africa, life changed during a period of climatic cooling 6 to 7 million years ago. In the humid forest of western Africa, the genus *Pan* arose to comprise two species: chimpanzees and bonobos. In the drier ecology of eastern Africa, great apes learned to walk on two legs.[1] It is now argued that the eastern African ecological environment did not become open grassland but a park-like one of trees and grassland. In any base, bipedal great apes arose in this region, and have come to be known as hominins.

This chapter surveys the genetic, phenotypical, and cultural evolution of bipedal hominin species from australopithecine beginnings, 4 million years ago, to the moment when certain groups among *Homo sapiens* stood, 70,000 years ago, at the brink of a worldwide population explosion. The point of the chapter is to assemble in one place a discussion of issues that are not commonly presented together. It begins with the evolution of the phenotype of hominin species up to 200,000 years ago, including the effects of climate on biological evolution, processes of life-course development (epigenetics), and the documentation of hominin species through genomic analysis.[2] Then the analysis turns to the expanding capabilities of hominin species in the era of expanding brain capacity, from 2 million to 100,000 years ago. These growing capabilities included the emerging processes of learning through cultural evolution and multilevel selection, growth in hominin group size, types of communication by gesture or by sound, expanded capability for reasoning, underlying language abilities, and new knowledge about the character of human emotions and motivations.[3] In the third section, the chapter compares the communities of Neanderthals, Denisovans, and *Homo sapiens* as of 100,000 years ago. In sum, the chapter displays the new thinking that goes beyond skeletal phenotype to account for biological and cultural evolution of capabilities that were to enable learning, deeper logic, communication, and a wider range of emotions. Such extensions in analysis of individual-level behavior have brought an understanding that hominin life in mid-Pleistocene times was more elaborate and advanced than previously understood. This lifestyle provided a platform for a quite different sort of social and cultural change – to be explored in Chapter 3 – on the emergence of spoken, articulated language among *Homo sapiens*.

Evolving Hominin Phenotype: Pliocene and Early Pleistocene

We begin with the biological evolution of bipedal primates or *hominin* – a term that includes the genus *Australopithecus* in the Pliocene and the genus *Homo* in the Pleistocene.[4] The various species of the genus *Australopithecus*, all of them bipedal, lived in eastern and southern Africa during the Pliocene geological epoch (5 million to 2.6 million years ago). The succeeding Pleistocene epoch began 2.6 million years ago; at much the same time, the first species of the genus *Homo* arose: *Homo habilis*. In the next few paragraphs I provide phenotypical descriptions of successive hominin species, discuss the effects of climate and other environmental changes on hominin evolution, and then turn to the genotypical changes underlying what can be observed in the record of paleontology.

The genus *Australopithecus*, which emerged some 4 million years ago, was the first to become bipedal, though australopithecines did not walk as smoothly as the humans of today. The most famous skeletal remains are those of Lucy, a young female *Australopithecus afarensis* who lived 3.2 million years ago. Her skeleton shows clearly that she walked upright, and that her brain capacity was 380 to 430 cubic centimeters (cc). A recent comparison of australopithecine remains shows that their sexual dimorphism was relatively modest. That is, males exceeded females in weight by about 15% – much the same as for humans today – in contrast with the much greater disparities between males and females in other primate species. This result suggests that the main mating pattern among australopithecines (and for virtually all subsequent hominins) has been monogamy.[5] Males averaged heights of 1.2 to 1.5 meters (3.9–4.9 ft) and weighed between 30 and 55 kg (66–121 lb). *Australopithecus* continued to about 1.4 million years ago.[6] The 1960 archaeological discoveries by Louis and Mary Leakey brought to light both skeletal remains and tools of *Homo habilis*. The Oldowan or pebble tools of these small beings remain the oldest confirmed type of tools. Further discoveries now suggest that *Homo habilis* lived from 2.4 million years ago to 1.4 million years ago, in East and Southern Africa, with male dimensions averaging 1.00 to 1.35 meters in height, 32 kg in weight, and a brain capacity of 550 to 687 cc. Given this brain size, researchers believe that *Homo habilis* had no more social learning skills than apes living today. Tool knowledge, therefore, was transmitted not by imitation but by other learning mechanisms.

As more individual hominin remains have been recovered, and as they have shown to be widely varied according to several criteria, a controversy has persisted among paleontologists as to whether they should be classified in broad or specific categories. At one limit, the late Louis Leakey argued that broad categories should be employed, so that the small individuals who lived up to 2 million years ago should be classified within the genus *Homo*, as *Homo habilis*. At an opposite limit are scholars who prefer to identify numerous categories in both genus and especially species for the various hominin remains. These discrepancies in labeling human remains make it difficult for readers to trace the broad evolutionary patterns. I have tried to adopt terms in the middle range of those used.[7]

Processes of Biological Evolution. Biological evolution for humans today is, in its biochemical process, the same as it was for australopithecine individuals including the famous Lucy. The elements of Darwin's theory hold across time. One of the three pillars in Darwinian analysis is natural selection: the varying fortunes of individuals within their environment help determine which have the most numerous offspring to carry forth their genome (modified by sexual reproduction) into the next generation. A second pillar is variation: in the time of Lucy as in our own, there has been substantial variation in the characteristics of individuals (at levels from the genetic to the phenotypical) arising from genetic inheritance, genetic mutation, and the interaction of individuals and their environment. The third pillar is inheritance: at conception, each human individual receives a genome combined from the genomes of her or his parents, and the genes on those 23 chromosomes guide the individual throughout life. A further point in Darwin's analysis is that the individuals with the greatest "fitness," especially as measured by the number of their surviving offspring, will be able to sustain the lineage and will pass on their individual characteristics.

From our distance in time and space, we are unable as yet to see the biological details of the australopithecine genome or its processes, but we may have information on the *phenotype* – information on the physical type and behavior of the individual. Paleontologists, looking at skeletal remains and their surroundings, have focused on identifying such characteristics as height; weight; skeletal structure; teeth, skull shape; brain capacity; and information about diet, tools, and activities. The process of selection among variant individuals can take place at several levels – an incorrectly coded gene might fail to reproduce, or incorrectly coded proteins might not perform their metabolic processes, or the phenotypical function might be performed at a level that does not match up to that of competing individuals.

Modern laboratory work tells us that DNA, with copies in the nucleus of cells throughout the body, serves as the archive of evolutionary information. The DNA molecule can reproduce itself, whenever a cell reproduces. And the DNA is able to dispatch its information to cellular processes by serving as a template to create RNA, which moves from the nucleus to the cytoplasm of the cell and then creates specific proteins for each of the many metabolic processes. The extension of this work means that researchers have advanced steadily in documentation of human DNA and its history. Rebecca L. Cann, Mark Stoneking, and Allan C. Wilson led with a 1987 study by relying on mitochondrial DNA to document the African origins of modern humans through the female line. Later work expanded to analyzing polymorphisms of Y-chromosomes to trace the male line and somatic DNA, and most recently to whole-genome analysis of ancient DNA.[8] In addition, some of the proteins participate in epigenetic processes, alternately encouraging and restricting the production of RNA and new proteins. The interplay of these factors adds up, eventually, to the phenotype of an individual who lives from infancy through childhood and adulthood, where that phenotype constitutes the physical characteristics and the behavior patterns of that individual.[9]

In another long-term environmental pressure, the East African System of monsoons (in which summer heat on the continent forced winds north and south, alternating from one side of the equator to the other) brought alternations, every few thousand years, in the relative temperature and humidity of northeast and southeast Africa. This recurring pattern may have encouraged periodic hominin migrations back and forth to the most fertile region – a pattern that may have built into hominin species a proclivity to migrate.[10]

Genus* Homo*: Early Days to 500,000 Years Ago. The earliest known remains of the genus *Homo*, from the Afar region of northeastern Africa, are dated to 2.3 million years ago. More complete fossils that are now labeled as *Homo ergaster* date to 1.8 million years ago, especially in eastern and southern Africa. The most recent remains of *H. ergaster* are dated to 1.4 million years ago. It is generally argued that this period of relatively rapid growth in brain size was a time at which hunting and consumption of meat had expanded significantly. The rate of development from infancy to adulthood for *Homo ergaster* was similar to that for modern apes, slower than for australopithecines, and much faster than for modern humans. Early specimens of *Homo ergaster* were associated with Oldowan tools.

Homo ergaster, in its relatively brief history, appears to have undergone a number of phenotypical and cultural changes. The brain size of known African *Homo erectus* individuals averaged 700 cc.[11] (Figure 2.1 portrays the distribution of skulls for which the brain capacity is known, showing age and brain capacity over 3 million years; the top graphic shows the decline of temperature and its increasing fluctuation with time.) High-yield foods were necessary to achieve brain growth and, at the same time, reduce the size of the gut, where digestion took place. Fishing contributed to a high-protein diet, and it is possible that fire began to be controlled in this era. Then from 1.6 to 1.4 million years ago, the industry of Acheulian tools appeared in Africa, presumably among African *Homo erectus*. This industry, centering on tear-shaped hand-axes, continued in Africa and also in western Eurasia for the next million years. Gradual modifications of the Acheulian tools, though small, have been documented. In East Asia, however, Oldowan tools continued to be used. Further, at some point in hominin evolution, most body hair disappeared. While there is no direct evidence, recent work suggests that, as brain size grew, it became advantageous for hominins to lose most of their body hair at about 1.7 million years ago, in the era of African *Homo erectus*. Evidence for the timing of this change comes from genomic analysis of body lice and pubic lice.[12]

In addition to Darwinian genetic evolution, epigenetic changes are now argued to have been significant in the rapid phenotypical changes of *Homo ergaster*. Ian Tattersall makes a detailed argument for this interpretation, based especially on a single skeleton: the remains of Nariokotome Boy, the nearly complete skeleton of a youth who lived 1.6 million years ago at

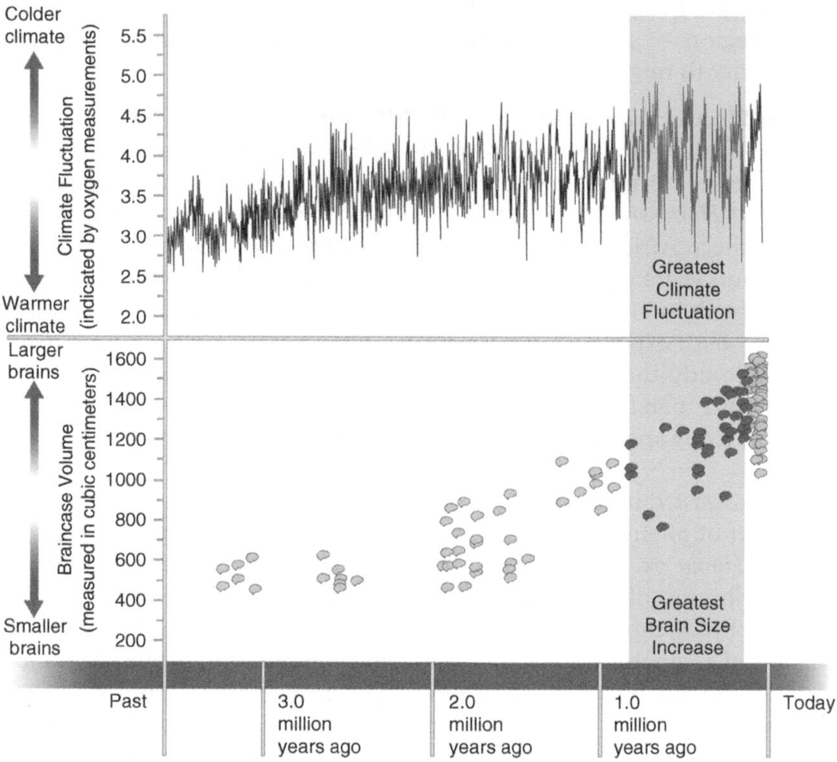

Figure 2.1 Range of brain sizes in hominin specimens. (Courtesy of the Human Origins Program, Smithsonian Institution.)

Lake Turkana in Kenya. This individual, with a height of 1.6 meters and now estimated to have been 8 years old at death, was "a hominid that grew up fast but physically was like nothing we know from earlier in time, and a creature that was clearly at ease away from the ancestral forest" (105).[13] The long legs imply adaptation to grasslands and an ability to carry out some hunting activities. Tattersall suggests that the sudden appearance of this significantly different phenotype may have resulted from epigenetic change, in which "a minor mutation that occurred in the Boy's lineage that, through altering gene timing and expression had radically changed its possessor's morphology – and had, entirely accidentally, opened new adaptive avenues to them."[14] Based on this example, Tattersall argues that the size of family groups or bands changed with the function of a species in the food chain. That is, earlier australopithecines, as foragers with little skill in aggression, had benefited from large groups, while those similar to the Nariokotome Boy, carrying out hunting, benefited from smaller groups so as not to eliminate all the prey, as their range size was limited by the mobility of females, whose infants were helpless.[15]

As of 1.4 million years ago, *Homo erectus* became the dominant hominin species. This tall and lanky species apparently began in Africa but spread widely throughout the Old World. Recent discoveries of skeletons and tools in the Caucasus date remains of *Homo erectus* in that region to as early as 1.8 million years ago. The physical height and body type of *Homo erectus* was very similar to that of people today, though skull shape was different and brain size was commonly 900 to 950 cc. Climate fluctuation became increasingly severe in the time of *Homo erectus*, and it is now argued that the steady increase in brain size was an adaptation to climate instability.[16] The larger brain meant that the process of development to adulthood was slower than for previous hominin species. *Homo erectus* was associated with Acheulian hand-axes in Africa and elsewhere. In Africa, *Homo erectus* ceased to be the dominant species as of about 500,000 years ago, though this species persisted in eastern and Southeast Asia, in some cases to as recently as 50,000 years ago.

Beginning some 700,000 years ago in Africa, *Homo heidelbergensis* arose as a modified offspring of *Homo erectus*. This species expanded in population and, as of about 500,000 years ago, migrated into the western and central parts of Eurasia. Dimensions for males averaged 1.75 meters in height and 62 kg in weight; for females, 1.57 meters in height and 51 kg in weight; and with brain capacity ranging from 1200 to 1300 cc. This species thrived for some time, using Acheulian tools. *Homo heidelbergensis* is best documented between 600,000 and 300,000 years ago; the African fossils tend to be older than those from Europe.[17] The individuals were tall with strongly built bodies and their faces were relatively flat but they still did not have the pointed chins typical of modern humans. They lived and worked in cooperative groups, hunted large animals, and made a variety of tools including stone hand-axes and wooden spears set with stone spearheads.[18]

***Offspring of* Homo heidelbergensis.** While the details of time and space are not known with precision, *Homo heidelbergensis* gave rise to three successor populations. These were *Homo neanderthalensis* living mostly in Europe, starting about 400,000 years ago; *Homo sapiens* in Africa, at least 200,000 and more likely 300,000 years ago; and the Denisovan order in Asia, perhaps 300,000 years ago. Neanderthals have been studied in great detail: their remains were located near the European centers of study, though sites are now known in West and Central Asia. Neanderthal mitochondrial DNA suggests an origin of 500,000 years ago; skeletal remains go back 300,000 years in Europe and reveal a descent from *Homo heidelbergensis*. There are "anticipatory fossils" at Sima de los Huesos at Atapuerca in Spain, 600,000 years old, that convey the gradual development of the Neanderthal phenotype.[19] Neanderthals developed and used the innovative toolkit known as Mousterian from 200,000 years ago to 40,000 years ago. From a flint core, many small flakes were chipped off, preparing blades of various sizes and shapes, smaller than the preceding Acheulian hand-axes. Some of these blades were attached to a haft or handle. Neanderthals had large brain capacity, of 1500 to 1600 cc, perhaps in part because of the need for adjustments to life in a cool environment. Neanderthal families appear to have been constituted as consanguineal bands, including couples and children of perhaps three generations. These bands communicated and exchanged with one another, especially in seeking partners for mating.[20]

A Central Asian hominin population, known as Denisovan, is known almost exclusively from genetic evidence. The initial skeletal evidence was a single finger bone from Denisova Cave in the Altai Mountains. The individual was a juvenile female who lived 41,000 years ago. Based on the full range of genetic evidence, it is now argued that the Denisovans were a sister group to the Neanderthals, who branched off from *Homo heidelbergensis* perhaps 600,000 years ago.[21]

Homo sapiens arose in Africa 200,000 or more years ago, according to a combination of genetic and skeletal data. The preceding *Homo heidelbergensis* sites in Africa show the development of modified stone tool technologies.[22] Current data suggest that *Homo sapiens* evolved from *Homo heidelbergensis* primarily in East Africa. A 195,000-year-old fossil from the Omo 1 site in Ethiopia shows the beginnings of the skull changes that we associate with people today, including a rounded skull case and possibly a projecting chin. A skull of 160,000 years ago (from the Herto site in the Middle Awash area of Ethiopia) also seems to be at the early stages of this transition: it has the rounded skull case but retains the large brow ridges of *Homo heidelbergensis*. Somewhat more advanced transitional forms have been found at Laetoli in Tanzania dating to about 120,000 years ago.[23] The average brain capacity of these individuals was roughly 1500 cc.

Nevertheless, for *Homo sapiens*, there is so far no indication of "anticipatory fossils" parallel to those that have shown the gradual development of Neanderthal phenotype. Tattersall treats this pattern as opening the possibility that our species "originated in the kind of systemwide genetic regulatory event" that he proposed as well for *Homo ergaster*.[24] As he argues, "this suggests that the physical origin of our species lay in a short-term event of major developmental reorganization, even if that event was likely driven by a rather minor structural innovation at the DNA level."[25] Overall, the Eastern Hemisphere, as of about 200,000 years ago, included various hominin populations, with at least some of them undergoing both physical and cultural-technical changes. They included *Homo sapiens* in northeast Africa, *Homo heidelbergensis* or successor species in other parts of Africa, *Homo neanderthalensis* from Europe to Central Asia, Denisovans in Central Asia and perhaps regions to the south, *Homo erectus* in eastern and southeastern Asia, and even the diminutive *Homo floresiensis* in island Southeast Asia.[26]

Advances in Hominin Capabilities

The foregoing picture of hominin evolution up to 200,000 years ago draws especially on paleontology, which conveys its increasingly clear picture of the physical evolution of hominin species – especially the growth in brain capacity – based on physical remains but also on growing understanding of epigenetics. We turn now to the strong likelihood that expanded capabilities arose along with the size of the human brain. Four more areas of advances in research need to be added into this still-speculative overview of hominin evolution in the times before language and speech. Two of these areas, the fields of cultural evolution and evolutionary linguistics, have expanded impressively since 1980. Three other areas of analysis – visual communication, the size of hominin social groups, and emotions – have shown themselves to be worthy of deeper investigation. Altogether, these research efforts show that the expansion in hominin neocortex had complex and interconnected implications for hominin capabilities.

Cultural Evolution. Cultural evolution has become a discipline linking social science, ecology and biology, hypothesizing a "dual inheritance" of biological and cultural change to "clarify the logical relationships between cultural transmission and other Darwinian processes."[27] Scholars in cultural evolution assume significant advances in human capabilities, somewhere in the time from the rise of *Homo erectus* 1.4 million years ago to the disappearance of *Homo heidelbergensis* 200,000 years ago. Robert Boyd, Peter J. Richerson, and their colleagues trace the steps by which, from this base, social learning took place, expanding to cultural learning and the coevolutionary development of population properties. They use the term "culture" to refer to any imitation, teaching, and learning that transmits behavior from one generation to the next. (As a matter of world-historical scale, I label this as "individual-level culture" to distinguish it from "group-level culture."[28]) Their theory includes identifying ways in which individuals within small groups developed patterns of cooperation, the basis for all future development of the processes in cultural evolution. The initial vision of social learning came from the psychologist Albert Bandura.[29] The process of cognition "in a social context" enables the individual to learn in a context of exposure to a variety of behaviors. At a more complex level, that of cultural evolution, the individual has the option of explicitly imitating a model behavior, likely but not necessarily that of a parent. The social-learning effects of cultural evolution are changes in behavior, not any changes in the physical form of the organisms. As a result, researchers have modeled changes in cultural evolution in a search for logically consistent hypotheses of changes. Researchers in cultural evolution turn to archaeology in an effort to trace the use of physical techniques, such as in tool production, that could have come only from social learning. The process of social learning arguably continues in the present; studies for recent times focus on cooperation.[30]

Such cultural evolution is a process for creating and inheriting learning. It emphasizes learning from one another, and there is no doubt that it could proceed, at least in part, without spoken language. The mechanisms of cultural evolution are assumed to be adaptive for fitness that will maximize the number of the individual's offspring. The evolving characteristics, in the case of cultural evolution, are a type of behavior, such as a technique for making a stone tool. According to the theory, when the individual successfully learns, the result is encoded in that individual's brain – the brain thus becomes the archive for the characteristic (in contrast to the biological genome). In this case, since learning goes to the brain, there is no distinction parallel to that of phenotype versus genotype. The characteristic must be heritable, but the inheritance takes place through the individual learning of the characteristic by persons in the next generation – most obviously from parent to child, though other paths of transmission could just as well be followed. The characteristic must be variable – in the example, there must be varying techniques for tool making that can be compared in effectiveness.

Further, because cultural evolution takes place at the level of the individual organism, it is evidently in competition with genetic evolution, which also focuses on the individual level. What is the consequence of competing evolutionary processes? The theorists of cultural evolution have conducted modeling to demonstrate that the two processes are opposed in a zero-sum game.[31] Ideally, the result is an "equilibrium" of cultural and genetic fitness – thus, a stable and balanced degree of genetic and cultural characteristics, which must work out even though the two evolutionary processes are quite different. For instance, the timing of transmission is different: genetic character is received from both parents by the individual only at birth; cultural character is contributed over the course of a lifetime, the contributions of parents can vary, and persons other than parents may contribute. Overall, however, cultural evolution can be modeled so that it will not permit individual-level culture to reduce individual genetic fitness.[32]

Three key analytical points in cultural evolution are known as inclusive fitness, the Price Equation, and multilevel selection. W. D. Hamilton argued that in "inclusive fitness," the genetic success of an individual can be treated not only as an organism's direct descendants but also includes the descendants of siblings and other close relatives.[33] This raised the possibility of inheritance of altruistic behavior among individuals. The Price Equation, based on the work of George Price and John Maynard Smith, extended this reasoning to groups of similar individuals, focusing on the coevolution of the frequency of a characteristic with the fitness of the individual.[34] The concept of multilevel selection is that selection can be assessed at the personal level, the level of inclusive fitness, and as multilevel selection. This facilitates analysis of such cases as sickle cell anemia and the inheritance of altruism, showing how it could be that inheritance of altruism would reduce the fitness of each individual yet increase the fitness of a group of such individuals.[35]

Further, it is argued that cultural transmission of this sort can become cumulative and can become a population characteristic.[36] Is cultural transmission really a population variable? This is important for whether the mechanism can indeed work for larger populations. Joseph Henrich argues that social learning began as early as 1.8 million years ago with *Homo erectus*, in the form of individual imitation. An individual develops one innovation; another person (likely a child) imitates it. Then the imitator arranges for the next generation to imitate it. In this way, each innovation gradually becomes more frequent in the population. If it can be dependably passed on, and if one can teach the innovations one has imitated plus one new innovation, overall technology advances. Henrich uses the term "Crossing the Rubicon" to refer to the tipping point at which the cumulation of innovations becomes self-reinforcing.[37] Henrich argues that there was cumulative cultural evolution at least for a while around 750,000 years ago at the site of Gesher Benot Ya'aqov in Israel, then argues that at about 450,000 years ago, *Homo heidelbergensis* in general had achieved that level of cumulative cultural evolution.[38]

In multidisciplinary work, the issue of "reductionism," discussed throughout the sciences, eventually came to studies of human evolution. The general strategy of reductionism is to use a simplified model, often taken from a work at a lower level of aggregation, to explain an apparently complex phenomenon. Thus, modeling from physics was important in the rise of molecular biology; here, biological modeling is used to explain social phenomena and the question is whether human group behavior can be reduced to equivalent individual behavior. The limits on reductionism are reached when new dynamics appear that cannot be accounted for in the reduced theory. That is, as I argue, the individual-level analysis of cultural evolution, while it can explain some group cooperation, is not able to account for the dynamics of group social behavior.[39]

Linguistics, Reasoning and Internal Language. The identification of the *FOXP2* gene, known to affect the ability to speak, initially raised hopes that a gene for speech had been discovered. But it came to be understood that the full complex of language and speech requires the interplay of numerous genes, so that the single *FOXP2* gene was not the key to language.[40] For the evolution of language, Berwick and Chomsky trace steps parallel to those proposed by the cultural evolution theories, from a parallel base, in rather explicit detail. They label the capability for syntax of an internal language as a "CPU" which computes the unspoken internal language (i-language), drawing on an available *lexicon* of word-like meanings, and linking them to an *interface* with the system of thought, which carries out the processes of reasoning, inference, and planning. This is their statement of the initial conditions, the basis from which their analysis of language begins – all within the mind of the individual. In addition, they allow for the existence of a "sensorimotor" system that connects the individual to the rest of the world. The CPU thus connects with two interfaces: the internal interface with thought, which combines to create internal language, and the interface with the sensorimotor system, which has the potential to include spoken language and which also is the interface through which learning takes place.[41] Berwick and Chomsky then propose a second biological change: a small-scale mutation that adds to the power of the CPU. This is the Merge function, which provides a simple way to combine words into hierarchical groupings. As a result, the interaction of CPU, lexicon, and the system of thought became far more sophisticated. They assume that the Merge function, once it emerged, spread rapidly throughout the population by natural selection. They emphasize that the benefits of Merge for the internal thinking processes (i-language) were greater than the benefits of the later development of speech.[42] Nonetheless, under this reasoning, the later emergence of e-language, syntactical speech, would have enabled the exchange of higher-level thinking among individuals, leading to creation of new sorts of group behavior.

Berwick and Chomsky assume that the advanced i-language, relying on Merge, persisted and expanded among hominins for at least several thousand years without leading to the invention of speech. They assume that the higher skills in hierarchical reasoning gave great advantages to those who had that capability, and that they grew and spread in influence. Yet once the Merge function had established itself, they assume that speech was sure to follow. In time frame, Berwick and Chomsky estimate that the Merge function arose about 80,000 years ago, and that the development of spoken language had taken place before 60,000 years ago, when groups of *Homo sapiens* began migrating from Africa into Asia. They choose 80,000 years ago because of the archaeological investigations of Blombos Cave in South Africa, which show illustrations drawn on the wall that suggest an unprecedented skill in representation.[43] So I emphasize the importance of the emergence and spread of the Merge function, as described here, and I follow up in Chapter 3 with analysis of the next step, the invention of speech.[44]

Visual Communication. Evolutionary psychologist Michael Tomasello drew on the turn to epigenetics, emphasizing ontogeny or development of the young. Tomasello had already begun work on "cultural learning."[45] Starting in 1998, he conducted detailed comparisons of development of young primates: humans, chimpanzees, and bonobos, with attention to visual communication. He concluded that perception of the physical world developed in parallel for all three species but that perception of social relations developed to a much higher level in humans. He hypothesized that early hominins began food sharing, which opened the door to cooperation; that 400,000 years ago humans began to work closely in pairs (especially as mates); and that population rose 150,000 years ago, requiring humans to work in groups.[46] While Tomasello's work paralleled that of Boyd and Richerson in some ways, the two groups pursued different agendas. Boyd and Richerson emphasized dual heritage and group selection, yielding genetically supported advance in cooperative outlook. In this model, groups and tribes grew slowly but at expanding rates. Tomasello emphasized ontogenic processes of development, giving more attention to intimate social interaction.

Groups of Hominins. Efforts to document the size of hominin social groups focus, in part, on brain size. Leslie Aiello and Robin Dunbar studied primates of all sizes to compare their brain capacity with the size of common social groups for each species; they concluded that group sizes increased linearly with brain size. Projecting this relationship to the brain size of *Homo sapiens*, they proposed that human group size should be 150 members.[47] Dunbar's later cross-sectional studies of humans found that groups of 150 are common in human society today. This result has a relationship to human speech in that anthropological observations suggest that about 150 speakers is the minimum number necessary to sustain a language. Meanwhile, Aiello and Dunbar found a similar correlation among primates in brain size and time spent in exchanging grooming. But grooming time reaches a maximum beyond the size of chimpanzees, so human group size could not have moved smoothly up to 150. Perhaps, then, humans were frustrated with too-small groups until they developed speech – the increase in efficiency of communication by speech would then have allowed a rapid growth in group size up to 150.

Archaeological data give little information on the size of human groups: archaeologists find remnants of individuals and sometimes of households, but it is rarely feasible to know how many people participated regularly in larger social groups. While we await results, I offer the following assumptions on the size of human groups. I assume that before speech, family groups included 15 or more members of all ages, regardless of brain size. With the rise of speech, as I argue in Chapter 3, I assume that speaking communities rose rapidly in size to 150 members. If groups reached a size near 300 members, I assume they split into separate groups. For times after speaking was well established among humans, as described in later chapters, I will propose that groups expanded by a different logic, relying on multiples of the basic 150.[48]

Emotions. Another of the fields of study reinforced by the neo-Darwinian synthesis and microbiology was ethology, the study of animal and eventually human behavior in biological context.[49] The model of emotions advanced by Ralph Adolphs and David J. Anderson, working within the field of ethology, has now produced a comprehensive model, beginning with a functional definition of emotions, in which an emotion is defined by the bodily state that it reinforces. Inputs to each emotional state include various stimuli and the context of those stimuli; in addition, there is some volitional control by an organism over its emotional states. The consequences of emotional states include behavior of the organism, somatic responses, cognitive changes, psychophysiology, and, for humans, subjective reports on emotions or feelings.[50] The analysis of Adolphs and Anderson focuses on a wide range of animals, including humans. They assume that a typology of specific emotional states can be developed through research, but they decline to speculate on a specific list. They do, however, propose a "provisional list" of the properties of emotion states.[51] In sum, they argue that emotional states exist generally for animals, but they are open to the argument that emotional states may lead to distinctive behavior in humans and that human perceptions and reports on emotions may be different from other animals, especially because of the higher cognitive power of humans. In a very different approach to emotions, Lisa Feldman Barrett, working with assumptions more like those of Tomasello, focuses on the construction, perception, and expression of emotions. In place of the emotional states of Adolphs and Anderson, she sees a set of physiological responses to stimuli reaching an individual human. The "emotions" are constructed in the interplay among humans. In what she calls "a new view of human nature," Barrett criticizes Darwin's "essentialist view," with its implication that emotions are genetically coded and inherited from early animal species. She emphasizes continuing interpenetration of the mind and the external world – that is, she relies heavily on the new discoveries in epigenetics as a framework for her analysis.[52] The two research groups, each pursuing their approach with detailed experiments, appear to be advancing the study of emotions in promising ways. Neither group, however, has got to the point of proposing the type or the time frame of any specific evolutionary changes in human emotions or their expression.

To summarize these varying analyses of hominin change, my impression is that the parallels among approaches are sufficient to support an overall interpretation in which the expansion of the human brain without creating specific new suborgans, created a space for additional computation that enabled the emergence of new functions for learning, group behavior, reasoning, visual communication, and expansion of emotional expression.[53] Later on, with the creation of syntactic language and detailed verbal communication, it became possible to utilize each of these functions at a much higher level – relying on Merge-based reasoning to construct and parse sentences, expanding learning through verbal guidance, supplementing visual communication with an expanding verbal lexicon, and articulating emotional responses with the spoken word.

The World of Genus *Homo*, 100,000–70,000 Years Ago

What was the full range of hominin populations, in Africa and Eurasia, as of 100,000 years ago? Did *Homo heidelbergensis* persist in certain parts of Africa or did additional hominin species develop in Africa as offspring of *Homo heidelbergensis*? Did *Homo sapiens* migrate from East Africa to other parts of the continent before 100,000 years ago? If so, whom did they encounter as they spread? The one well-documented case of expansion and possible encounter is that of Qafzeh Cave in Israel. As of 115,000 years ago – at a humid time when the Sahara was easily crossed – *Homo sapiens* settlers occupied the cave. But they were unable to sustain their occupation of the region: Neanderthals occupied the cave at a later and cooler time. Yet there were other hominins, whose remains were left in Moroccan caves 110,000 years ago, who used a technology known as Aterian. For a slightly later time, remnants were found at Qafzeh Cave and also at the nearby site of Skhul, which might have been hybrids of Neanderthals and *Homo sapiens*, or which might have been hybrids of Aterians and *Homo sapiens*.[54] In these and other remains of North Africa, South Africa, Israel, and Congo, indications are found of the gradual development of decorative material culture, including the piercing of marine snail shells, the heating of pigments to deepen their color, the creation of bone harpoons and, in the most striking item, the ochre plaques, geometrically engraved, found in Blombos Cave of South Africa. Such evidence supports the argument by Sally McBrearty and Alison Brooks for early and gradual development of decorative material culture in Africa.[55]

The lifestyles of the widely dispersed hominin populations must have been broadly similar. The populations lived particularly along waterways, produced and used stone tools with the Middle Stone Age technology, and had a foraging lifestyle – gathering vegetable and animal matter from land and water, gaining meat primarily by scavenging, but also hunting. They shared ways of learning, though speech was not central to their life.[56] The bodies of the various species were virtually identical; the brain size and cranial shape varied somewhat among them. Each population, in order to sustain itself, required at least several thousand individuals, in the view of population geneticists. In addition, these hominin populations found themselves in competition for resources with other medium- to large-sized mammals.

What was the nature of the psychology and behavior of these hominin individuals and families? While we have no direct observations on them, what we know about modern humans and other animal species enables us to assume that they had individual personalities and temperaments, and that they were active agents in the lives they lived. The families in which they lived, mostly small groups that were rarely as large as 30, interacted according to patterns that provided roles for family members by age and sex; mating was commonly monogamous.[57] The balance of violence, collaboration, submission, love, and ambition cannot be known for sure, but all of these and other motivations were surely in play.

The research on cultural evolution, evolutionary linguistics, and other behavior gives us additional questions to ask about these hominin species. According to recent research, we are left with parallel possibilities for each of these issues: that someplace in the time from the emergence of *Homo erectus* to *Homo sapiens*, as brain size grew rapidly, human mutations or developmental shifts may have provided these beings with additional capability for social learning, for enhanced emotional expression, for communication (via gesture, music, or basic lexicon), for internal language that heightened logical abilities, and for variations in the size of their communal groups. Which of these species had achieved the level of basic capability for cultural evolution and for internal language? Which of them had advanced in processes of cultural evolution to the point where "cumulative cultural evolutionary products were driving the genetic evolution of our genus, shaping our feet, legs, guts, teeth, and brains"?[58] Which of them had benefited from the mutation that brought the Merge function, leading to advances in hierarchical construction of sentences in an internal language? Which of them gained the capability of developing further behavioral predispositions, such as the need to symbolize, to explain, or to act, that would influence the evolution of human behavior? To the degree that any of these changes took place in the era of *Homo heidelbergensis*, up to 500,000 years ago, these advanced capabilities would have been passed on to all of the hominin species. To the degree that these changes took place after 200,000 years ago, only individuals among *Homo sapiens* had access to these new capabilities. Further analysis of available information should help sort out the possibilities. For instance, the remains of *Homo sapiens* at Qafzeh, 115,000 years ago, and their disappearance thereafter, may indicate that they were then lacking some capabilities that had been gained by 60,000 years ago.

3 Speech and Social Evolution

This chapter presents my hypothesis of the sudden rise of spoken, syntactic language in Northeast Africa, between 70,000 and 65,000 years ago. It follows up with the resulting process of social evolution; the formation of the Human System; and the initial coevolution of the social, cultural, and environmental aspects of human life. At the conclusion of the preceding chapter, we left *Homo sapiens* as one of three or perhaps more foraging hominin populations. Today, *Homo sapiens* alone has become the dominant species on Earth. Was this transformation sudden or gradual? In an important 2001 article entitled "The Revolution that Wasn't," archaeologists Sally McBrearty and Alison Brooks argued for slow and steady transformation within Africa's *Homo sapiens* population.[1] Most forcefully, they argue against the thesis of a "human revolution" that took place about 40,000 years ago, during which humans might suddenly have gained advanced conceptual skills (reflected, for instance, in cave paintings). Specifically within African hominin populations, McBrearty and Brooks argue for continuing development in technology, behavior, and skeletal structure from 250,000 years ago, rather than a long wait before a "human revolution" 40,000 years ago. Instead, McBrearty and Brooks rely on the continuous African fossil record to document a succession of developments beginning with Middle Stone Age technology and including the early development of decorative practices relying on ochre for coloring and on ostrich shells for making beads.[2] Since publication of their 2001 article, archaeological advances have confirmed the recurring changes and developments in African social practices.[3]

In this chapter, however, I offer a picture of sudden and revolutionary change, though I argue that such change took place within the framework proposed by McBrearty and Brooks. Change did take place at a revolutionary pace – not with the arrival of *Homo sapiens* in Europe 40,000 years ago but with the emergence of speech in Northeast Africa at about 70,000 years ago. As a result, *Homo sapiens* expanded to occupy all of the Eastern Hemisphere by 25,000 years ago – by which time all other hominin species had disappeared as independent communities (though with individuals absorbed into the expanding *Homo sapiens* population). My approach is to embrace the results reported by McBrearty and Brooks, but to argue that revolutionary change still took place. I emphasize articulation of speech rather than the emergence of cognition; and I hypothesize complex relationships between an initial community of speaking humans and the other humans with whom they interacted. I argue that the initial commitment to creation of a speech community laid the groundwork for other instances of explicit cooperation. Such cooperation – distinctive from earlier forms of cooperation because it involved an explicit, voiced commitment to a group – yielded the formation of social institutions, beginning with language, community, rituals, and migration, and more thereafter.

Language and Speech

"The Founders" is the label that I give to the initial community of speaking humans.[4] "Speech" is the term I give to the syntactic language that the Founders created.[5] The apparently sudden changes and expansion of this founding community, occurring from about 70,000 years ago, set the pattern for human expansion ever since.[6] Physically, the Founders were virtually the same as humans today – their average height and the size of their brains were very much like those of today and similar to those of other hominins in their time. They were good at walking and running – foragers who lived in savannas (grasslands) and did some hunting. They relied on a standard set of Middle Stone Age tools, of a type that had been in use in Africa by 300,000 years ago.[7] In such a speaking community, its members commit themselves to the group through the effort of learning their shared language.

We lack direct evidence to confirm the dating I propose for this transformation: 70,000 years ago.[8] The timing of my hypothesis relies on a triangulation of genetic, archaeological, and linguistic data plus the assumption that the documented human demographic and migratory expansion began soon after the rise of speech communities. Specifically, I propose that the rise of syntactic speech communities was as early as 70,000 years ago and as late as 65,000 years ago. My overall hypothesis proposes a process by which speech, representation, community, and migration combined to create a process of social evolution. I then argue that social evolution brought about the emergence of the Human System, encompassing all speaking humans, the dynamics of their community life, and the processes of their expansion and change. The preexisting processes of individual-level, cultural evolution persisted and contributed to overall coevolution of human behavior. But new processes of group-level evolution emerged along with the construction of syntactic language. From this point on, the overall coevolution of humanity was to depend increasingly on group behavior, focused on the evolution of institutions.

In linguistics, I rely on work documenting elements of a proto-human spoken language to support the argument that syntactic speech arose within a single community and did so rapidly rather than slowly. Continuing research, notably by Christopher Ehret, reaffirms the argument that a proto-human language arose in a single community, and that all languages of today are descended from it. The language has by no means been reconstructed, but Ehret has made advances in showing substantial commonality in basic vocabulary for such terms as pronouns (first and second person singular), words for mother and grandmother, and for *deictics,* terms such as "this" or "that." Thus, he reports the term "**mai" for first-person singular, the terms "**wai" and "*mue" for second-person singular, and the terms "*ina" and "*aya" for mother. Ehret's data on vocabulary of proto-human language are reconstructions, drawn from languages all over the world. These words give evidence that all languages today are descended from a single stock.[9]

Relying on this assumed creation of spoken language by the Founders, I assume that, at some point, a small and localized group of people – perhaps a mix of adults and children – began sharing their thoughts by assembling vocalized sentences to represent specific and even complex meanings. I assume, as I have emphasized, that the Founder population lived in Northeast Africa.[10] Yet there were prerequisites to be satisfied before the Founders could gather to undertake their path-breaking verbalization. To begin with, they required the conceptual capability to speak. Where did the logic of speech come from? Was spoken language necessary for survival or for thriving? Linguists and philosophers have considered this question in depth for the past 50 years. Noam Chomsky long led the discussion, arguing that the complex logic of language was somehow inbuilt into the human mind, waiting for the practical opportunity to apply it. No one could find a specific organ prepared to handle language: the debates were ferocious, but research advanced only slowly.[11] Scholars of collective learning came to suggest that, even without special suborgans within the brain, tool-making hominins had been gradually expanding their logical capabilities. As described in the previous chapter, linguists Berwick and Chomsky have developed the notion of the Merge capability – a simple but crucial logical step, enabled by a genetic shift, that enabled individuals to link two ideas into a conceptual bundle and then link the bundle to a third idea, in a process that could be extended for several steps. This enhanced capability for reasoning took place within individual minds and was not necessarily linked to communication.[12] Berwick and Chomsky propose that Merge arose 80,000 years ago.

A second prerequisite was the physical capability for speech. The biological argument emphasizes a shift in the position of the larynx that may have developed in the mid-Pleistocene, making it so that the passage between the mouth and the lungs was enlarged. This shift had the advantage of permitting the creation of a wider range of sounds; it had the disadvantage of allowing food to fall into the lungs. (Study continues on the shape of the tongue, lips, lungs, and of the varying types of sounds that could be made.)[13] Lieberman's review of the physical capability for speech centers on *Homo sapiens*; the question thus remains as to whether the crucial physical changes took place only among *Homo sapiens*, or whether Neanderthals and Denisovans also experienced the same changes.

A third prerequisite was that of a social demand for speech or, alternately phrased, the logical demand for speech. To be careful in our modeling, we must make sure to distinguish the processes of reasoning, communication, and vocalization, recognizing that they are separate though related issues. Tomasello, for instance, has explored the possibilities of communication through hand signs and other body language, regardless of the level of vocal communication.[14] Berwick and Chomsky argue firmly that the principal function of language was not communication but the articulation of internal thought.[15] So we have a debate as to whether a need for improved communication was a prerequisite for speech.

A fourth prerequisite was the social capability for speech. Here we encounter the limiting fact that, among humans today, children learn to speak readily but those who reach adulthood without having spoken cannot learn to speak. Yet the initial creation of language must surely have been the work of adults or at least of adolescents, because comprehensive knowledge and sophistication must have been essential, at the earliest stages of speech, in making language useful. To bridge this gap, I hypothesize that children of 10–15 years of age began speaking, drew other children in, and eventually populated the adult ages.[16] Thus, I propose that adolescent children created spoken language as one of the games of creativity that appear among the playful people of that age, and that the innovation somehow survived and spread. Perhaps the young people sought to teach their parents their new form of communication, or perhaps the creation of speech caused a great social breach between parents and children. But with time, when the first speakers had children – perhaps 5 or 10 years after inventing language – they were able to teach their children to speak, making language more likely to persist and spread.

(Dunbar, however, has argued that adult women and female bonding were the main progenitors of language.)[17]

The institution of *syntactic speech* thus emerged as the first social institution.[18] It centered on devotion to a collective practice: those who shared in speech had to devote years to developing words and syntax, exchanging them with others. The common objectives included describing feelings, responses, and objectives, combined with the related objectives of listening and learning in a new way – not to mention the shared frustrations of failing to find the words. According to patterns known for recent times, a language must have well over a hundred speakers in order to sustain itself over time.[19] Speakers would have recruited more people to their group, attracting them through the excitement of speech and the accompanying innovations. The emergence of speech was surely a situation of social upheaval, as young people established the new order. Perhaps the rhetoric enunciated by senior figures set the standards for language, or perhaps the innovation of teens set the standards. In the course of these processes, speaking people gradually rose to senior status among their compatriots and were able to begin creating more institutions besides language.

Yet the riddle of Universal Grammar persists: what accounts for the complexity of syntax? To explain the dynamics of the institution of language, we must return to the details of language itself, and to the problem raised by Chomsky as to how infants could conceptualize language and participate in the logic of word order and other grammatical structures. I offer three main points. Ultimately, Berwick and Chomsky identified the Merge capability as an elegant simplification of this problem, by formulating the reasoning of bundling prior reasoning and adding new concepts incrementally. The Merge capability, once created, had long persisted among nonverbal individuals; it worked effectively, but in part because it was interior to each person – one could avoid the specificity that would be necessary with spoken language, where one had to be understood by other people.

A second response to the riddle, I argue, is to be found by looking elsewhere, beyond the brain. Much of the logic of language, rather than sitting encoded and awaiting in the brain, lay within language itself, and it gradually unfolded through the practice of speakers as they spoke. Here are some examples that tend to sustain this thesis. Language has various dynamics of inherent change: from generation to generation, pronunciations change, grammar changes, and new terms are "borrowed" from one language into another. Children in every generation innovate and revise the language of their parents. As populations separate, their common language diverges steadily until, eventually, one requires a translator to connect the two speech communities. These properties of language arise not from some grand design, but from the interplay of the sounds and meanings in the practice of speaking. The inherent characteristics of language generated dynamics within speaking communities: we will see parallels to this pattern in the dynamics of other social institutions. My third point is that, however speech got started, it unleashed discourse, social reorganization, and innovation. Which words were created first? The labeling of family roles and the naming of individuals are good candidates. For instance, naming and categorizing family members meant being able to speak explicitly about hierarchy and interplay of male and female, of inside and outside the family. In addition, the naming of parts of the body and the most basic elements of material culture are also good candidates: hands, feet, eyes; food and water. All of these are nouns.[20] Further, the most common actions of humans are also good candidates for early words. Along with the invention of nouns and verbs, people would soon have developed modifying terms for what is big or small, fast or slow, hot or cold. The invention of language was a unique step in itself; it also set a paradigm for social evolution.

Language itself, however, was not enough to sustain the promise that it held out. One must hypothesize that the institution of *community*, including some 100–300 people, soon arose as a second social institution, with its members speaking a single, original language. Once spoken language was created, it could persist only if a community provided social structures to sustain the shared words and meanings, preserving and expanding them. Such a community of speaking individuals, once it formed, was a new experience.[21] It must have appeared as a massive and formidable social grouping, immensely outsizing the family groups that surrounded it, each with perhaps 20 nonspeaking persons.[22] Family units, of course, continued to exist within the new and larger language community, yet the nature of family life must have undergone transformation because of the ongoing connections through speech to members of other families within the new community. I argue that a speaking population, with its language, community, and rituals, acts according to the plan of the group as a whole. That is, speakers of a language agree to accept pronunciation and meanings of words that are used by the group as a whole. In contrast, informal groups of hunters may pursue similar objectives, but individuals within the group may also take their own path.[23]

The formalized community, providing a common identity to the group whose members spoke the initial syntactic language, still required additional devices with which to give coherence to the expanded community. Thus, a third institution was required – the institution of *ritual* or customs in which community members participated to reinforce their communal sentiments. Language required commitment to the shared practice of speaking; community required commitment to an expanded social identity; ritual required participation in events that reaffirmed, rationalized, and celebrated the community. As I argue, however, the central element of an institution is a group of people who share an objective. The institution of rituals, therefore, was either the full group that acted out the ritual or a smaller group that planned and directed the ritual. The dynamics of rituals were the movements and sounds of dance and song and the emotions of solidarity evoked by ritual.

In defining institutions, I will offer this type of reasoning at multiple reprises in this book. Other major innovations, large and small, created new institutions in human experience, and the internal logic of each institution brought into play the dynamics necessary to sustain the character of activity in that new institution. For agriculture, the nature of the agricultural calendar arose from the needs for planting, weeding, harvesting, and storing. For ceramics, the need to control both clay and fire developed specific skills of modeling and firing that generated a recurring desire for creativity in design. For migration, as we will see, the combination of the human life cycle and the availability of different language communities both encouraged and required migration and learning among young adults. The distinctive dynamics of speech, agriculture, ceramics, and migration were neither inherited nor invented by the innovators. Instead, they were encountered by those who invented the new institution, as inherent and inescapable elements of this newly formed aspect of life. Such was the case, later on, for libraries, the state, war, and science. Thus, for speech and other big steps in social evolution, human agency launched the innovation but did not determine the details of its dynamics.

From the moment that syntactic speech emerged, there arose the question of whether speaking humans were able to teach language to others. Was the capability to learn to speak determined by biological ancestry or by social learning? Of course, parents taught language to children, as is done today. The question is whether speaking adults (or children) were able to teach language to nonspeaking adults (or children). At least within the earliest days in Northeast Africa, it seems that language could have spread rapidly only by teaching language to new groups. This experience of teaching and learning of language, once adopted at the start, tended to continue. Speaking humans grew in number and, at the frontiers of their communities, other humans learned to speak. Thus, the early days of language necessarily involved processes of both convergence and divergence: people spoke at all levels of proficiency, with varying accents, so that pidgins and creoles emerged along with the development of standard languages, the divergence of dialects, the rise of new languages, and at the edge of speaking communities.[24] This complexity has characterized the history of language ever since.

Nevertheless, teaching nonspeakers to speak may have required that the learners had already inherited the Merge capability and that their sensorimotor capabilities were adequate for speaking. There is no reason to doubt that sensorimotor capabilities were similar throughout hominin populations, but the Merge capability had not initially been inherited by many. Could those without Merge learn to speak, even at a very basic level? At worst, they simply could not learn, and the spread of language was limited by the pace of biological evolution – for hominins, a rate of about 35 generations per millennium. At best, there were no limits on learning to speak, in which case speech spread as the large communities of speaking people expanded and incorporated people from nonspeaking groups. In between, there is the possibility that those without Merge could learn an inferior sort of speech and could be incorporated as subordinates, inferior in their reasoning and their social capability. The recent attention to interbreeding of hominin groups, however, suggests that learning language might have been possible. An alternative is that early speaking communities included some people who could speak and others who could not. Any such patterns ultimately disappeared as Merge spread through the population, but the inequalities of early days may have instilled certain inegalitarian values and practices that were influential over the longer run.

Social Evolution

Social evolution, as defined here, centers on the creation and change of social institutions, where institutions are defined as structures formed and maintained by members to achieve agreed-upon purposes. These paired definitions of social evolution and social institutions are central to my analysis. The term "institution," however, has two principal meanings in common usage. I adopt the definition commonly used by anthropologists, in which institutions are groups of people and their behavior, rather than the approach of sociologists and economists, who treat institutions as society-wide norms and rules.[25] The advantage of the group-focused anthropological view is that it treats group-based structures as elements of society that are created, transformed, and eliminated according to their strengths and weaknesses. These institutions, while not exactly parallel to the genes within genetic evolution, reproduce themselves and change gradually according to social and environmental pressures; they contribute their behavior to the functioning of a larger Human System. In contrast, the norm-focused, sociological approach is so broad and abstract that it never considers group behavior within institutions; instead, it assumes that institutions rely only on individual behavior.[26] In using the smaller-scale anthropological approach, one cannot escape addressing the role of groups in society or the overlap of groups and individuals in the functioning of institutions.

Types of Groups and Group Behavior. Here I explore the logic of members of a speaking community. The central concept is that of "collective intentionality," developed by philosophers John Searle and Raimo Tuomela. This concept describes and analyzes the nature of agreed-upon human groups, in contrast to the logic of individualistic persons on their own or in groups of other individualists. I begin by distinguishing an "I-group" from a "we-group" in the behavior of humans in groups. In I-mode, each person, whether alone or in a group of people, thinks of himself or herself as "I," making decisions based on his or her own interests and information. In we-mode, several individuals agree to think in terms of "we" and a shared, common interest. An "I-group" consists of several individuals – participants – whose individual opinions or actions happen to be closely linked. A "we-group" includes several individuals – members – who have agreed on a common objective and who agree to cooperate in reaching that goal. These basic terms are then combined in several ways.[27] To change the direction of a we-group, communication within the group would result in a decision as to the direction that all would take. To change direction in an I-group, all the individuals would decide on their direction – and might go in several directions. Therefore, in comparison to I-groups composed of participants making individual-level decisions, we-groups have advantages in solving collective problems through collaborative agreement of members. To be more specific, the three elements required for forming a we-group are the existence of a reason for the group, the understanding of members that they all share a common interest, and the practice of members in committing themselves to the common effort. Satisfaction of these conditions can create a group agent with clear decision-making powers that make its choices simpler than is the case for an I-group.[28] We-group networks each have a membership, and the members agree to follow the norms adopted by the group. Yet there are overlaps of the categories: within a we-group network, there may be participants who follow their own independent thinking. Within an I-group, there may be subgroups acting as we-groups.

Along with the scholars who developed this reasoning, I define a "social institution" or an "institution" as a we-group created by members who communicate through speech, to serve a selected social purpose. My assumption in this group-based analysis is that institutions serve as the basic elements of social evolution. I seek to show how these elements have evolved on their own through a process of social evolution that has important parallels to Darwinian evolution of the biological order. For early speaking communities, I seek to develop and apply logically consistent models of social evolution – including their interaction with processes of biological and cultural evolution. Through direct observation today and through study of the written, anthropological, and archaeological record, we may hope to build appropriate models by identifying the social changes that have taken place, fitting them into an explanation of the process of social evolution. A key point is that the effects of social evolution are changes in behavior, not changes in human physical makeup. Nevertheless, the effects of social evolution have tangible material and social results.[29]

Processes of Institutional Evolution. Darwin, in his analysis of biological evolution, underscored three aspects of the process: *natural selection*, the recognition that individual organisms may not survive the struggle for existence; *variation* in the characteristics of individual organisms; and *inheritance* of changes that arise in organisms. With time, attention has grown for a fourth element of Darwinian theory, the *fitness* of variations – whether subsequent generations of organisms bearing a certain variation will survive and thrive. Indeed, there are two more elements of Darwinian thinking that have also gained attention with time. One is that inheritance is carried forth across generations by an archive of information on the organism – an archive that we now know to be self-reproducing strands of DNA. The other is the point, emphasized forcefully by Ernst Mayr, that Darwin relied implicitly on population thinking, which later came to be formalized.[30]

For social evolution, I seek to identify analogs to these criteria for variations in social institutions. Explicit group behavior and communication via speech, however, require an amplification of Mayr's emphasis on population thinking.[31] Social institutions arose as the principal implementation of group behavior, so they are the key element of social evolutionary analysis. The nature of the archive for social evolution – the preservation of information on each institution that is necessary for its reproduction – is not fully clear. I assume, however, that there exist processes, distributed throughout the human community, by which each constituent archive stores information, reproduces itself, and reproduces the social institution.[32] The result, conceptually, is an overall social archive that reproduces itself for the next generation and also reproduces the relevant social institutions. My theory proposes that institutions are created and sustained through aspects of social evolution including selection, variation, inheritance, and social fitness. The following paragraphs discuss each of these aspects, introducing the processes and the additional terminology necessary to explain their dynamics.

Selection among Institutions. Real-world experience will give information on which social practices are favorable for the social order or individuals and which are not. The continuing shifts in the natural environment may make existing human behavior appear to be either beneficial or harmful. Thus, out of the interaction of human behavior and the natural environment, a process of social selection emerges to provide feedback to the social archive. In social evolution, selections take place at varying levels to determine which innovations will continue to be supported and which are rejected and no longer reproduced. At the endogenous level of "micro-selection," innovations that are simply inconsistent with the prevailing structures of the social archive are not selected. At the larger-scale level of "macro-selection," innovations that undermine normal social practice may be added to the archive but are unlikely to be reinforced in future generations. At the level of the environment, exogenous environmental changes that affect the human social order will have an effect on whether social innovations lead to benefit or harm for the social order.[33] Social selection is a mix of processes – unconscious, conscious, and environmental – that combine to provide feedback to the social archive, indirectly affecting the character of social behavior that the archive generates for the coming generation. The unconscious part of social selection is linked closely to the biologically based behavior of humans that survives from our hominin ancestry – and, arguably, linked as well to behavior that has arisen from cultural evolution. The conscious part of social selection is different. For instance, in some cases there arises a community desire to eliminate certain behaviors that are agreed to be negative or dangerous. In effect, this means creating a new innovation to eliminate an old practice.[34] Discourse in communities could lead to consensus but also to extended debate between viewpoints: for instance, categorization could lead to creation of spurious categories, yielding innovations of value to special interests but not to the community as a whole. Overall, social selection commonly led to a consensus in favor of adopting and preserving innovations. Further, achievement of a consensus that a certain practice ought to be ended is not the same thing as modifying the archive to halt the reproduction of that practice. Further, the changing environment has an influence in social selection roughly parallel to either unconscious patterns of behavior or the conscious development of social consensus or even ideological unity.

Variation: Change through Representation. Variation in the human social order arises especially as a result of representation. By representation is meant the portrayal, modeling, conceptualization, translation, or recapitulation of a phenomenon, often projecting the portrayal from one arena to another. Speech was not the first human venture into representation – the earlier practices of dance, music, and personal decoration qualified as acts of representing the world. But speech provided a tool for representation that was so general in its breadth, so open to specificity, that it surely encouraged early speakers to expand the range and the depth of their representation in speech and in all other directions. Once people represented their world by coining thousands of words, they could turn more readily to picturing aspects of their world in other media – such as with the representation of animals in visual art. This growing discourse revealed new questions in logic and knowledge. In these discourses of categorization and representation, people communicated meanings with a mix of clarity and error, agreement and disagreement, information and misinformation. The result, overall, was that individuals and groups repeatedly proposed innovations in conceptualization, social structures, and material life.

Vocabulary, as soon as it was invented, led to *discourse*. That is, people's interaction through speech brought interpersonal discourse, the exchange of information and sentiments. Discourse gave specific names to the elements of the world (nouns), to actions that can be taken (verbs), and qualifications of these (adverbs and adjectives). While people found more and more to talk about, they did not necessarily agree on every point. Speech opened up arguments over the meaning of words. It led to understanding and misunderstanding, so that truth and lying came to have extended meanings, once they were expressed in words. That is, discourse led to *debate*. In these discussions, the difference between what was known and what was not known could be debated. For instance, the ability to put into words the problem of the unknown opened a new area of thought and activity. In this discourse, speakers engaged in specific *categorization* and general *representation* of their world. Through *categorization*, people selected terms and assigned meaning to them, thereby constructing knowledge about the sociocultural order, the natural world, and any subject imaginable. Categorization brought comparison, which then developed into analogies and metaphors. With every act of labeling a thing or action in their world, members of the early speech community were undertaking representation of the world in speech. Out of these discourses arose *ideology*, sets of ideas that served either to sustain or undermine a social consensus, often reflecting specific interests within a community. Representation and discourse led ultimately to action – to innovation.

In social evolution, the mechanism for change begins with representation inside human minds: reflection and modeling of the world as it is and as it could be. From such representations arise suggestions for action: conscious innovations that amount to new types of sociocultural behavior and that may result in construction of new social institutions. These innovations may be thought of as exogenous to the social system, providing surprises that may turn out to benefit the social order. But they are equally well considered to be endogenous results of feedback, in which representation of the social situation as it is and as it could be leads to proposals for changes. The creation of any social institution required discussion to formulate the institution. The initiative may come from an individual, but an institution is significant only when it is serving a group. Any new institution, such as migration or other sorts of rituals, will have planned benefits but will also have unplanned advantages and disadvantages.[35] In the earliest days of social evolution, community meetings and community rituals served to support and reproduce the institutions. Preservation of each institution might require training of personnel to sustain it and perform its functions. With time, newly initiated institutions will undergo some degree of regulation – they may be reaffirmed and strengthened, or perhaps limited, or dropped as of no interest or as actually harmful. Thus, I have chosen to argue that social innovations are *constructed.* As a result, the processes for generating change in the archive are very different for social evolution than for biological or cultural evolution.

Inheritance: Reproduction and Regulation. For a newly formed social group, it must rapidly identify its social institutions, preserve them, and become able to replicate them for future generations. I present this step as parallel, in social evolution, to Darwin's step of "inheritance." A social archive dependably preserves information that can generate social behavior. Given the materials one has to work with – human consciousness and human institutions – there is no way for the archive to have the unity or precise replicability of biochemical DNA strands. Yet in parallel to the biological archive, in which DNA molecules are distributed throughout the cells and organisms of humanity, the sociocultural archive is distributed throughout the organisms, institutions, and material resources of humanity. The archive includes both materiality and behavior: it includes brain cells, parental practices, the wisdom of senior persons, institutions, rituals, and doubtless more.[36]

There are two stages to inheritance or reproduction of social institutions: the archive must reproduce itself in order to function for each generation; then it must reproduce each social institution. The social archive does not necessarily direct social behavior itself – it reproduces the institutions which, in turn, produce the social behavior of communities.[37] Sheer individual genius must have had its effects, but only for a single lifetime unless these effects were carried forth in an organized fashion. The actual preservation of the archive presumably begins as the conscious, collective memory of founding community members, reinforced by the structure of social institutions. Replication of the social archive for the next generation has required expanded processes – child socialization; schooling of many sorts; and construction of libraries in oral, text, and now digital format.

Reproduction of each social institution required recruiting the next generation of group members, instructing them in duties and outlook. Peer pressure, verbal coaching, rituals, and traditions all became elements of reproducing the institution. The archive and its preservation of an institution could expand but could also contract as minds and voices were lost. Since there was significant migration, one can imagine the difficulties of young migrants hiving off on their own, having to work without the wisdom of the ages. On the other hand, in-migration to a given community might bring valuable and positive contributions to the local archive. One key point of my hypothesized view of reproducing institutions is that it does not enable innovations to bring an instant leap ahead to a changed social order; instead, a step is required to update the social archive, which can then act to make the innovation available to the next generation.

Further, *regulation* is a dimension of inheritance or reproduction that is distinctive for social evolution. While selection is a yes-or-no choice, the process of inheritance receives conscious feedback from the assessment of fitness. This conscious feedback creates space for an institution to undergo regulation from social pressures, in the course of reproduction, so that a flawed institution may benefit from inheritance "with modification" rather than be eliminated. For instance, the institution of language, shared and preserved by all in the community, undergoes steady modification in its reproduction through the input of all speakers. Regulation and reproduction of social institutions, as a combined process, may be difficult.

Social Fitness. The "fitness" of a social institution is the degree to which the institution yields improvements in the social welfare of a beneficiary population. As a default, the beneficiary population should be the community (or, at a higher scale, the society) within which the institution is located. Here one must ask: Which results? Which group? The members of an institution might choose to focus its benefits on themselves rather than on the wider community, thus leading to possible disputes in assessment of the institution.[38] The measurement of social fitness could be that it brings more population to the community, or that it brings more resources, or that it brings a mix that can be called general welfare of the community. These ambiguities in assessment of social fitness are inherent and must be worked out by social compromise or by campaigns for regulation of the institution. The selection of institutions takes place in the struggle for existence in a natural and social context, with feedback effects determining whether the institution survives or not.

Dynamics and Functioning of Social Institutions. For several early institutions, I offer further details on their functioning, in parallel to the preceding discussion of the dynamics within the social institution of syntactic language. I focus sequentially on the reformulation of family groups, which had long existed, once they became incorporated into larger communities; the creation of the community to sustain the nascent group of speakers; and the formulation of two practices of migration. Family groups, based on the small-scale bands of up to 20 individuals that had long existed, persisted as speech developed. Small bands (or nuclear families) presumably continued as residence groups, but periodic connections and growing discourse among them created a larger identity – acted out at occasional celebrations of life, death, the seasons, or other events.[39] As the speaking communities took on more roles, the place of the family or band within it underwent adjustment. Speech began to take place within the family as well as in the emerging community, thus changing family relations. In addition, at both levels, the degrees of equality, specialization, and dominance in gender and family relations were open to renegotiation.

The community, as it took form to sustain the close communication of speakers of articulated language, might have required four or five generations to coalesce as a coherent institution. The speech communities, now encompassing discourse and debate on a growing number of issues, found themselves in the position of developing their organization and regulation, including ways of preserving the social innovations that began to be created out of discussions among community members. Thus, the institution of ritual came to have a new significance once the young community came to have senior members. In rituals and practices, communities could restate to each generation the rules that were being evolved for behavior. The separate communities became distinct and specialized in their habitat – they developed their own rituals as well as language specific to their situation. Among the many issues that arose was the effect of speech on violence: Did speech both limit and expand violence? Was the result verbal peace-making or rhetoric that created larger-scale violence?[40]

The institution of *migration* took two principal forms among early speaking communities.[41] First was the pattern of colonization, in which a whole community or many of its members – young and old – moved to a new habitat and started a new life. The community might seek to replicate its old life in a new place, thereby spreading the terrain of occupation by speaking humans. Human groups may have inherited some migratory tendencies from their deep ancestry – including the ancient climatic pressures for migration north and south within eastern Africa. With time, migratory communities became distinct from each other, as their languages and accompanying rituals evolved. Once colonizing new regions, such migrant groups could at once develop distinctiveness from other communities yet retain some contact with the activities and developments of their parent communities. The other form of migration was cross-community migration.[42] In it, small numbers of people, most commonly young adults, moved from their home community to join an existing community some distance away. Migrants who moved from community to community needed to learn a new language, could also learn new rituals and social and technological devices, and might return to their home community with the new ideas. Migration of young people from one community to another was functional in that it shared ideas from one community to the next, brought together different experiences for the creation of further innovations, and built the practice of learning languages and rituals as a common human activity. Migration itself was not the institution. Instead the institution included the migrants themselves and those who facilitated migration or who facilitated and hosted migrants. Thus, migratory subgroups existed within early communities; migrants moved to another community and joined another migratory subgroup including migrants and facilitators.

Further, the reinforcement and benefit of cross-community migration brought an important biological result. Migration of young adults brought interbreeding and genomic exchange among human communities, which preserved broad genetic unity, limiting the development of highly differentiated local populations. Humans, in comparison to other mammalian species, have a relatively high degree of exogamy, in which mates are couples with origins in different communities. Thus, while individual humans have significant genetic differences from each other, these differences are spread widely because of cross-community migration, so that the genetic differences *between* communities are smaller than the genetic differences *within* communities. In sum, cross-community migration became an institution within human social practice that generated inherent dynamics building certain practices into human life: the common migration of young people across community lines, the resultant facilitation of learning and innovation, and reinforcement of genetic unity throughout humankind.

The Human System and Its Scales

The initial Human System, as I have described it so far, was small: it consisted of the community of the Founders, their language, their community, the families within the community, their rituals and migration practices. This pioneer system expanded through population growth and immigration from surrounding groups, until it underwent its first division, becoming two communities, followed by later such divisions.

Interpretation of the Human System can be expanded by including the perspective of *Living Systems*, as set forth by James G. Miller.[43] Miller's volume, at once encyclopedic and analytical, applies systems thinking to the biological order, emphasizing the multiple scales of living systems – from the cellular level to that of global societies – and the remarkable parallels in the functions performed in living structures at each scale. Miller's levels of analysis were the cell, organ, organism, group, organization, society, and supranational system. His principal objectives were to show the parallels in the various levels of life, to treat each level in systemic terms and – especially – to emphasize the functional subsystems within each level of life. He identified 19 subsystems in any living system, and listed the *name* he assigned to each subsystem, its *purpose* (or function), and the *agents* (or *organs*) fulfilling the purpose.

Miller's 19 subsystems fall into three groups: (1) for the unit as a whole, two subsystems, focusing on reproduction and on boundaries; (2) for the treatment of matter and energy, eight subsystems; and (3) for the treatment of information, nine subsystems. A remarkable point that arises immediately is that the number of subsystems focused on information is as large as that for matter and energy – and this for living systems at every scale. The implication of this framework for early human social evolution seems unmistakable. That is, for the period from 70,000 to 25,000 years ago, during which expanding human communities implemented many new and practical decisions, it appears that establishing subsystems in information – that is, conceptual and social changes, not just technological innovations – must have been of high priority. For instance, the social practice of migration enabled both the gene pool and the cultural archive to be widely exchanged among all human communities; creation of effective ritual for reproducing the community must also have been a high priority. Based on these and other cultural and social advances, human reconfiguration of the material world would come later.

Miller's framework also clarifies the relations among biological, cultural, and social evolutionary processes, as well as interaction with the environment. Biological evolution, with its long predominance, directs the evolution of the cellular level, the level of organs, the individual organism, and even the family. Further, the rise of cultural evolution brought changes to characteristics of individual human organisms and to the family level, because of its interactions in learning, so that biological and cultural processes compete to bring change and continuity to the human organism and family.[44] In that case, the subsystems at the individual and family level, organized in biological and cultural terms, may have archives and systems of natural selection that are duplicates or at least supplements to each other. Social evolution, meanwhile, is defined for the level of the community and, eventually, also for the higher levels of society. Of course, all of these factors influence each other indirectly through coevolution: thus the community must interact with the biological and cultural structures of the organism and the family.

Human Social Subsystems. The Human System, created in the late Pleistocene with the rise of language, was initially composed of overlapping and interacting communities at levels of family, language group, and ethnicity. In the Holocene epoch, communities expanded to form societies that took the form of towns, states, and empires. These social groups, to survive and reproduce themselves, required *subsystems* functioning to perform the tasks of reproduction, maintaining boundaries, processing of matter and energy, and processing of information.[45]

What were the subsystems and institutions of the early, constructed social systems that stretched in a thin layer over such a huge terrain? In the early days of social evolution, humans needed implicitly to create their 19 subsystems to sustain their process of social evolution – and then those subsystems expanded along with the Human System overall. What functions did these subsystems serve? Did they benefit the whole system or just specific social groups? The Human System had no central brain, though many conscious individuals and communities were able to share information and develop consensus behavior. The biological systems and subsystems of human individuals and groups (families) continued to function much as they had before, but were supplemented by social systems and subsystems, expanding and subdividing their functions.

Two subsystems address the needs of the system (in this case the community) as a whole: the reproducer and the boundary. The *reproducer* subsystem includes the archive, which preserves the information necessary to generate the behavior of the community system and actually implement it. The reproduction of the individual members of the community is handled by the biological reproducer at the organism level, but the behavior of those individuals at the community level is handled by the reproducer. The *boundary* is most likely a social boundary, identifying which individuals are members of the community. Eight subsystems process matter and energy for the community. The *ingest* brings in food, water, firewood, and raw materials for tools; the *distributor* shares these inputs among community members. The *convertor* transforms firewood into heat or water into drink. The *producer* creates food out of meats and vegetables; the *extruder* removes the waste material, its sight and smell. The *motor* is the ability of the community to move; the *supporter* can be seen as the habitat in which the community lives.

Nine subsystems transfer and transform information in various ways. The *input transducer* brings information from beyond the community; the *distributor* spreads information within the community, by means of the *channel* and *net.* The *decoder, associator, memory,* and *decider* work together to make sense of information from within or beyond the community and made decisions on actions. The *encoder* conveys the decision to destinations within the community; the *output transducer* conveys decisions beyond the community. The information sub-systems are striking in their differentiation and in their importance at the community level, and indeed at all systemic levels. How is information collected and prepared for use? How is information processed and stored, yielding decisions? How does information become output? Who or what receives and processes the information that is output? The responses to these questions are not entirely clear, but the questions are shown to be significant, and it becomes clear that issues of conceptualization and organization of information in the early days of the Human System set important and lasting patterns.

Institutions within Subsystems. New institutions, while constructed by human agency, commonly function through dynamics that are inherent to the logic of the institution. As I have argued earlier, language inherently includes elements that we label as vocabulary, parts of speech, grammar, phonology, plus patterns of gradual change over time. In another example, the function of the *producer* subsystem is to synthesize materials for growth and repair. When agriculture arose, it became a distinct institution, governed by the seasons and the characteristics of the various crops. A parallel institution within the *producer* system developed for animal husbandry, with its patterns governed by the breeding, pasturing, and exploitation of the animals. Much later, libraries arose as an institution within the *memory* subsystem: library dynamics rely on the logic of classification and access to resources. (Libraries and other parts of the *memory* subsystem must also make the contribution to the *reproducer* subsystem that produces group-level culture for the next generation.) Each institution, through its novel contribution, generated an appropriate sort of human behavior that facilitated the task of its subsystem, such as with weeding in agriculture and reshelving in libraries. In the case of each institution, whether it addressed information or matter and energy, the relevant *institutional logic* could be encountered only by entering each of these new domains. The growing system of human knowledge developed many discrete disciplines of knowledge about the inherent character of many types of activity.

How have these numerous elements and subelements of the Human System sustained the system as a whole? One big question inherent in Miller's scheme is about the functioning of the *decider*, the subsystem that is to make decisions based on available information. The answer, consistent with Miller's observations, is that the mission of the community's *decider* subsystem can be performed by downwardly dispersed subsystems.[46] That is, communities disperse decisions to families and individuals, societies disperse decisions to communities and organizations, and so forth. So it is with decisions in political, economic, cultural, and social arenas – the decisions are taken in a distributed fashion among overlapping communities.

In those situations, contending views are expressed through *ideology*, in a continuing debate on social priorities, especially the degree to which they should be decentralized or centralized. Yet there must exist some pattern (or institution) by which the dispersed decisions are coordinated for the functioning of the community or societal system.[47] Humans everywhere, in all their communities and societies, preserve three common archives: their genetic, cultural, and social archives. Persistence of these archives is reinforced by migration, enabling both the gene pool and the social archive to be widely exchanged among all human communities. Together, these archives provide a platform for further social evolution. Even today, with superpowers and international organizations, we are far from having a unique "decider" to answer all the big questions.

Networks. The concept of the "network" provides a way to expand the scale of analysis beyond the level of discrete units by modeling the links among units.[48] For the "social networks" that I discuss here, network analysis makes it possible to expand the logic of individual and group behavior to the spatial distribution of this behavior. For network theory, I have relied on the analysis of Paul McLean, in which he uses networks to explore cultural behavior. McLean begins with networks as the spatial linkage of nodes, where the nodes may be individual persons, then expands the logic to various cultural interactions.[49] I further extend McLean's analysis in two directions, relying in each case on the logic of collective intentionality. First, I assume that some or all nodes in a network may be groups rather than individuals; second, groups among the nodes may be either I-groups or we-groups. As a result of the second added assumption, the entire network may be a we-group. Thus, the constituent nodes of a social network may be individuals, I-groups, or we-groups; a network overall may therefore be characterized as an "I-group network" or a "we-group network." In the historical narrative to follow, I will identify I-group networks and we-group networks in various social situations: I argue that I-group networks have been relatively easy to form and to modify, but that we-group networks, once formed by agreement as to their objective, become influential in transforming their world.

Communities of speaking humans, organizing themselves as systems and as we-groups, communicated at one level while functioning within a social group and communicated at a different level when in separate groups. Communities diverged from each other as they grew in size, scattered through migration, and became different in language.[50] Yet these communities, as they diverged, remained linked in I-group networks among themselves, through cross-community migration, other forms of exchange, and through acknowledgment of their similarities and common origins. In the aggregate, these linked language groups became a wide social network of communities.

In the time from 65,000 to 50,000 years ago, these communities and their diversifying subsystems expanded across the Old World tropics. By retaining their inherent capabilities and remaining in contact through migration, they came to be linked into a set of overlapping networks of social systems – an expanded version of the Human System, stretched in a thin but growing collection of communities across land and littoral from the African Atlantic to the South Pacific. An individual community, once created by language, benefited from the creation of additional social institutions, each with its own inherent dynamic, each bringing a different sort of change. These included rituals (to sustain the community – including dance, song, feast, initiation rituals, funerals, recognition of migrants entering and leaving); migration (a process that benefited technical and cultural knowledge and broadened the genome); artisanal workshops (to create works of representation); institutions of collective governance; and eventually, agriculture and states.[51] It may be noted that communities, while they play an important part in this narrative of the early life of speaking humans, are not well known from the archaeological record. Instead, archaeology documents the remnants of individual humans and of small groups for the period before 10,000 years ago. Nonetheless, while there are few material remains of human communities in this era, we have strong if indirect evidence of their existence from research on human group size and from the knowledge that a language needs a group to sustain it.[52]

Coevolution and Human Nature

The expanded coevolution within East African human life yielded the organization of speaking communities; they were to expand and develop for many thousands of years. This is the historical and analytical point, therefore, for reviewing the interaction of the elements introduced so far. The review focuses on the concept of coevolution as it is applied among as many as five orders of social change – biological, cultural, social, environmental, and systemic – and the practical results for the early days of human expansion.[53]

In genetic terms, the early speaking communities were likely the subject of genetic drift – the initially small groups had a genetic composition that was distinctive from that of the overall regional population; their distinctiveness might be reinforced if the groups grew autonomously. On the other hand, the steady incorporation of nonspeaking people into the speaking communities limited the degree of drift and brought the speaking populations steadily toward being representative of the general African population of *Homo sapiens*. This is an initial example of coevolution of biological and social patterns of evolution.[54]

Environmental, cultural, and social coevolution took place as humans settled into new homes. The relatively rapid expansion of speaking communities throughout the diverse landscape of Northeast Africa meant that individual communities, as they moved and incorporated new members, found themselves with the need to adapt to new habitats. In a sort of rehearsal of a pattern that would spread throughout the world, they settled into their new situation, adjusted their technology and social relations to the needs of the local habitat, but also brought about modifications in the land, the flora, and the fauna of their new home. Thus, the movement of a single community from the hillsides to a valley habitat would likely bring about elements of biological, cultural, and social evolution in interaction with each other. Biological evolution adjusted the physical phenotype – and perhaps the basic emotions. The mechanisms of cultural evolution, in which individuals learned new techniques by imitation and through instruction, adjusted individual behavior and patterns of cooperation. Social evolution adjusted group behavior, such as migration and community formation, as well as network formation and interactions.[55] The expansion of the Human System was not just a matter of extending the frontiers of humanity to all the shores of the Old World; it was also a process of filling in many distinctive niches.

Such identification of these elements of the Human System can facilitate the discussion of the character of their various interactions, including cases of overlap and possible duplication. One area that calls for further study, within this framework, is the possible overlap of subsystems at the level of the family and community, especially in the early days of social evolution. The family continues as a biologically and culturally governed system; the community comes into existence through processes of social evolution yet must rely initially on biologically and culturally governed processes; as the community solidifies, its socially governed processes remodel the family system. Is it adequate, under these circumstances, to keep the model simple by assuming that the family is biologically governed and the community is socially governed?

"Human Nature" and Emotions in Institutions. The term "human nature," used in general parlance to refer to the presumptively dependable patterns of human behavior, is occasionally the topic of more specific analysis. Human nature is understood variously to refer to behavior that is instinctive, predictable, or unavoidable – in any case, human nature is thought to be closely related to the emotions. One way to phrase the question of the fixity of human nature is: how much of human nature is determined by biological processes and how much is conditioned by social evolution or cultural evolution? In my opinion, emotions are programmed most basically in biology; they are expressed in individual behavior in interaction with stimuli received from other individuals and the environment. With feedback among these elements, individuals can influence emotions through their own volition.[56]

Yet here I add two dimensions to the consideration of emotions and human nature. First, the initial discourse on emotions and behavior, as it arose in the early speaking community, gave rise to questions and conceptualization about human nature. Human nature is a fundamentally social concept, and it arose as a topic of discussion and social problem in the initial days of social evolution. As a result, in this volume, I speak of emotions throughout the analysis, but I speak of human nature only for the period in which groups of humans shared syntactic language. Second, I argue that social institutions draw on and reinforce certain individual behaviors and associated emotions, linking them to group behavior and outlook for each institution. For social institutions, I propose giving attention to the specific dynamics of each, especially the ways they draw on biological emotions and styles of cultural learning, so that given institutions might be characterized by different emotional outlooks and behavior. An institution sustaining migration might draw on individuals with an adventurous outlook – thus, institutions of charity and of commerce might reinforce quite different outlooks. Coevolution is likely to mean that institutions, increasingly numerous with time, each build outlooks and give openings to various emotions or "natures." As a result, even if the biological nature of humans were static, the social nature of humans and their institutions would change with time and situation. To this degree, "human nature" is open to conscious change. More broadly, the balance of institutions and resulting outlooks in a community might create an overall *zeitgeist*, as one or perhaps more outlooks gain influence over a community with significant internal diversity. Group-emotional characteristics are reaffirmed through song, dance, ritual, and performance of duties; these group emotions interact with the individual-level emotions of group members. Details of these processes and their links are not well known. A zeitgeist appropriate to early years of speaking human communities might have focused on creativity, as there was so much work to do in developing the new institutions of language, community, and ritual. In any case, these links among emotions at individual and group levels, in biological and social-evolutionary processes, are potentially central to the overall links of biological and social processes in humanity.[57]

As a result, I offer the argument that social processes have become steadily more important in human behavior, and that the socially based dimension of behavior can be described as the accumulation of institutional behavior. For instance, the creation and teaching of language are best seen as social rather than biological processes. Both language and its teaching arose through conscious agency, although it may be that teaching and learning of language have now existed for long enough that there have been biological adaptations that facilitate them. The creation of a speech community, further, must have been a conscious act. Once the speech community was created, there began a coevolution of emerging communities and existing families. According to this hypothesis, there should be an expansion of research on communities as a part of the creation of language, and further research on the later emergence of multicommunity societies and social organizations.

Cultural Evolution in the Process of Social Evolution. One more important element in the coevolution of human life is that the processes of cultural evolution necessarily continued to function during the era of speech and social evolution. Peter Turchin has offered the fullest model of this issue, in three stages. First, he clarifies an important difference between biological and cultural evolution, defining the functioning of cultural evolution in the Pleistocene. Then he makes a two-stage historical argument as to how cultural evolution brought major changes during the Holocene epoch, to which I shall return in Chapter 6. The hypothesis of cultural evolution, however, added a new factor. Social learning, which developed new skills and built them into a population-wide genome through the dual heritage of cultural and biological evolution, enabled populations to become relatively homogeneous and distinct from each other. Thus, a basis arose for variation among groups (or "tribes"), and the tribes with the highest degree of cooperation won in competition among tribes, thus advancing cooperative behavior for the species.[58] To lay out his model in the first stage, Turchin points to the work of George C. Williams, who argued that groups of animals did not become a base for biological evolution because the groups would have had to be internally homogeneous and distinct from each other to provide the variety necessary for selection among them. But groups of animals were too diverse, notably because of migration among them, so that biological evolution inevitably fell back to reliance on the variety among individual organisms – basic biological evolution.

The hypothesis of social evolution works quite differently, as I have argued earlier. Groups form by conscious decision of the members.[59] Variety is found in the innovations within groups and innovations within institutions, which add to the strength of the community (or tribe). The contrast is that, in cultural evolution, variety within tribes is kept at a minimum by social learning, to enable selection among tribes – it leads to elimination of the weak tribes. In social evolution, the variety is among institutions and their innovations, and the selection is among institutions – the process leads to elimination of the weak institutions. Migration provides a clear example of the difference between these mechanisms: migration is dangerous for cultural evolution, as it adds variety within tribes; but migration is beneficial for social evolution, as it adds variety and innovative power to institutions. We are left to explore the implications of these contending hypotheses through their internal logic and their consistency with historical evidence. It is conceivable that the two mechanisms operated in competition through the late Pleistocene – with elimination of weak tribes in some cases and elimination of weak institutions in others. The contrasts of these two models may be kept in mind as we proceed through this narrative of human history and human evolution.

In analyzing coevolution among multiple influences, one must also consider the specific character of each component.[60] The list of influences includes biological process and evolution (at levels from the cell to the social group), cultural process and evolution (at levels from the individual to the social group), social process and evolution (at the level of the community and the higher levels erected upon communities), the combination of these in the Human System, and the environment of Earth with which the Human System interacts. Consideration of these categories raises the question of the process of self-organizing adaptation at the level of the Human System. Does there exist an additional mechanism of evolution at the aggregate system level? Or is it simply the resultant of all the processes at the lower level? In my opinion it may ultimately be possible to identify an overall dynamic of change of the Human System, but this question should be revisited when we discuss later times in history and a broader range of human experience.

4 Systemic Expansion

My narrative now extends to the expansion of the Human System, from 65,000 years ago to 1000 years ago. The narrative presents, in three chapters, an expanded combination of data from genomics, language, archaeology, climate, geography, technology, and social change. In it I seek to integrate the efforts of contemporary scholars to combine the new data and outlooks from multiple fields of study. Of course, this tale of human expansion cannot be definitive, because it is concise and especially because new evidence, arising from rapidly advancing research, will soon require updates in any interpretation. Nevertheless, because I am attempting a comprehensive narrative of human expansion, I hope that it will set a standard of cross-disciplinary breadth, so that subsequent analyses will make their revisions to a steadily more comprehensive picture of human expansion and evolution. Chapter 4 (from 65,000 years ago to 25,000 years ago) and Chapter 5 (from 25,000 years ago to 12,000 years ago) combine to explore the migration and social change of humans in the late Pleistocene epoch; Chapter 6 continues with stories of migration and social evolution through most of the Holocene epoch, to 1000 CE. Language classifications, as I employ them, provide information on the location and movement of language groups in the Pleistocene as well as in later times, though migrations are difficult to date. Genetic data include newly available results of whole-genome analysis of ancient DNA, for which dates if not locations are precise. Climate data stem from worldwide samples and simulations; anthropological and archaeological data, steadily growing in quantity and quality, confirm both local and general patterns. My theory of social evolution, relying on other recent theories, assembles an interpretation from these data.

The emergence of the Founders and their initial speech community, as described in the previous chapter, launched the process of social evolution. This chapter begins with the crystallization of the Human System in small communities of Northeast Africa. The narrative then emphasizes the differing character of human expansion along three main axes of human expansion – the African tropics, the Asian tropics, and temperate

Eurasia. As I argue, the transformations in group-level culture, technology, the social archive, and sociocultural institutions took place in different ways along each of these migratory trajectories. Expansion of human population was like no other expansion of a large mammal. While we have increasing knowledge of widespread migration of hominin species in earlier times, as shown in Chapter 2, there was nothing to even approach the sudden expansion of a single population from a clearly defined homeland to varying habitats throughout Africa and Eurasia. This explosion of population resulted not from some massive shift in human biological or conceptual makeup, but from a specific new capability – a mental capability, a technology, or some mixture – and I have assumed that the new capability centered on the rapid rise of spoken, syntactic language.[1]

Previous analyses of human expansion have tended to separate the expansion within Africa from that into Eurasia, often neglecting the African expansion entirely. In contrast, the expansions in Africa and in Eurasia are best seen as contemporaneous with each other and similar in many respects.[2] The amount of territory occupied by speaking humans in the period from 65,000 to 25,000 years ago was about equal for Africa and Eurasia; the time period required for this occupation may have been much the same on the two continents. Both continents were also the home to other hominins up until about 25,000 years ago, so that each has its stories of interbreeding. While the Founders spread south and west from their Northeast African homeland – including southward along Africa's Indian Ocean littoral – they also crossed the Red Sea and began migrating eastward along the Indian Ocean littoral of Asia. The geography of Africa made it easier to expand inland than was the case for Asia, where mountains and deserts held the migrants near to the coast, at least for some time.

I seek to document the path and timing of this global migration through four types of evidence – archaeological, linguistic, genetic, and climatological. In archaeology, new sites continue to be discovered throughout the Eastern Hemisphere for this period, and they are documented with steadily improving estimates of the dates of each site. For linguistic data, by analyzing the earliest and other principal subdivisions within a major language family, it is possible to estimate the geographic region in which the language family originated – its homeland. In genetics, steady expansion in the analysis of DNA of living populations, linked by assumptions on how to link today's population to past populations, has yielded migratory paths through analysis of several genetic constituents: Y-chromosomes, mitochondrial DNA, and somatic chromosomes; important new genetic information comes from whole-genome analysis of ancient DNA.[3] Because of advancing technology, genetic data are updated frequently, yielding more cases, better efforts to link present and past populations, and better estimates on the timing of past genetic changes. Both linguistic and genetic data contain a full stratigraphy, with information on stages from early to recent times: they, along with archaeology, are worthy of detailed and continuing analysis. The distinction between linguistic and genetic stratigraphy is important and valuable. Each language has a single parent, though it may have many sisters – every language has a single line of ancestry. Genetic descent is bisexual: each organism has a male and a female parent – a single line of female ancestry and of male ancestry, though overall ancestry is highly complex. Combining the logic of linguistic and genetic descent should amplify our information on details of migration patterns. For instance, language evidence gives the best idea on where speaking communities emerged and on their principal movements; genetic evidence gives the best ideas on the ancestry of the Founders and on the cross-breeding of subsequent communities.

The Founders in Their Homeland

It was in the dry eastern half of Northeast Africa that bipedal hominids first arose and where they first learned to use stone tools. Here too was the main habitat for *Homo sapiens*, as the species emerged between 300,000 and 200,000 years ago. Within this species in the same region, 70,000 years ago, arose a community that began to speak an articulated language.[4] The initial population of speaking humans was small. Geneticists have argued that the ancestors of today's human population experienced a genetic bottleneck – an extreme narrowing of the size of the breeding population – some 70,000 years ago. What interpretation are we to give this evidence? Stanley Ambrose speculated that the bottleneck was the result of an ecological crisis after the gigantic Toba volcanic explosion (in Sumatra, 75,000 years ago) – arguing that the atmospheric dust caused a cold snap and famine that reduced the *Homo sapiens* population to a very low level. Others have expressed doubt about whether such a cold snap could have reduced population of all *Homo sapiens* – or other species – to such a degree.[5] As an alternative, rather than think of the apparent genetic bottleneck as indicating the decimation of the whole hominin population, Christopher Ehret has suggested instead that this narrowing reflects self-selection by the initial speaking population.[6] Thus, while Ambrose proposed that the bottleneck came first for natural reasons, and that human innovation came later, Ehret suggests that the innovation of speech came first, and that the bottleneck reflects the self-selection by those who could speak – a population that subsequently expanded dramatically. The latter is the hypothesis I adopt in tracing the emergence of social evolution.

The early Northeast African communities of speaking humans lived in a region of ecological variety: they had the option of living out their lives in highland or lowland habitats and were able to encounter sharply differing plants and animals and differing access to water and firewood.[7] These communities grew, once they formed, through two social dynamics. First, small family groups steadily joined each other to form communities of those sharing a language, thus reorganizing their existing population into multifamily communities. Second, the advances in adaptation of these communities – better information on available resources and, perhaps, better tools – enabled them to survive at a greater rate and to grow in number.

The existence of speech at the community level provided a platform for new types of representation, such as ideas about the known and the unknown – life, the afterlife, the skies, the animals, and the possible existence of spirits. This was, in effect, the institution of *religion*, which now formed as a body of thought and practice. It was an institution in that there was a group of people who shared discussions, developed interpretations of these big issues, and gave names to the concepts they created and shared. Flannery and Marcus extend this logic, contrasting human and primate social orders: they argue that the dominant alpha male in the primate hierarchy is replaced, among humans, by supernatural figures that form part of "a natural/supernatural dominance hierarchy."[8] Syntactic language permitted articulation of the world in a way that had not previously been possible. One may therefore ask whether humans, as they began to speak, immediately began to consider themselves as a species apart, with a special relationship to the supernatural. The institution of religion relied on groups of people at two levels: religious believers as the participants in religious practice; and creators of religious thought as the core group that formulated its ideas. The dynamics of religion included developing the concepts, questioning them, assessing their implications, and articulating relationships of the religious beliefs and social practice.

To trace the expansion of the Founders and their offspring across Africa and beyond, I give particular attention to evidence on language distribution, as it is reinforced by evidence on genomics, archaeology, and climate. For this and subsequent summaries of language data on migration, I provide information at three levels.[9] First, in the text, I summarize the results of the analysis, as I will do in subsequent paragraphs for the case of the Founders and further on for later migrants. Second, in an associated online resource I provide maps for each of 14 *language phyla* or major language groups worldwide, showing their extent and the pathways of their principal migrations, along with supplementary information. I define a *language phylum* (pl. *phyla*) as an ancestral language family that gives strong evidence of having existed for at least 15,000 years, with families (or subfamilies) within it that formed later. (The language-mapping site is within the Resources tab at www.cambridge.org/Humanity.)[10] Third, in this book's concluding appendix on Frameworks for Analysis section on migration, I include additional methodological discussion of historical analysis of languages.[11]

My analysis of language and human migration follows on the tradition of Joseph H. Greenberg, who classified the languages of Africa and went on to lead in study of language classification worldwide, analysis of language universals, and exploration of the thesis that a single, founding language could be documented.[12] Greenberg's approach shows how language data – backed up by paleontology and genomics – confirm that the Founders population of speaking humans arose in Northeast Africa.

While historical linguists working on some regions – the Americas, Melanesia, and northeastern Asia – have sought to refute Greenberg's large-scale classification of languages, I will show in this book how language classifications in the tradition of Greenberg's analysis fit closely with genetic data on human occupation of Earth. The consistency of linguistic and genetic evidence on migration tends to confirm both that the Greenberg classifications are largely valid and, especially, that the distribution of languages known today provides clear information on the location and migration of languages during settlement of Africa, Eurasia, and the Americas.[13]

As the earliest languages diverged, so also did communities. We do not have direct evidence on the original language, "proto-human," except for a few basic words that have been reconstructed. Yet of the four language phyla now known in Africa, all of them have their homeland in Northeast Africa, suggesting that they are relatively closely related to the founding or "proto-human" language. Three of these language homelands were in the middle Nile valley; the fourth group, the Khoesan languages, was based in northern Tanzania. Location of these language homelands is shown in Map 4.1. The initial rituals that held together the founding community, unusually large in comparison to nonspeaking families or bands, had to change with the early divisions among communities, now accounting for a different sort of specificity. Out of these experiences of representing specific communities came the emergence of cultural practices that were new because of their scale and because they were articulated verbally. In the process, humans may have found ways to incorporate the music and dance that had existed since earlier times. At each step, new vocabulary terms were developed and spread through the communities. Naming plants, animals, and places and their characteristics was but one dimension of the expanding verbal

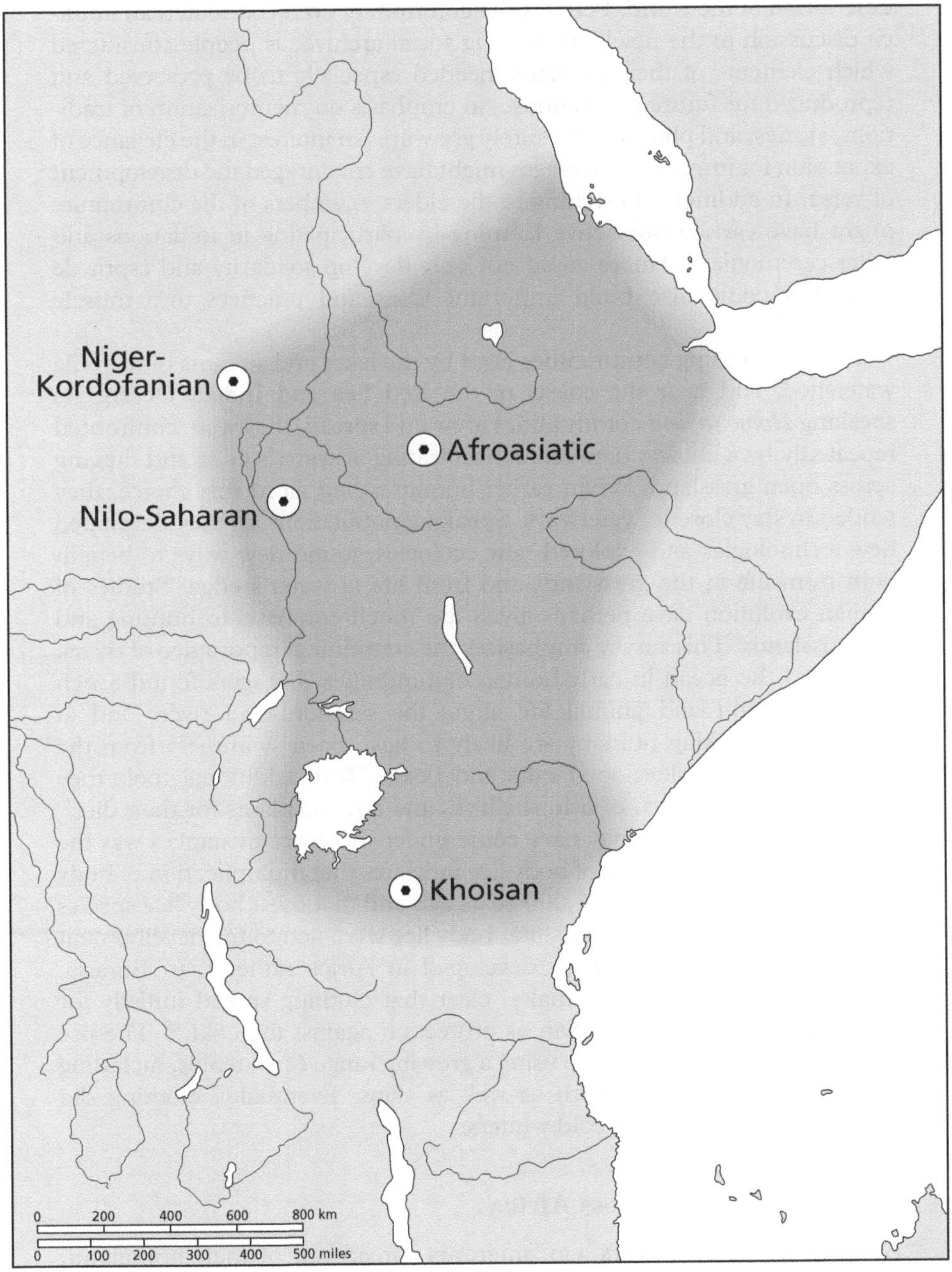

Map 4.1 African homeland and language families. (Redrawn based on original image, courtesy of the University Library System, University of Pittsburgh; generated by Daniel Andrus, 2019.)

description of the world. Perhaps the community even conducted an implicit discussion of the newly developing social archive, as people considered which elements of their existence needed especially to be preserved and reproduced for future generations. An emphasis on memorization of traditions, stories, and philosophies surely grew up. An interest in the elegance of expression for important messages might have encouraged the development of verse. In addition to listening to the elders, members of the community might have carried out active learning by participating in initiations and other ceremonies. Dances could not only develop solidarity and esprit de corps but could also build important ideas and practices into muscle memory.[14]

These expanding communities lived by the lakes and streams of the Nile watershed, and near the coasts of the Red Sea and Indian Ocean. As speaking *Homo sapiens* communities grew and spread, they were confronted repeatedly by a choice: between concentrating at water's edge and ranging across open grassland. When earlier hominins had faced this choice, they tended to stay close to waterways. Speaking populations, as they developed new technologies and explored new ecologies, found new ways to benefit both from life in the grasslands and from life at water's edge. Studies of human evolution have perhaps given too much emphasis to hunting and the grasslands. This survey emphasizes the continuing importance of rivers, lakes, and the ocean in early human communities. Foragers found a rich variety of plant and animal life along the seashore and rivers and at lakeside.[15] Speaking humans are likely to have been swimmers from the first, and to have developed rafts and boats. With additional tools they advanced their retrieval of fish, shellfish, and aquatic plants for their diet.

One innovation that may have come under these circumstances was the creation of clothing. Study of body lice indicates that multiplication of body lice species began roughly 70,000 years ago and that most body lice species remain in Africa. This suggests that body lice were tied to the development of clothing, and that clothing developed in Africa rather than Eurasia. Further, this African origin makes clear that clothing spread initially for decorative purposes rather than as protection against the cold.[16] The use of clothing opened the door to using a growing range of materials, including leaves, grasses, and barkcloth as well as skins. Eventually, clothing did provide protection during cold winters.

Expansion across Africa

As of about 65,000 years ago, migrants spread out of their homeland, headed in all directions. The best summary of evidence on the expansion of Founder communities across Africa, by Christopher Ehret in the

Cambridge World History, gives a skillful account of archaeological and linguistic evidence.[17] According to Ehret, migrants from Northeast Africa moved first to South Africa. There they encountered preexisting populations; for a considerable time, the two populations coexisted. Next, migrants moved west from Northeast Africa into the forest region of West Central Africa, where they encountered hominins whose material culture is known as the Lupemban culture; the pygmy peoples of Central Africa may be descended in part from them. For West Africa, migrants from Northeast Africa entered the region early on but were not able to gain a foothold, as the preexisting culture and population persisted with little change. In all three regions, the immigrant populations ultimately gained predominance, though elements of the preexisting culture – and presumably of their contribution to the gene pool – persisted.[18] By 22,000 years ago, the "modern" population occupied virtually all of sub-Saharan Africa.

Temperature and humidity declined during this era of 40,000 years, with a high point at 60,000 years ago and low points at 40,000 and 25,000 years ago – the Sahara became steadily dryer, forested regions contracted, and grasslands dominated. For the path of Pleistocene migration, we can rely on the evidence of linguistic distribution, as reinforced by genetic evidence. The persistence in northern Tanzania of Khoesan speakers of sharply different languages reaffirms that this was the homeland from which speaking humans moved into southern Africa; genetic evidence tends to confirm that view. (On Khoesan languages and migration, see www.cambridge.org/Humanity, resource text file 1.Khoesan.)[19] Speakers of languages related to Afroasiatic expanded north to the Red Sea hills, east to the Indian Ocean coast, and south to the headwaters of the Nile. (On Afroasiatic languages and migrations, see www.cambridge.org/Humanity, resource text file 2.Afroasiatic.)[20] Speakers of languages that became known as Nilo-Saharan expanded southwest in the savanna and, later, northwest into the Sahara. (On Nilo-Saharan languages and migrations, see www.cambridge.org/Humanity, resource text file 3.Nilo-Saharan.) The homeland of languages ancestral to the Niger–Kordofanian languages, in the Kordofan Hills near the middle Niger, still has populations of the ancestral group. (On Niger–Kordofanian languages and migrations, see www.cambridge.org/Humanity, resource text file 4.Niger-Kordofanian.)[21] It appears that members of this group made early and unsuccessful efforts to migrate to the west. In the late Pleistocene, however, a group moved far west to the upper Niger Valley, developing what are now the Mande languages, followed by related groups. Thereafter, migrants speaking Afroasiatic languages moved west along the savanna to the basin of Lake Chad and developed dense populations in that region. Some Nilo-Saharan speakers moved west to the middle Niger Valley while some Afroasiatic-speakers settled and expanded in the Lake Chad region. The overall pattern seems to be creation of a base area, followed by Pleistocene-era migrations in multiple directions.[22] This pattern is repeated in the language groups of regions outside Africa. For the Holocene epoch, as we will see, further sets of migrations changed these patterns significantly.

Social, technological, and conceptual changes took place during the course of these migrations. In social structure, the institution of *marriage* emerged during this era. While the mating of men and women, usually in monogamous unions, was nothing new, the formalization of relations linking the couple and their families was definitely something new. As families and rules of exogamy became formally defined and bounded through language, the marriage of a couple could formalize responsibility of the two families to each other.[23] The institution included the couple but also the families linked through the couple; more broadly, the institution of marriage included all those who participated in the marital order. The dynamics of marriage came to include age at marriage, childbirth, and child care; duties owed to family members; and relations among senior and junior persons linked through marriage. As marriage became formalized in speaking communities, the bilateral families or lineages surrounding and overlapping marriage also became formalized.

At the same time, there was a persistent though minority tendency for mating and perhaps marriage across community lines.[24] Mating across community lines has had two important biological results. First, it has increased the variety within the gene pool of each localized community. Second, because unique genes are spread widely, it means that, despite the genetic variety in each community, the average genetic composition of human communities remains similar within large regions and even across the continents. In this sense, genetic differences *within* human communities are greater than genetic differences *between* communities. In technological change, we now have evidence that the technology of bow and arrow was developed in southern Africa roughly 64,000 years ago: this technology gradually spread and underwent further development worldwide.[25] One would also like to know more about the migrants' skills with watercraft for venturing on lakes, rivers, and along the Indian Ocean coast. These could have been rafts, but more likely would have been reed craft.[26]

What was the interaction of speaking *Homo sapiens* with people who did not speak? Was it possible to teach them to speak, perhaps only at a very basic level? If speech was not possible, was it possible for individuals or groups from nonspeaking populations to join and participate in the communities of speaking people? This issue was biggest in Africa, as the largest nonspeaking hominin populations were in Africa. Within speaking communities, there surely arose efforts to develop excellence in rhetoric, in which the speakers spoke not simply to command but to entertain and to explore. Stories of past experience surely developed expressions of variations in human behavior and understandings of human nature.

Networks formed as communities expanded, growing in both explicit and implicit forms. The explicit institution of a we-group *network* became an institution in that its members were linked by agreement, for instance as a cross-community network through marriage or for individuals who shared names and gift-exchange relationships.[27] More commonly, networks were implicit groups (I-group networks) of participants who served as nodes in a network of interaction but without an explicit agreement. Networks also took the form of we-groups in institutional structures, as in the direction of rituals. Equally importantly, the question arose on how to link speaking communities to each other. Here again, both sorts of linkages arose. At the level of a community, cross-community migration was an institution, as each community organized its handling of migrants who departed or entered the community. But at the level of several communities that were exchanging migrants, they had formed an implicit group, in which the communities cooperated without a formal structure. In a more elaborate level of networking, selected families from several communities might agree to set up a we-group of collaborating families in exchanging candidates for marriage. Communities naturally differentiated themselves from each other – certainly in language, but perhaps also in dress, in cuisine, and perhaps in the details of social organization. With these differences, the benefits of a network that formally linked communities became more evident. The steady exchange of cross-community migrants spread innovations among communities, though not rapidly enough to prevent distinctive community rituals and behavior from expanding. The network of implicit and explicit communities – differentiating yet maintaining contact for matters of marriage, migration, exchange, healing, and the understanding of sacred power – was the initial extension of the Human System to the multicommunity level. To the degree that the institutions of human communities, in this long era of expansion, sustained an overall zeitgeist encompassing their priorities, it might have emphasized settlement – the establishment of new communities and strengthening of ties to the habitat.

Tropical Asian Migrations

As migrants steadily occupied the various regions of Africa, some crossed over to Asia. Relying on water's-edge technology – necessarily including the use of boats – they crossed the Bab al-Mandab (less than 20 km wide at the time) and expanded eastward. This crossing into Asia took place as early as early as 65,000 years ago or as late as 50,000 years ago. In no more than a few thousand years some migrants had reached Australia and New Guinea.

New research in genomic analysis argues firmly that there was a single migratory movement of humans from Africa to Asia, rather than multiple migrations into Asia.[28] This means that all settlers in Europe, northern Asia, tropical Asia, and Oceania owed their existence to that single group of initial migrants, and to their later differentiation. This important finding shapes and limits the potential paths of the great eastward migration. It also adds new value to language distribution as evidence for the path of migration, as the location of language-phylum homelands fits closely with genetic evidence on migration. That is, the relative consistency of genetic and linguistic data confirms that language homelands have been persistent in their location, and that genetic and linguistic data can be used to check each other. Relying on the interplay of language and genome, we can now ask, with greater clarity: which African groups were the source of migrants to Asia? The candidate groups are Nilo-Saharan, Afroasiatic, and Niger–Kordofanian languages, and future research may enable us to select among them.[29] Further, while the migrants spread through tropical Asia from a single initial stream, they appear to have met and interbred with Denisovans along the way: my impression is that this process took place in India, and that all those who moved further east carried their Denisovan DNA with them. This model is made more likely by the growing evidence that there were two centers of Denisovan population, one in Central Asia and one in regions to the south.[30]

Climate fluctuations governed the path and timing of human occupation of the Asian and Oceanic tropics. The fluctuations were regionally specific, but the overall pattern was set by levels of insolation – the varying force of the sun's radiation on Earth (see Figure 4.1). Northern Hemisphere insolation hit a low point at about 74,000 years ago, bringing cool temperatures, low humidity, and low sea levels. By 60,000 years ago, insolation had reached a relative peak, bringing warm and humid times and higher sea levels. The next and more modest trough in insolation was 48,000 years ago, followed by a still more modest peak 37,000 years ago. Overall, times of warmth and humidity (as shown in the shaded portions of Figure 4.1, above the threshold level of insolation) were the easiest times for migrants to move to new lands. In times of lowest insolation (74,000 and 48,000 years ago and the Last Glacial Maximum of 22,000 years ago), arid regions became relatively impassable. The initial migrants into Asia must have passed through Arabia, but debate continues as to

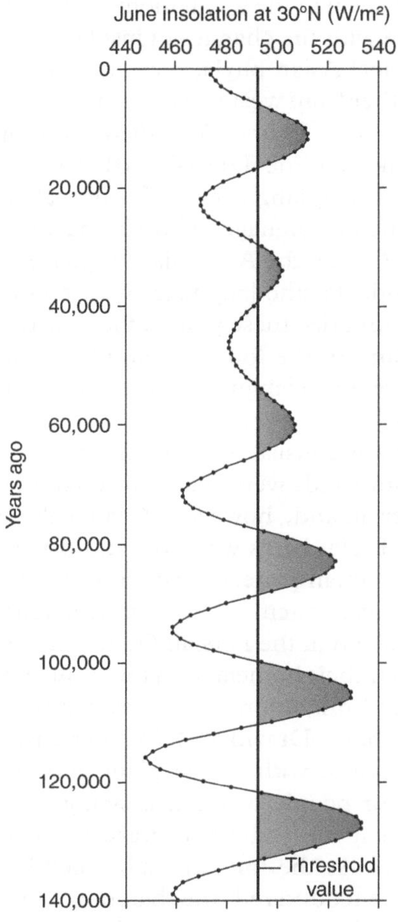

Figure 4.1 Fluctuations in insolation, mid-northern latitude. (Courtesy of Aubrey Hillman.)

whether their Arabian settlements were along the coast or in the interior: at 60,000 years ago, the interior would have been accessible. In general, I prefer to emphasize human reliance on seafood and the ecological similarity of the Indian Ocean coast all the way from East Africa to South China and Oceania. The Himalayas, however, always had some well-watered regions, as was the case for the mountains of Ethiopia, so mountainous regions always attracted settlers. The great land masses of Africa, Central Asia, and China heated and cooled with the seasons, forcing monsoon winds to blow north and south, which brought precipitation from the Indian Ocean in the warm months. The strength of the monsoons varied over time with the changes in insolation.[31]

The homelands of five tropical Asian phyla, created by early settlers, were established one after another from west to east, relatively close to the coast. In order, these were the Elamo-Dravidian languages with a homeland in what is now Pakistan; the Trans-Himalayan (also known as Sino-Tibetan or Dene-Caucasian) languages with a homeland in what is now southwest China; the Austric languages with a homeland at today's border of Vietnam, Laos, and China; the Australian languages, centered in north Australia; and the Indo-Pacific languages with a homeland in what is now Indonesia. To summarize these points, the dots on Map 4.2 indicate the estimated homeland of the four African phyla and the five tropical Asian phyla. All of these Asian language homelands are in regions with rich natural endowments.[32]

The habitat of tropical Asia was consistent along the coastline, as flora and fauna of the ocean and shorelands were shared through the rhythms of the Indian Ocean. Further inland, however, Asian habitats varied dramatically. From Arabia to India, lands were low and dry, with highlands only in Yemen and the Iranian plateau, and with only the Tigris–Euphrates and Indus as major river systems. The Elamite languages (now extinct) established their homeland in the Persian Gulf, while the related Dravidian languages established their homeland in the Indus Valley. (On Elamo-Dravidian languages and migrations, see www.cambridge.org/Humanity, in resource text 5.Elamo-Dravidian.) Moving eastward, the coast of India was a fertile region, with ample plant and animal life. Along this coast it was easy for people to build growing communities and to expand inland; the Dravidian languages spread through most of India in the Pleistocene epoch. As the migrants advanced beyond the mouth of the Ganges, they encountered the bamboo forests along the coastal plains of Southeast Asia, watered by major rivers.[33] Inland (to the north), the valleys became steep though still fertile, as several great rivers flowed south and east from the snowy Himalayas: the Brahmaputra, Salween, and the Irrawaddy to the west; the Mekong and the Red to the east.[34] The Trans-Himalayan languages developed a homeland in the mountainous valleys of western Southeast Asia and spread widely from that base. (On Trans-Himalayan languages and migrations, see www.cambridge.org/Humanity, in resource text 8.Trans-Himalayan.)[35] The Austric languages arose nearby in the valleys of the Mekong and Red Rivers; the four main subgroups of Austric overlapped geographically with each other and with certain Trans-Himalayan languages. (On Austric languages and migrations, see www.cambridge.org/Humanity, in resource text 9.Austric.)[36]

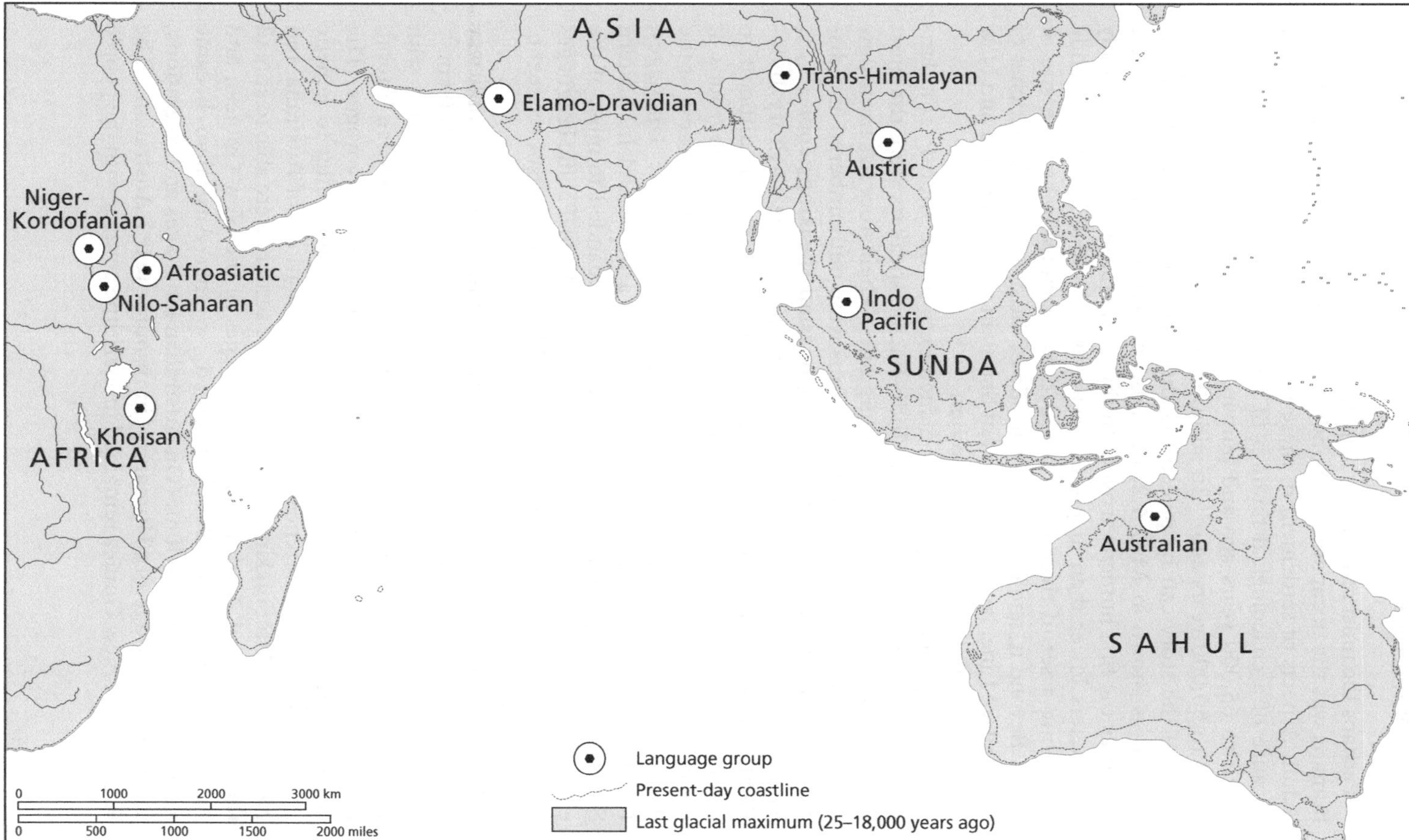

Map 4.2 Homelands of tropical language phyla. (Redrawn based on original image, courtesy of the University Library System, University of Pittsburgh; generated by Daniel Andrus, 2019.)

Perhaps the most remarkable step in the tropical expansion of modern humans was the occupation of Sunda and the movement from Sunda to Sahul. Aspects of settlement in these regions are well known, especially because of archaeological findings of human remains in Australia now dated to 65,000 years ago and human remains in Papua New Guinea dated to 50,000 years ago.[37] In addition to these high points, however, more of the story can be pieced together. The geographical context is central: as shown in Map 4.2, the subcontinent of Sunda had a huge extent as the first humans arrived. In this cool era, sea level was roughly 100 meters lower than at present, so that the subcontinent of Sunda (including a set of rivers) was revealed as a peninsula half the size of India. Beyond it, and on the eastern side of the deep oceanic trenches of the Wallace Line, lay the continent of Sahul, in which Australia and New Guinea were then linked to each other.

The story focuses especially on members of the Indo-Pacific language family: it appears that the Indo-Pacific speakers were the most energetic mariners of early human migrants. Their story has been slow to reveal itself, because their populations were outnumbered and displaced by later migrants. That is, Austronesian-speaking peoples – the Malays of today – become dominant in most of the territory Indo-Pacific speakers had long since settled. As a result, it is difficult to estimate the homeland, the distribution, and the initial subgroups of the original Indo-Pacific–speaking settlers. Because inhabitants of the Andaman Islands, west of Sunda, speak Indo-Pacific languages, I assume that Indo-Pacific speakers formed a homeland in western Sunda, as they entered the region, and then expanded to most of the extensive coastland and interior of Sunda. (On Indo-Pacific languages and migrations, see www.cambridge.org/Humanity, in resource text 6. Indo-Pacific.)[38]

Further, the Indo-Pacific speakers settled Sahul as well. The only way to get to Sahul was to cross stretches of open ocean of at up to 100 km.[39] We can be sure that human migrants had developed their skill with watercraft from the Bab al-Mandab all along the Indian Ocean littoral, but the crossing to Sahul was an extraordinary set of voyages. The crossing was made not once but several times, according to genetic evidence showing differences within the populations of Australia and New Guinea. After making this crossing, the settlers were able to spread throughout Sahul. With time, the languages of the New Guinean and Australian portions of Sahul diverged, so that Australian became a separate language phylum. (On Australian languages and migrations, see www.cambridge.org/Humanity, in resource text 7.Australian.) The languages of New Guinea remained in maritime contact with the people of Sunda, so that the Indo-Pacific phylum encompassed both regions – the greatest extent of early languages.

The continuing maritime energies of Indo-Pacific speakers are revealed in their migration east from New Guinea to the Solomon Islands, north into the islands of Micronesia, and south, as they appear to have voyaged along the east coast of Australia to Tasmania, where remnants of their language were recorded in the nineteenth century. We do not know much about the watercraft of Indo-Pacific speakers. Watercraft made of bundled reeds are known from Tasmania; dugout canoes are another possibility, as they are used throughout the South Pacific today.[40]

Up to this point, my analysis of migration into Asia has focused on documenting and interpreting the migratory steps and the material life of eastward-expanding humans. Yet there are reasons why we should attempt to reconstruct not only migration and material life, but also the conceptualizations of the migrants. To pursue our focus on language, we should note that migration and the passage of time necessarily brought the coining of many new terms to describe the lands and the experiences of people as they spread steadily further from the Northeast African homeland.

While reconstructing most early words is beyond our current competence, identifying works of visual art is turning out to be quite feasible. Visual art is now being discovered throughout the Old World tropics, some of it created more than 40,000 years ago, thus confirming the universal human impulse toward visual representation. The cave paintings dated to 35,000 years ago in temperate Chauvet-Pont-d'Arc Cave, in France, are now balanced with the recent discovery of figurative paintings created at least 40,000 years ago in the cave of Lubang Jerji Saléh, in the highlands of tropical Kalimantan, then part of Sunda.[41] These discoveries confirm that Indo-Pacific speakers occupied interior as well as coastal regions from the beginning.

An encounter with these works gives rise to the hypothesis that *workshops* served as institutions, small but valuable, for facilitating the creative process. Workshops included small numbers of people who joined together for a common purpose, though individuals performed different roles in achieving the purpose. Typically, a workshop had its leader, its creative genius, while others provided support and learned more about the task. Workshops were valuable in conceptual and visual representation, in foraging, and in productive work. In practical and material terms, the preparation of food, containers, physical decorations, and ceremony may all have advanced according to the logic of the workshop, in which a leading figure conceptualized the task to be performed; designed the product (conceptual, representational, or material); and gathered assistants to collect materials for the task, perform subordinate tasks, and, for those who showed skill, become apprentices who gradually learn to contribute to completion of the main task. Still, workshops commonly lasted only as long as the completion of a single task or the working life of the founder. Other forms of expressive culture – dance, music, dress, poetry, oratory, history, epics, jewelry, hair dressing, the painting and scarring of bodies – may have benefited from workshops rather than resulting simply from individual creativity.

Migration to Temperate Lands, 45,000–25,000 Years Ago

Northward migration – from the tropical latitudes into the grasslands and forests of the temperate zone – was no easy matter for communities of *Homo sapiens*. There appears to have been a delay of as much as 10,000 years and perhaps more between the arrival of humans in Australia and the first settlement of *Homo sapiens* in temperate Eurasia.[42] From 45,000 years ago, however, numerous remains of *Homo sapiens* are known for central Eurasia, southeastern Europe, Anatolia, and northeastern Asia, suggesting a settlement that was rapid once it got started. Yet launching this migration required the satisfaction of at least three criteria: a densely populated tropical region from which migrants could make repeated advances into new territory, a route northward that was free of geographical obstacles and furnished with adequate food and resources for the transit, and a technology and social order that would be productive in the areas of settlement. Temperate lands were quite different from the tropical habitat that humans knew. Even the most favorable environments in the temperate zone were very different from the tropics in temperature, seasonality, fauna, and flora.[43]

One must begin with the geographical and climatological obstacles. By 45,000 years ago or more, humans had established themselves across the tropics from the Atlantic to the Pacific, and their long record of migration would appear to have made them ready to move to the north. Yet the moment was hardly auspicious. This was a time of especially cool and dry climate (see Figure 4.1). A great belt of desert stretched from Africa's Atlantic shore, across the Sahara, the Arabian Peninsula, and Central Asia to the Gobi Desert with only a few interruptions. While the lands of Iran, the Levant, Arabia, and the Sahara each experienced humid times, these were limited to periods of a few thousand years' duration, centering on 130,000, 105,000, and 10,000 years ago.[44] And for India and Southeast Asia, the Himalayas blocked movement northward, except for the plains of China and the western Pacific coast.

Of the potential routes from the topics to the temperate zone, only two regions gave much promise. The first of these was a path from the Indian Ocean to Central Asia, along a northward route chosen from possibilities running east of the Caspian Sea and west of the Himalayas. The second was a path from Southeast Asia to northeastern Asia, along the lowlands of eastern China or the Pacific coastline. Other possible routes must be ranked lower for geographical and climatological reasons. For instance, it has been tempting for researchers to assume that *Homo sapiens* moved directly and without difficulty from Africa to Europe via the Nile Valley and the Mediterranean shore of the Levant and Anatolia: many arrows trace this route on published maps of human expansion. But the Sahara, Arabia, and Levant were exceedingly dry after 55,000 years ago, with only a modest and late increase in temperature and humidity at 35,000 years ago – so that recent advances in the genomic and climatic records give little evidence of migration along this route until much later times. The Atlantic coastal route from Africa to Europe was similarly desiccated.

The most likely route north, in my opinion, was the Khyber Pass route, leading from the Indus Valley along the western flank of the Himalayas – hilly but dependably well-watered – and continuing to the west of the Altai Mountains up to the grasslands of the Central Asian steppe. The migrants would encounter herds of large mammals for hunting, along with nourishing fruits in certain valleys. From this beachhead, migrants were able to expand to the west and the east, rapidly making their imprint in temperate Eurasia. If this was the route, migrants needed to be prepared to follow it. Fortuitously, a technological advance may have helped. The recent discovery of an atlatl or spear-thrower at a site in the Levant, with a date of roughly 46,000 years ago, suggests that people from this region might have been prepared to enter the temperate lands and hunt the large animals there.[45]

From the standpoint of mythology, the work of folklorist Michael Witzel raises the possibility that a shift in ideology may have accompanied the migrants who reached the temperate zone. Witzel, after tabulating and mapping the mythologies of people throughout the world, concluded that they fall into three main categories: the African-based "Pan-Gaean" mythology, ancestral to the others; a continuing "Gondwana" mythology among the peoples of the Old World tropics that tells the tale of a recurring cycle of life; and a continuing "Laurasian" mythology for northerly regions of the world that assumes the creation of the Earth by supernatural beings, its transformation and eventual end and collapse.[46] That is, one might hypothesize that Laurasian mythology was a new view of the world, invented among those who moved, some 45,000 years ago, northward from the Indian Ocean littoral and into Central Asia.[47]

One more question arises as we consider the departure of the migrants who set off for temperate lands: which tropical language group undertook whole-community migration into the temperate zone? The choices are arguably limited to the three groups closest to the Khyber Pass: Elamite from the Persian Gulf, Dravidians from the Indus Valley, and Trans-Himalayan speakers from southeast Asia. The latter group, though the most distant, is in my opinion the best candidate because its members had the option of migrating dependably along the well-watered Himalayan foothills, reaching Khyber Pass and continuing by the same process to the steppes. This approach is supported by the more recent migration of Trans-Himalayan groups along the same foothills in Holocene times, which included their occupation of the Tibetan plateau. Dravidian speakers might equally well have formed the migrant parties. Elamite migrants are less likely to have migrated north because the lands of present-day Iran and Iraq were then very dry.

Once migrants had made the transit to the steppe and learned to live with the climate, flora, and fauna of Central Eurasia, they were able to form areas of settlement and language homelands. The new arrivals moved both west and east along the grasslands: westward to the Caucasus and eastward to what are now Kirghizstan, Tajikistan, and northeast to the slopes of the Altai Mountains. These *Homo sapiens* from the south, as I have argued, had already interbred with Denisovans in the tropics. As they entered the temperate zone, they encountered and interbred with Neanderthals, and perhaps also with northerly Denisovans.[48] Genomic studies show greater Neanderthal ancestry among humans in northeastern Asia than in Central Asia or in Europe. There is not yet evidence that tells us whether the interbreeding was peaceable or violent, though we may assume that it was some of both. This documentation of interbreeding confirms the persistent if minority practice of exogamy in human social relations.

Two groups of settlers found the Caucasus to be a strategically located, productive, mountainous region, and their descendants remain in the region. The North Caucasian languages are to the north (on North Caucasian and Basque languages and migrations, see www.cambridge.org/Humanity, in resource text 10.North Caucasian). The Kartvelian languages are spoken in the central Caucasus (on Kartvelian languages and migrations, see www.cambridge.org/Humanity, in resource text 11.Kartvelian).[49] North Caucasian languages are reported to be related to Trans-Caucasian languages; Kartvelian languages might therefore be descended from Elamo-Dravidian languages. Archaeological sites suggest that, roughly 45,000 years ago, migrants moved along the northern and western shores of the Black Sea, then divided into one group moving up the Danube Valley into Central Europe and another group settling the Mediterranean littoral all the way to Iberia. The Basque languages of Iberia, likely related to North Caucasian languages, provide another indication of this initial *Homo sapiens* settlement of Europe.[50] (Many thousands of years later, in the early Holocene, agricultural populations speaking Indo-European languages moved into the same regions and their languages became dominant.)

Populations that moved eastward from Central Asia moved along the steppes until they reached northeastern Asia. There two great river valleys provided a region rich in resources – with forests in the Amur Valley to the north and grassland in the Yellow River basin to the south. The ancient DNA retrieved from remains 40,000 years old at Tianyuan Cave, near Beijing, reveals a *Homo sapiens* individual with traces of Neanderthal ancestry.[51] This tends to confirm that the population including this person had passed through Central Asia, where Neanderthals lived in sparse populations. Settlers in this region – most likely in the Amur Valley – developed the Eurasiatic phylum, which was later to expand substantially. (On Eurasiatic languages and migrations, see www.cambridge.org/Humanity, in resource text 12.Eurasiatic.)

Two further migrations into northeastern Asia are indicated by linguistic and genomic data. Yeniseian languages, spoken in the Yenisei Valley in central Siberia, appear to result from a later migration of Trans-Himalayan languages.[52] The remaining migration to northeastern Asia was of migrants who moved northward along the western Pacific coast. Genetic analysis of northeastern Asian populations reveals a certain portion of ancestry that is closely related to the genomes of people living in Papua New Guinea, who speak Indo-Pacific languages.[53] This raises the hypotheses that the experienced mariners among Indo-Pacific-speakers worked their way northward along the coast and the coastal archipelagos from Sunda to Taiwan, the Ryukyus, Japan, and areas to its north.[54] Whether their languages survived in this new territory, it is likely that their skills in fishing and watercraft became important additions to the regional economy. The specific new technology one may hypothesize is the invention of skin boats – light boats, made of wooden frames with hides sewn and stretched across them, floating high in the water and keeping the passengers dry. These boats were effective in the cold northern oceans and were equally effective in the cold and rapidly running northern rivers.[55]

In this same northeastern Asian region, humans and dogs encountered each other at roughly 30,000 years ago.[56] The two species formed a bond: dogs apparently joined human communities readily. This association gave humans what may have been their first experience with breeding.

With time, dogs spread through human communities on all the continents, thus revealing the I-group social networks of continuous contact among human populations. The pace of diffusion of dogs among humans is not clear, but the fact of diffusion reinforces the notion that cross-community migration linked human populations everywhere, maintaining similarities in the genome as well as spreading sociocultural innovations.

Humans, as they grew in numbers, began to exert pressures on the ecology within which they lived. An outstanding example of how a small human population can modify an environment is that of Australia, where the settlers (at an undetermined time) took up the practice of burning much of the land to assist them in hunting. Ultimately the whole landscape was modified, and the species of plants and animals that thrived were those that could recover from periodic fires. In contrast, migrating humans were themselves modified by their ecology. After numerous generations in very cold regions, heads and bodies became rounder. Skins became lighter in color at high latitudes, as the reduced pigment allowed more vitamin D to be produced through sunlight. (In the tropics, dark skin minimizes the effects of ultraviolet radiation, which destroys folic acid.) In general, despite the tight genetic unity of all the humans who migrated into new areas, it was their external physical features – their "phenotype" – that changed the most, in response to the varying temperate regions in which they lived. Phenotypes of eastern and western Eurasia differentiated further, in response to local conditions. For all of those who remained in the tropics, local physical differences also developed, although they were much smaller in scale because the ecological differences were smaller. Later on – mostly in the Holocene – migrants from the temperate zone expanded into tropical and subtropical regions of the Old World, and the mixing of these distinctive phenotypes accelerated.

The clearest record of material culture from the years of human expansion is in the stone tools that have been retrieved from sites of residence and work: choppers, arrowheads, fishing weights and hooks, grinding stones, and more. Less well documented are the numerous tools of bone and wood, the vines, ropes, and other perishable tools. The bows and arrows, ultimately from Africa, the atlatl or spear-thrower, ultimately from Eurasia – each spread to every continent, and they were modified and improved endlessly. Containers provide a good example of mundane but essential developments in human life. Containers made of wood and leaves were gradually supplemented by mats, baskets, and animal skins. Techniques of sewing – documented for Central Asia and relying on awls and needles to make holes for the twine – arose to link skins and mats, while the weaving of baskets and mats eventually led to textiles.[57] We are left to imagine about the innovations in cooking, but one may be sure that the taste buds of our ancestors were well developed and that there was a priority, at least in good times, on the creation of good-tasting food.

5 Production and Confederation

In this chapter, I emphasize the importance to human history of the period from 25,000 to 12,000 years ago. Too often, interpretations of the big changes in human history draw a heavy line, somewhere from 7000 to 10,000 years ago, and label it as the Agricultural Revolution. Such an interpretive boundary makes the claim that history began at that time, with a great change in productive systems that provoked the rise of chiefdoms, towns, social hierarchy, cities, and civilization itself. The terms Agricultural Revolution and Urban Revolution, popularized by the distinguished archaeologist V. Gordon Childe in his 1942 book *What Happened in History*, have been taken to stand as the productive launching pad for all the complexities of history.[1] Childe himself actually wrote with considerable nuance and detail, yet the symbolic transformations became fixed in many minds. Newly available data now require that we take a different approach. As a result, I focus in this chapter on arguing that, during the 10,000 years before agriculture took its full form, major changes unfolded in material and social life. From 25,000 years ago, the effects of severe cold followed by rapid warming accelerated the inherent path of human innovation. Humans met the climatic challenge of the Last Glacial Maximum by developing an economy of production to supplement their practice of foraging, and by grouping their communities into larger confederations in order to manage their expanding economic complexity.

The 40,000-year process of settling the Eastern Hemisphere took place during a time of gradual cooling in climate. From the peak in warmth, 120,000 years ago, temperatures slowly declined, humidity gradually decreased, and sea level declined. Individual humans would have observed only the short-term fluctuations in climate – though such changes were commonly severe – and not the long-term change. But the cooling gradually restricted the range of habitat for plants and animals, so that competition for resources among all species became more serious.

Crisis: The Last Glacial Maximum

At the end of this long cooling phase came a sudden, deep freeze: the Last Glacial Maximum hit its lowest temperature 21,500 years ago. Temperatures fell enough to cause a great expansion in glaciers in the Northern and Southern Hemispheres, so that almost no life forms existed in the glaciated areas. So much of the world's water was in the form of ice that sea level fell to 100 meters below its previous peak. Great amounts of land were exposed: Britain was united with Europe, Japan was united with Asia, and Alaska was united with Siberia. The subcontinent of Sunda and the continent of Sahul had long been revealed; their dimensions expanded as sea level fell further.

Most humans lived in the tropics of Africa and Asia, thus avoiding the coldest temperatures. Nevertheless, they too felt declines in temperature and especially in humidity, the shifts in annual weather patterns, the change in habitat for themselves and for the plants and animals on which they relied. Great storms and catastrophic droughts were not out of the question. As a result, migration and reorganization of life style were prevalent in the tropics, though less so than in the temperate zone.[2] The era of the Last Glacial Maximum was a difficult time for all. In the multimillion-year perspective of Gaia, this may have been but one more fluctuation of climate within the well-established limits of global temperature and humidity.[3] The plants and animals within Gaia's planetary system, governed by their biological archives, had been through earlier such episodes and were in some sense prepared for the adjustments that needed to be made. For the Human System, a newly formed network of speaking communities, it was the greatest climatological challenge yet encountered. As some human habitats became unlivable, people found themselves in conflict with neighbors over diminishing resources or had to move in search of better spaces. Families were doubtless broken up by the cold; survivors' best hope was to be adopted into other families. Scarce resources could lead to infanticide, to giving away children, and perhaps to abandonment of the aged or infirm.

The crisis put pressure on families and communities to maintain such valuable resources as hunting grounds, fishing gear, sources of salt and other minerals, groves of fruit trees, and implements of various sorts – and even control over the lives of people. For personal implements, individuals who had created them or received them as gifts chose, perhaps, to pass them on to children. Communities asserted their right to hunting or gathering at certain lands or fisheries, and out of these came a sense of property rights. As the range and extent of foraging and production increased, the new types of work brought new types of family relations.

Division of labor by sex and age developed in different ways. Ceremonies of initiation advanced the children to adulthood and to new responsibilities in work; the ceremonies were led by knowledgeable adults. Overall, humans explored the production of material goods, rather than simply foraging.

Of particular importance was the growing tendency of humans to find and create permanent shelter. While archaeologists have identified some examples of shelters built by hominin groups well before the emergence of *Homo sapiens*, it was in the aftermath of the Last Glacial Maximum that human communities systematically began to find or build shelters.[4] Among the most remarkable were Eurasian dwellings made of the tusks of mammoths.[5] An example is the Mal'ta–Buret culture of Siberia, just west of Lake Baikal and dated to 24,000 to 15,000 years ago. The population of this site remained in place through the worst of the Last Glacial Maximum, with subterranean dwellings, reinforced by mammoth bones, revealing the impressive adaptive skills that this hunting and foraging population had developed.[6] More broadly, rock shelters and caves were used where available, but most families had to create their own shelter. These foraging communities were mobile in their search for resources but were not entirely "nomadic" – most had a home territory and a home base. In these home bases, people developed constructions relying on such materials as earthen walls, stone walls, and wooden constructions, including some on stilts.

Sequel to Crisis: Production and Confederation

The cycle of change that brought the Last Glacial Maximum turned to bring a period of warming, especially from 19,000 years ago. Polar ice caps melted, the levels of the oceans rose by tens of meters, and rainfall increased, leading to new vegetation everywhere. A detailed and precise video, "Deglaciation of North America," provides valuable simulation of the retreat of the Laurentide Ice Sheet from its peak at 21,500 years ago to 5000 years ago.[7] This video shows the geographical framework for social change: as glaciers melted, plant species expanded and so did populations of humans and other animal species. Plant and animal species moved from their concentration in equatorial zones, colonizing lands at higher altitudes and higher latitudes, though with reverses. Humans too expanded. Most likely, because of their improving technology and knowledge of the land, they expanded at the expense of other species, such as by diverting watercourses or as with Australian burning of terrains to concentrate the animals to be hunted. Meanwhile, megafauna became extinct in the Americas in the aftermath of the temperature reversal.[8] During these 12,000 years at the end of the Pleistocene, humans faced the need for recurring changes in their lives. To the degree that their shifting institutions gave rise to a zeitgeist reflecting their dominant outlook and values, it is likely to have been a zeitgeist of flexibility.

Overall, the sequel to the Last Glacial Maximum brought two great transformations in human society: expansion of productive activity and formation of expanded social groups. These paired transformations arose through ramifications at all social levels. An excellent illustration of the gradual but profound changes of this era awaits the reader of anthropologist Steven Mithen's *After the Ice: A Global Human History, 20,000–5000 BC.*[9] As Mithen shows, members of human communities expanded their productive activities during this era of erratic climate, both in artisanal work and in food production and preparation. Artisanal work included the construction of housing out of wood, bamboo, stone, mud brick, skins, and other materials; work in weaving that created textiles as well as mats and basketry; use of fire in advanced methods of cooking and in creation of ceramics; as well as creation of sculpture in wood, bone, and other materials. Watercraft and fishing tools had been produced in earlier times; advances in these implements may have occurred in this time.

I'll illustrate changes first in productive activities and then in social organization, though I emphasize that the two types of change took place simultaneously and interactively. But it is instructive to begin with ceramics because of the range of talents required for that work. *Ceramics* arose in some regions in late Pleistocene times; eventually the production of pottery developed in most areas of the world. The institution of ceramics production required three sorts of mastery: control of fire, control of clay, and control of linking them in a kiln. Individual ceramic pieces, including sculpture, have been found in archaeological remains dating as far as 30,000 years ago. Pottery was invented independently in many regions of the world, but at quite different times. The systematic production of clay pots began about 18,000 years ago, notably in North China and Japan.[10] The ceramicists known as the Jomon culture, whose work in Japan began at least 12,500 years ago, were hunter-gatherers who did not adopt agriculture for several thousand years but made deep ceramic pots, eventually with elegant decorations. Other regions of early development of pottery were the West African savanna and the eastern Sahara, the latter an area where cattle tending and gathering of millet began early. There and in the middle Nile Valley, traditions of pottery manufacture continued in an unbroken line of stylistic development from 9000 years ago to the present.[11]

The expanded level of production was carried out not only by individuals but especially in workshops. The institution of ceramics required a leader to direct the design and production of the ceramic works, but also to direct the preparation of the clay and the creation of the kiln. Less skilled yet dependable assistants were required to collect and stack the firewood in the kiln, to tend the fire and its temperature, and to assist in bringing the firing to an end.[12] For construction of homes, creation of mats, creation of fishing equipment – and the preparation of music and dance for ceremonies – equivalent sorts of workshops convened as needed. Leadership decisions, including leadership in workshops, became more varied and perhaps more important along with the increasing differentiation in human economic and social life. For migrating populations, it was the choices on the timing and direction of movement. In gathering grains and roots, it was the timing of harvesting and the allocation of labor during the harvest.

In some regions, the intensive gathering of food developed by stages. One such intensification was the harvesting of tubers, portions of which were replaced in the ground to sprout again: this practice arose in West Africa, Southeast Asia, and New Guinea. Another intensification was the harvesting of wild grains, especially in western Asia, northeastern Africa, and Southeast Asia. In another example, some of the communities of northeastern Africa and western Asia lived in close association with wild cattle and sheep. A fourth intensification in food gathering occurred in the area from southern Mexico to northern South America, where various types of squash were consumed.[13] An additional factor – ecological – facilitated the development of agricultural techniques in several regions. Mountainous regions – more specifically, small regions with sharp variations in altitude – brought a significant advantage. The range of altitudes meant a range of climates, so that neighboring groups could rely on a range of different and complementary vegetable species.

Visual representation, already an ancient human practice, conveyed complex conceptualization, pictorial skill, and collaborative creation during the late Pleistocene. The images of Lascaux Cave, masterpieces created about 17,000 years ago, were arguably preceded by sketches, as the artist developed the idea to be portrayed, and may well have been created by workshops, as assistants gathered materials and filled in certain details of the work within the outline set by the master artist. For example, within the extraordinarily continuous and well-preserved record of rock paintings in Arnhem Land of north Australia, one may find the image shown in Figure 5.1, with a woman shown holding a stone axe, dated stylistically to 20,000 years ago. It provides a rare image of an axe in use, showing it to be hafted with a handle, and confirming that women used

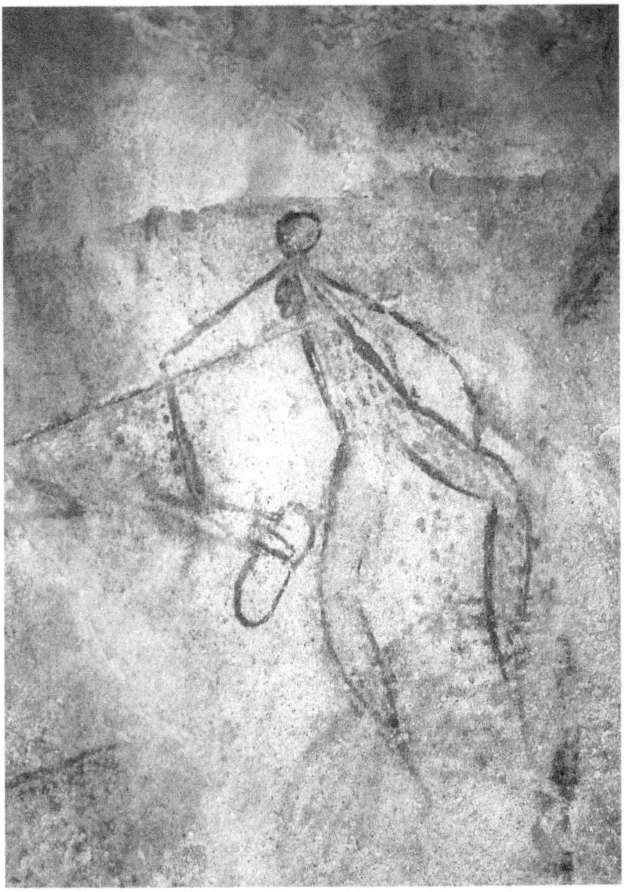

Figure 5.1 Woman with stone axe. Burungkuy site complex, Kakadu National Park, Australia. [Photo by George Chaloupka, *Journey in Time: The World's Longest Continuing Art Tradition* (Chatswood, AU: Reed, 1993), 120. © Copyright George Chaloupka 1993. Courtesy of Pina Giuliani (photo) and the Traditional Owners of Kakadu National Park (image).]

these tools. Other images that are being documented for Africa, tropical Asia, and temperate Eurasia may be analyzed for what they reveal of the material life and conceptualization of the artists.

In social organization, I hypothesize the late-Pleistocene formation of *confederations* as institutions in which previously autonomous communities agreed to join into larger groups. The action was to transform informal networks of communities into explicit confederations as institutions, so that the new confederations were explicit, we-group networks. These were the largest-scale acts of cooperation for humans to this time. At present, however, our knowledge of confederations is more a matter of social-evolutionary theory than of empirical social science documentation. In this discussion, therefore, I begin by asserting the theoretical case for confederations, and then explore the available information to see whether it sustains the argument.

The expansion of material production in the late Pleistocene epoch has been documented clearly in the archaeological record. Logically, this innovation required social coordination. I hypothesize that, to coordinate the new work of production and to adapt to rapid climate change, closely related communities formed *confederations* enabling larger-scale sharing of resources.[14] I argue that these changes took place from the peak of the Last Glacial Maximum and up to 12,000 years ago, in advance of the Holocene epoch. Confirming the formation of confederations in the late Pleistocene will be difficult to document, however, because the archaeological record is strongest in documenting life at the family and community levels rather than aggregates of these.[15] Nevertheless, the logic of social evolution of a living Human System suggests that such a transformation ought to have taken place along with growing material complexity. A confederation, formed out of the union of four or five independent communities, might have had a total population of 600 to 800, composed of several residential groups spread across a common territory. It may be that evidence from historical linguistics and comparative anthropology will ultimately add documentation to address the size of these social groups. Forming confederations as larger groups to address the complexities of production, however, might not have been an easy transformation. Confederation may have been rejected or revised on multiple occasions, as it required new practices in communication and decision-making and opened possibilities for abuse and confusion. That is, the innovation of confederation had to pass a social-evolutionary test of selection and regulation. For instance, if several communities with different languages sought to confederate, that act of union created pressure for selecting one of the constituent languages for public discourse, while the other languages were for home use only. This issue surely came up frequently with aggregation of confederations and, later, societies. Combining communities might have resulted from mutual agreement, from request for acceptance by a community suffering great misfortune, or even from conquest of one community by others. Nevertheless, the advantages of confederation permitted sharing of resources and coordination of various types of productive work.

Archaeological data on the late Pleistocene, supplemented by ethnographic data on more recent times, give indications of how communities made confederation feasible, building up networks of exchange and practices of food-sharing. The Magdalenian foragers of France, 15,000 years ago, relied on reindeer, horses, and smaller animals for food; they fished for salmon, pike, and trout; and they made flutes of bone. From ethnographic surveys we learn that ethnic groups of the Andaman Islands recognized informal headmen, who had to undergo long preparation and deep learning before accepting the position. Australian tribes, commonly divided into moieties, each with roughly six clans, combined into an elaborate system of classifying both close and distant relatives, who relied on an intertribal trading system in spears, feathers, beeswax, resin, ochre, and beads. The Tlingit people of the North Pacific coast exchanged goods regularly with the Athabascan peoples of the Canadian interior.[16] The Tlingit exchanges were I-group networks of exchange, in which people exchanging goods were linked by their transaction but not in any further ties. Even at this informal level, exchange links tied communities and confederations into larger I-group-networks.

Humans exert leadership in any of several fashions: the leadership of command associated with forceful military leaders, the leadership of persuasion associated with religious figures, and the leadership of invention associated with novel technology or social organization. In addition, there is the leadership of facilitation provided by those who attend to many small details, enabling others to do their best. All types of leaders may have gained recognition as elite or superior individuals. As confederations formed, their structures allowed for elite individuals and elite strata. But then leadership and elite status faced choices with every generation: what rules of inheritance for passing on resources and decision-making power? Funerals not only celebrated the life of the departed but also presented the occasion for passing on institutional leadership. Marriage, under these circumstances, was more than the choice of two young people to start a family. It could be an alliance between two existing families, with important implications for how resources would be passed to later generations. In addition, because property was not evenly distributed, hierarchies developed within and among families.

The formation of confederations required a social archive to document and reproduce the institutions of this new level of social structure, necessary to preserve the codification of law, allocation of properties, practices for passing on family and community leadership, and the negotiation of disputes among communities within the confederation and with competing confederations. The confederation also provided a framework for occasional mobilization, as for response to disaster, defense against attack, initiation of the young males and females, and construction of public works such as sites of worship or the diversion of water. Indeed, the full complexity of a confederation, now a social system, included a set of subsystems to perform its various tasks.[17]

Emergence of this more complex set of levels now divided the responsibilities of life among individuals, families, communities, and confederations.[18] What decisions were taken at each level? The initial formation of the community, as I have argued, was an automatic response to the rise of language. But the later coalescence of confederations was less obvious and more contingent. How did human nature and behavior change once communities added the institutional structures of confederations to their existence? Presumably a wider identity resulted for most within the confederation. Did a confederation give support to certain institutions and types of behavior that encouraged certain emotions and behaviors in community members? The range of institutions and responsibilities, even in a small community, was such that it encouraged a wide range of emotions and behaviors. Yet the development of certain overall priorities, ideology, and institutions might elicit the rise of a zeitgeist, a way of thinking maintained by consensus in the community as a whole.[19]

Old World Migration

Communities had contracted their habitat during the Last Glacial Maximum. In the long period of warming thereafter, they formed into expanded confederations, took up production and a settled lifestyle, yet also migrated to expand their habitat once plants and animals began to flourish in areas that had experienced desolation and cold. For thousands of years of recurrent warming, human populations everywhere expanded their terrains and their activities, though they always had to face fluctuations and bad times. Certain of these migratory expansions gained particular importance, as people from key centers of population were able to occupy large territories that had been almost devoid of human occupation during the Last Glacial Maximum.

For late Pleistocene Eurasia, the overall migratory pattern included two major trends. First is that northeastern Asia, notably the Amur and Yellow River valleys, became a center of population dispersal in all directions. Second is that, in both East Asia and West Asia, migrants moved both to the north and the south, as people from tropical zones moved to the north and people from the north moved south to subtropical zones. The exact timing and the relative order of these migrations is not known, but the timing stretched from the late Pleistocene into the early

Holocene. In exploring these migrations, one must remember that, as of 25,000 years ago, temperate Eurasia had been settled by humans for roughly 15,000 years – so people were comfortable at home except for the decline and then rise in temperature.

Northeast Asia is best taken as the launching point for many of these migrations. The largest population of Northeast Asia arguably spoke Eurasian languages. The languages of this group presumably arose from early settlers of the temperate north: I have argued that their ancestors spoke either Dravidian or Trans-Himalayan languages. The Eurasiatic languages of today include eight major subgroups: their distribution centers on the Amur Valley of the north Pacific coast of Asia. (On Eurasiatic languages and migrations, see www.cambridge.org/Humanity, in resource text 12.Eurasiatic.)[20] Moving to the west were migrant subgroups of Eurasiatic that became the Yukhagir and Uralic languages of the Arctic zone, the Altaic languages of the temperate grasslands, and Indo-European and Etruscan further west. Moving south along the Pacific shore were languages that became Japanese and Korean. Moving northeast were Chukotian languages – and groups who settled in the Americas.

In addition to the speakers of Eurasiatic languages, people from three more language groups became parts of the population of northeastern Asia: they joined in the dispersal in various directions from the Eurasiatic homeland. The second group, speakers of Yeniseian languages, had long lived in northeastern Asia; their descendants live today just west of Lake Baikal.[21] Their languages have been traced to the Trans-Himalayan languages. Third, mariners speaking Indo-Pacific languages appear to have moved to northeastern Asia by island-hopping from the Indonesian or Philippine archipelago. They are documented especially through genetic data.[22] In addition, their maritime skills may have been important in the development of skin boats among the Eurasiatic speakers among whom they settled. Eurasiatic speakers then moved westward along Siberian rivers, settling among and perhaps dominating the preceding populations. This "westward movement" reached the Caspian Sea, the Black Sea, and Europe. As has been observed long ago, the distribution of skin boats ranges all across temperate and arctic Eurasia and into North America.[23]

The fourth group of migrants in Northeast Asia consisted of people speaking languages ancestral to the Chinese of today. Their ancestral languages were centered in the highlands of Yunnan, along with many other Trans-Himalayan languages. Nonetheless, significant numbers of people speaking Chinese-related languages had settled in the Yellow River Valley, perhaps in the early Holocene. These settlers came to have a physical resemblance closer to the surrounding Eurasiatic speaking peoples than to the majority of Trans-Himalayan speakers to the south. It appears that, from the mid-Holocene epoch, the predominant migration in China was of Han Chinese from the north to the south.

A somewhat parallel pattern of migration developed at a similar time in West Asia. Migrants from the tropics, speaking Afroasiatic languages, moved north from their homeland in the middle Nile Valley in the late Pleistocene or early Holocene. Their languages became dominant in Egypt, in northwest Africa, in the Levant, and in Arabia; some genetic evidence from Y-chromosome analysis confirms these northward movements.[24] In the same or later times, according to genetic but not linguistic data, there appear to have been southward migrations from Anatolia and the Caucasus. As a result, people of Egypt, the Maghreb, and the Levant speak languages of African ancestry but have a physical resemblance closer to Eurasiatic speaking peoples than to more southerly speakers of Afroasiatic languages.

Late Pleistocene migrations of the tropics in the Old World, while less spectacular than those of the temperate zone, were nonetheless significant. Afroasiatic-speaking populations, in addition to moving north, moved west and built up dense populations in the savanna lands near Lake Chad. West African speakers of Niger–Kordofanian languages migrated to new lands in both savanna and forested regions.

The Americas, 25,000–12,000 Years Ago

Continuing efforts to link archaeological, climatic, genetic, and linguistic evidence are yielding a steadily clearer picture of the occupation of the Americas. The combination of available data – genomic, archaeological, linguistic, and climatic – makes it possible to propose a comprehensive picture of settlement of the Americas. I present this interpretation in four chapters: first, the settlement of Beringia, 25,000 years ago; second, the rapid maritime settlement of the eastern Pacific shore, from just south of the Laurentide ice sheet to South America, beginning 19,000 years ago, by speakers of Amerind languages; third, a second and more northerly maritime settlement, beginning 17,000 years ago, by speakers of Na–Dene languages who expanded into Alaska and Athabaska; fourth, a rapidly unfolding complex of North American events some 13,000 years ago – the proliferation of Clovis spear points, the sudden cooling of the Younger Dryas, the disappearance of North American megafauna, and the possible crash of a comet or meteorite in Greenland at the same moment. In methodology, the most important result of this analysis is that genomic and linguistic data reinforce each other significantly, so that even the differences between the two types of evidence help to clarify the overall picture.

The story of American settlement opens with Beringia, the "land bridge" linking Alaska and Siberia because of low sea levels during the Last Glacial Maximum. It is now widely assumed that migrants from northeastern Asia settled in Beringia about 25,000 years ago and remained there.[25] Their progress east was blocked by the great Laurentide ice sheet, an obstacle covering all of Canada, much of Alaska, and today's Great Lakes region. The ice sheet reached its maximum roughly 21,500 years ago and then began to contract, first revealing islands free of ice along the north Pacific coast.[26] Beringia is shown on Map 5.2.

The second chapter in occupation of the Americas was the occupation of coastal lands beyond the Laurentide ice sheet by maritime voyagers, starting about 19,000 years ago, and the rapid settlement of inland areas by their descendants. I discuss first the linguistic evidence and then the genetic evidence for this migration. In Map 5.1, the main map shows both the coastal settlement and the inland migrations of language families within the Amerind phylum; the inset to Map 5.1 displays genetic evidence of this same migration (see www.Cambridge.org/Humanity, in resource text 13.Amerind).[27] As a key issue in technology of migration, an essential support for the maritime voyagers who settled the Pacific coast was the "kelp highway," the great maritime forests of brown algae lying just off the coast. Marine archaeologist Jon Erlandson led in developing the notion of "kelp highway" as an explicit challenge to the "land bridge" that had long been emphasized as the path for settlement of the Americas.[28] Kelp grows all around the Pacific, and especially in cool waters from Japan to Alaska to Mexico. It provides an ecology that attracts numerous species of all sorts – shellfish, fish, sea mammals, and seabirds. Human populations that had become accustomed to living in the western Pacific were therefore equipped to exploit the oceanic fringe north to the Aleutians, and then along the similar waters to Baja California. (Similar kelp forests are found along the coast south of Ecuador.)

For watercraft moving south and past the eastern Pacific edge of the ice sheet, the first sizeable ice-free terrain was the Salish Sea, as the lands and waters from Vancouver Island to Puget Sound have recently been labeled. There, maritime voyagers from Asia or Alaska settled and built communities of Amerind speakers that have expanded to this day, beginning no later than 19,000 years ago. (On Amerind languages and migrations, see www.cambridge.org/Humanity, in resource text 13.Amerind.) Language classification data confirm that the Salish Sea region was the homeland

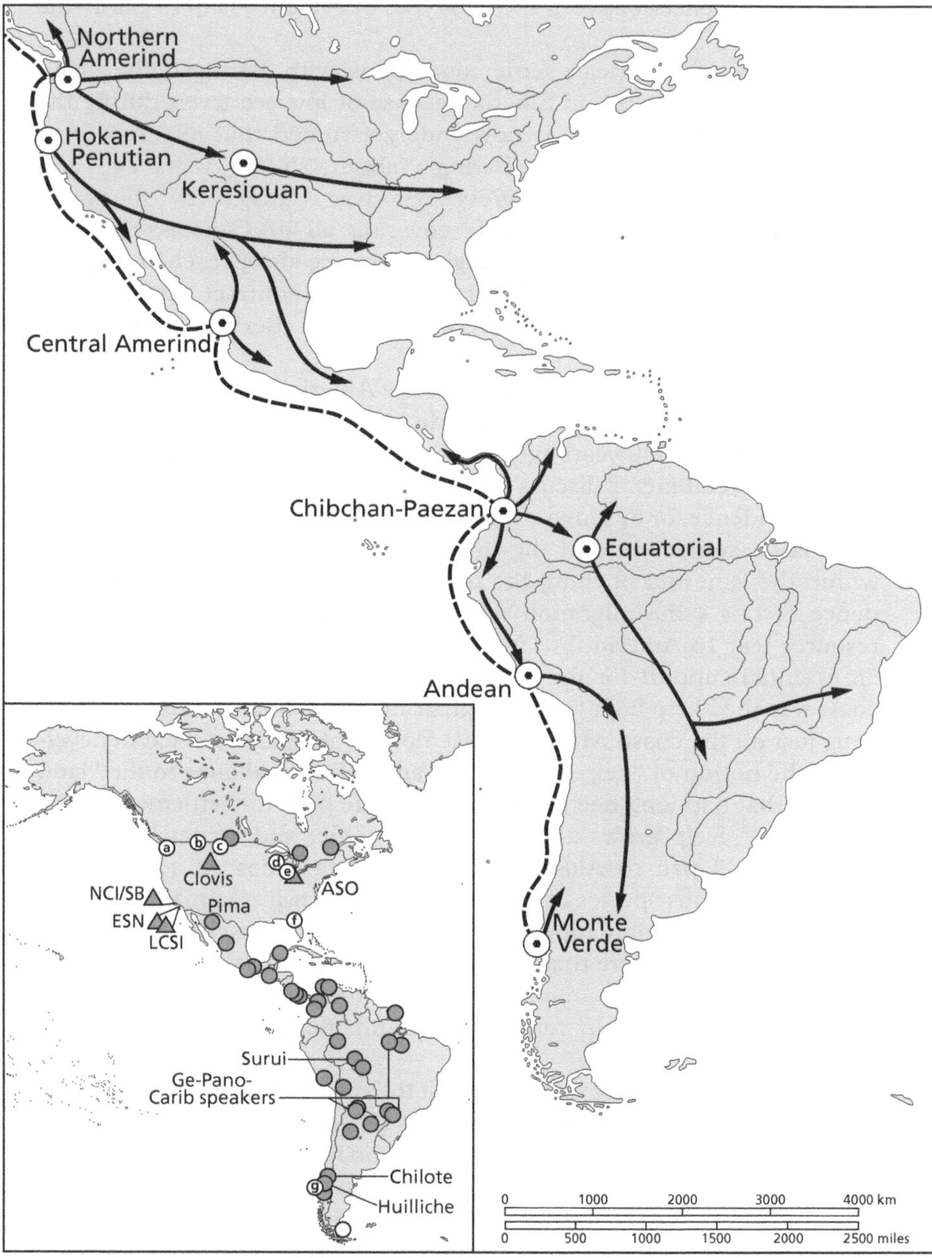

Map 5.1 Americas: genetic and linguistic evidence on settlement. (Main map: Redrawn based on original image, courtesy of the University Library System, University of Pittsburgh; generated by Daniel Andrus, 2019. Inset map: Redrawn based on original image, courtesy of Christiana Lyn Scheib et al., "Ancient Human Parallel Lineages within North America Contributed to a Coastal Expansion," *Science* 360 (2018): 1026.)

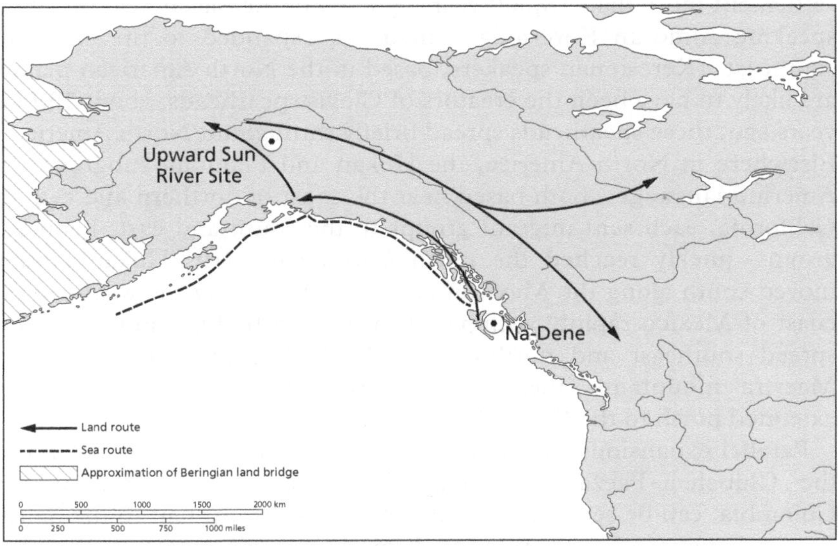

Map 5.2 Na–Dene languages and settlement. (Redrawn based on original image, courtesy of the University Library System, University of Pittsburgh; generated by Daniel Andrus, 2019.)

for all the Amerind languages of the New World.[29] The more southerly settlements, as shown on Map 5.1, were: (2) the coast of northern California: here the settlers became speakers of the Hokan and Penutian languages; (3) the coast of Mexico, in the region of Mazatlán, where settlers became speakers of Central Amerind languages; (4) the Colombian coast, in what is now Chocó province, where settlers became speakers of Chibchan-Paezan languages; (5) the Peruvian coast, near Lima, where settlers formed the Andean languages. Finally, the sixth and most southerly coastal settlement of Amerind-speakers has the advantage of being documented by an archaeological site: Monte Verde, on the southern coast of Chile, where a settlement has now been dated to 18,500 years ago.[30] Some of the languages spoken in this region today are of the Chibchan–Paezan group, which argues that the voyagers had close ties to the Chibchan–Paezan homeland in modern Colombia.

The populations in each of the six coastal areas of the Amerind-speaking settlement began to move inland immediately after their settlement more than 18,000 years ago. From the Salish Sea, migrants speaking Almosan–Keresiouan languages expanded to the east and southeast. (Keresiouan speakers, based in the North American plains, are likely to have been the creators of Clovis spearheads, about 13,000 years ago; these spearheads spread briefly throughout North America.) Elsewhere in North America, the Hokan and Penutian subgroups of Amerind languages, both based near the coast of northern and central California, each sent migrant groups to the south and east. Penutian groups quickly reached the north Caribbean coast; Hokan groups moved south along the Mexican coast. Settlers along the west central coast of Mexico rapidly acclimated to the nearby highland areas and spread southeast and northwest in the high country of the Sierra Maestra mountains: these Central Amerind languages eventually extended north to the Great Basin of Utah.

Parallel expansions took place in South America. The homeland of the Chibchan–Paezan languages, in Chocó province of today's Colombia, can be seen as the South American equivalent to the Salish Sea settlement in North America. From there, migrants moved northwest into Central America, northeast to the Caribbean coast, south to the coast of today's Ecuador, and southeast across the Andes and into the Amazon Valley. In the Amazon, the Equatorial languages took form, and settlers moved southward among the upper valleys of many rivers and eventually eastward and toward the Atlantic. Further south along the eastern Pacific, the Andean languages took form among settlers on the Peruvian coast. Their settlements spread north into the mountains of Colombia and southeast into the mountains of Bolivia; their ecological strategy was parallel to that of the Central Amerind speakers of North America. The southernmost coastal settlement, at Monte Verde, did not become demographically important, but the information it provides assists greatly in understanding overall patterns in occupation of the Americas. The early date of its settlement, 18,500 years ago, confirms the whole maritime pattern of migration.

A major genetic study by Christiana L. Scheib et al. provides remarkable confirmation for this pattern.[31] This research (as indicated in the inset map in Map 5.1) shows a broadly similar ancestry for all populations south of the glaciers, with a subdivision into two groups – northern and southern – and certain other exceptions. The northern genomic group is much the same as the Almosan–Keresiouan language family, while the southern genomic group includes virtually all remaining groups in the language classification. The genomic evidence indicates that a single ancestral population split into the northern and southern groups, which later rejoined through admixture. That is, most of the genomes of Central and South America trace ancestry to the southern group but show some admixture with the northern group.[32] Scheib et al. offer four different models of the split and admixture of the two groups. Language evidence gives strongest support to the first of those models, which shows that both separation and subsequent admixture took place mostly in the Pacific Northwest.[33]

A further and remarkable point within this second chapter, that of Amerind-language expansion, is that certain South American genomes show a measurable representation of DNA parallel to that of Papua New Guinea – that is, to speakers of Indo-Pacific languages.[34] This Indo-Pacific ancestry shows up at other places in South America and at some points in North America. The most obvious hypothesis is that communities and descendants of Indo-Pacific ancestry survived in Northeast Asia, participated in voyages to the Americas, and continued in voyages all the way to South America.

The third chapter in settlement of the Americas was the arrival of Na–Dene speakers. The Na–Dene languages, now centered in Alaska and northwest Canada, are related to the Yeniseian languages of central Siberia. (On Na–Dene languages and migrations, see www.cambridge.org/Humanity, in resource text 14.Na-Dene.)[35] As shown on Map 5.2 and based on a combination of linguistic and climate data, I believe that the Na–Dene speakers traveled by sea to reach Haida Gwaii Island, the largest of the Queen Charlotte Islands, about 17,000 years ago, after the island had become free of ice.[36] The Haida continue to live there today. As the glaciers receded further, Na–Dene-speaking populations expanded northward along the adjacent mainland, forming the Tlingit population, then westward to what became the Eyak population, then inland and northward to the Yukon Valley of central Alaska. Here the Na–Dene speakers reached the lands that had been settled earlier by the Beringians. As the inland glaciers melted, leaving a significant open space in Athabaska before the Younger Dryas event of 12,900 years ago, Na–Dene speakers moved up the Yukon, eastward and southward, occupying lands where their descendants still live (www.cambridge.org/Humanity, in resource text 14.Na-Dene.).[37]

Meanwhile – and from the same region – the genome of an infant girl has been retrieved and analyzed. Dated to 11,500 years ago, it is from Upward Sun River, in Central Alaska.[38] This individual was most likely a member of the Beringian community: her genome is related to others of North America and, more distantly to northeast Siberians, but it is argued to be separated from other North American genomes by roughly an additional 10,000 years. We do not have and are unlikely to find any information on the language of the Beringians. But the Na–Dene languages are related to Yeniseian languages, clearly distinct from the Eurasiatic languages that seem most closely related to Amerind languages. Further research should eventually make it possible to determine whether there were two or more populations in late Pleistocene Alaska.

The fourth major chapter of North American settlement opened some 6000 years after the initial settlements along the eastern Pacific coast: it linked the creation of Clovis spear points, the Younger Dryas or period of sudden cold from 12,900 to 11,500 years ago, and the decimation of megafauna. As of 13,500 years ago, artisans within the Keresiouan language family created the elegant spear points known as Clovis points, named after the New Mexico region in which archeologists first described them.[39] These points, of up to 10 cm in length, with sharp biface edges on each side, were affixed to spears for hunting of large animals. They became tremendously popular and spread rapidly, especially in what are now the eastern and southern United States. Study of ancient DNA has documented the remains of a boy believed to be associated with Clovis populations.[40] Yet production of Clovis points lasted for little more than 400 years, after which they were replaced by other spear points that were produced in smaller numbers. On one hand, it has been argued that the Clovis period came to an end because the megafauna populations had nearly been eliminated by hunting. On the other hand, the end of Clovis point production fits very closely with the initiation of the Younger Dryas.

The Younger Dryas episode was discovered at the end of the nineteenth century by geologists in Europe who noted succeeding shifts in plants and pollen that indicated sudden declines in temperature.[41] The effects were noted especially in North America and Europe and, to a lesser degree, in South America. The Younger Dryas episode is now dated with a sharp cooling 12,900 years ago and an equally sharp warming 11,500 years ago, followed by a slower warming. One explanation for this decline is that the very process of global warming brought a sudden and severe cooling. That is, the melting of North American glaciers in the previous era of warming brought a flood of cold waters to the North Atlantic and interrupted the northward flow of the warm Sargasso Sea, thus bringing cooler temperatures to all the oceans and to landmasses as well.

Quite a different interpretation has suggested an extraterrestrial impact – that a rocky meteor or an icy comet struck North America, with results that caused temperature decline. The thesis, proposed in 2007, has been debated steadily and inconclusively. The most extreme version suggests that the impact of a comet caused fires all over North America that killed animals and humans, and that a later return of the comet 11,500 years ago caused the sudden warming at that time.[42] The 2018 discovery of an immense crater in northwestern Greenland heightened interest in the impact thesis. Initial results suggested that the crater might have been created as recently as 12,000 years ago or as early as a million years ago.[43] If further research were to confirm an impact date anywhere near 12,900 years ago, it would become the center of further analysis.[44]

The impact thesis would account for the end of megafauna in northern America but not really for South America (at the same time), Eurasia (at much the same time), or Australia and New Guinea (earlier). That is, the recurring experience was of humans arriving in a region, encountering herds of large animals, followed by extinction of the megafauna several thousand years later.[45] Many scholars have argued that it is doubtful that human hunters could have killed off all the large animals, but it does seem that human presence was related in some way to the extinctions. Research continues on this issue. Further, the encounter of modern humans with other large mammals – that is, hominin populations such as Neanderthals and Denisovans – can be seen as a parallel example.

The history of migration throughout the Americas enables us to conceive of a remarkable sort of historical experiment in the significance of cross-community migration. Once in North America, the exploring and expanding human populations faced a problem different from any faced by their ancestors. Previously, the direction of migration had been from warm lands to cool. The first settlers of America, in contrast, were born and bred in cold regions. They and their migrant offspring had gradually to learn the techniques of life in warmer regions. They faced the task of learning new ecology after new ecology as they moved south and east, until they reached at last the limits of South America. Nevertheless, settlers of the Americas appear to have moved with amazing rapidity to populate all regions of the hemisphere. The range of Clovis sites indicates the density of settlement in North America, while archaeological sites and rock art have been documented in the Amazon valley for times 12,000 years ago. Three factors may be proposed to explain the rapidity of settlement. First, because temperature and humidity were rising throughout this period of settlement, despite fluctuations, it should not be surprising that populations would have expanded rapidly and that migration to form new settlements would have been facilitated. Second, one may ask rhetorically whether the

American settlers carried some sort of deep sociocultural memory that enabled them to retrieve the experience of tropically based ancestors many generations earlier, with which to solve the problems of life in the American tropics. Third and more practically, one may hypothesize that American migrants drew on the benefits of cross-community migration, in which various populations occupied different habitats, but exchanged information with nearby groups as they addressed new problems. American settlement did not rely on individuals pursuing a single path ahead, but on a broad front of migrating communities, exchanging information as they advanced in parallel.

The Achievements of Early Humankind

If we compare the initial Human System of Northeast Africa with the expanded Human System at the end of the Pleistocene – some 50,000 years later – the differences appear imposing. Still, at the beginning, the individual communities of speaking humans were formidable in comparison to nonspeaking families: the early network of a modest number of communities was sufficient to remake the human terrain. By the end of the Pleistocene, the combination of networks had stretched as a fragile web across huge terrains of six continents, though with the possibility that groups of communities might fall out of contact with it for a time.

Does this account of late Pleistocene history fit with my hypothesis on the rapid emergence of speech, some 70,000 years ago? As a method for evaluation of the hypothesis, I propose assessing available information on genome, population, language, the social order, technology, and migration patterns. Here I offer tentative results arguing that, across the disciplines, available empirical data and theoretical formulations are overwhelmingly consistent with the hypothesis of rapid emergence of syntactic language, 70,000 years ago. The genome shows rapid expansion from Northeast Africa, but with no clear such evidence before 65,000 years ago. Archaeology confirms the immense expansion in human population and range, both geographical and ecological. The rapid absorption of competing hominin communities is particularly striking. In language, the new work on proto-human vocabulary is consistent with the invention of language in a single community. The distribution of language phyla, as understood by those who accept their existence, gives a picture of human migration – outward in all directions from Northeast Africa – that is not only quite consistent with recent research in genomics and archaeology but also adds additional and plausible detail on migratory movements.[46] In the social order, we have the impression that the size of human groups expanded in the late Pleistocene, but documentation is difficult and the results are vague. Nonetheless, the theory of collective intentionality and group behavior argues that, once humans had language, they could form we-groups and build social institutions that could magnify human agency. In technology, we can document changes in visual representation, watercraft, weapons, shelter, and clothing – all were innovations but all were expansions of human activities before the great expansion. In migration, ongoing research shows a clear pattern of settlement throughout tropical Africa and tropical Asia by 50,000 years ago, followed by settlement of temperate Eurasia from about 45,000 years ago and widespread settlement of the Americas by 15,000 years ago. Each of these developments points to a release of extraordinary creative energies in Northeast Africa at a time not far from 70,000 years ago and to the steady elaboration of those initiatives. To return to the formulation of McBrearty and Brooks, however, all of these achievements could have been accomplished without the need for a "conceptual revolution." The logical powers, habits, and inclinations of humans could well have been continuous over a much longer time than the past 70,000 years. In sum, I argue that this compilation of available evidence strengthens substantially the argument that syntactic speech began, rather suddenly, some 70,000 years ago. In that case, the single step of creating syntactic speech, enabling people to communicate with greater specificity, made possible the development of a new and more conscious group behavior. With the addition of language, people were able to draw on their individual dreams and apply them in group projects.

The expansion of speaking populations of *Homo sapiens* can be traced distinctly in terms of biological, cultural, and social evolution. In biological terms, in addition to modest adjustments in teeth and skull, it may be that changes in the placement and shape of the larynx made our fully enunciated speech possible. Further, biological evolution can be traced in changing physical features induced by environmental factors. In particular, comparison of New World and Old World evidence on skin color clarifies the interpretation of "race" and "racial distinctions" for humans in general.[47] With time, those who lived in tropical American regions gained darker skins; those who moved further into South American temperate zones developed lighter skins.[48] Those who lived at high altitudes developed shorter stature. In the language of the twenty-first century, the term "race" is rarely if ever used to distinguish the physical appearance of indigenous populations within the Americas. Nevertheless, the emergence of significant phenotypical differences among peoples of North, Central, and South America is clearly observable, and is parallel to differences in appearance of people from various parts of the Old World. By this logic, the differences among people from various parts of the New World are smaller than those of the Old World largely because they had less time to develop.

In cultural evolution, patterns of reciprocity and social learning, including the growth of I-networks, must surely have been important in building learning and collaboration among human populations, though it is difficult to document such changes from this distance. At the level of implicit networks or I-group networks crossing the lines of community, archaeological evidence confirms the exchange of material items such as obsidian, the spread of dogs and the generalization of bows and errors as evidence of such informal networks.

In social evolution, the cascading transformations began with the creation of speaking communities bound by rituals. With time, migrations across community boundaries – whether voluntary or involuntary – required the learning of languages and customs and facilitated the broader sharing of local experiences.[49] Workshops, while small and transitory institutions, gave clear evidence of their existence in the quality of artistic creation and the remains of Pleistocene material culture. The hypothesized formation of confederations brought a conscious and beneficial expansion in the scale of social organization at a difficult time. In addition to such we-groups as workshops and language groups, social evolution led to spatially oriented we-group networks such as name associations including persons of shared names in different communities, who relied on each other in exchange of food or gifts.[50]

The Human System, by the end of the Pleistocene, linked members of the species into we-groups of languages and into wider and interconnected networks. Did the regions of Africa – both the Homeland in the northeast and the regions more recently settled – remain within the global system? Doubtless yes, as African population arguably persisted as the majority of human population until late in the Pleistocene and as that population maintained informal links across the continent through the "African Web," an I-group network.[51] Had each regional population become isolated, with its phenotype adjusting to the habitat in what came to be known as "race"? Yes, to a substantial degree, and this outlook is confirmed by the expansion of permanent settlements. But linguistic and genetic evidence tell us also of important movements of people and group-level culture from the South Seas to the Northeast Asian coastline; from Northeast Asia to Europe and the Americas; and from the Middle Nile to North Africa, the Levant, and Arabia. Does the experience of settling the Americas provide additional information on the patterns of human expansion? (Dogs came to North and South America, either with the initial settlers or soon thereafter.) While migrations between the Americas and the Old World were doubtless infrequent, roughly a half dozen such links have already been identified and there may well have been more.[52]

Settlement of the Americas recapitulates and extends most of the stories of the Old World expansion of humanity, with two major exceptions. The creation of language and communities, once accomplished, did not need to be reenacted; and human migrants to the Americas (as with Sahul) had the hemisphere to themselves – there were no preceding or competing hominin populations. But the tasks of physical movement and of accommodating to each new habitat – its flora, fauna, and geography – replayed themselves at every stage. The experience of surviving dramatic climate change reinforced flexibility in locating leadership for making difficult decisions in times of flood or drought. Thus, as people of the Eastern Hemisphere developed a new level of social organization in the course of the late Pleistocene – the confederation – those who came to the Americas either brought the confederation with them or developed it on arrival. Their populations were necessarily small and their lifestyle surely relied heavily on hunting, foraging, and fishing at the start. Yet they brought with them the knowledge of confederation-level coordination and productive processes. Perhaps the clearest example of their experience with production is the watercraft they built to transport themselves on long voyages along the eastern Pacific coast at a pace greatly exceeding that of Old World expansion. They completed a seaborne trilogy: the initial migration along the coast of the Indian Ocean from Africa to Sahul; the northward movement along mainland and island coastlines of the western Pacific from Sunda to the Arctic; and migration via the kelp highway along the eastern Pacific to South America.[53]

Given the number of human innovations, why have communities not become more widely different from one another? At least two factors narrowed the range of good choices in social evolution. The first were the limits of the natural world: we are constrained by the climate, plants, animals, and landscapes with which we live. Second, even in those areas of major human inventions, the newly invented institutions bring their own inherent dynamics that generate inescapable realities. Thus, the importance of fire in ceramics and metallurgy, the inherent structures of urban communities, and even the possibilities of human decoration, while allowing for much variety, are constrained by their inherent characteristics. Arguably, the story of human expansion was a positive achievement and an extraordinary success: perhaps the most rapid and extensive spread ever of a single species – certainly among mammals – throughout the world. The success of human expansion is more than the achievement of gifted individuals: it is a collective venture in which conscious collaboration through formation of social groups and institutions has expanded, preserved, and exchanged innovations. This Human System, however, just as surely achieved its expansion at the cost of significant conflict, oppression, and neglect within the growing communities.

Holocene Evolution

6 Society: Network versus Hierarchy

This chapter traces the innovative development of institutions from 12,000 years ago to 1000 years ago. The expanded social institution of *society* arose to encompass the communities and confederations of earlier times. With the rise in society populations to more than a thousand, this social consolidation brought the second great expansion in the human population groups (the first was the creation of communities). Networks and hierarchies expanded within and among societies: spectacular migrations and innovations in society and technology interacted to expand levels of cooperation but also the violence, oppression, and revolt that would continue to grow thereafter. The creation of societies thus brought a need to adjust the standard of social fitness for institutions, as the new institutions generated tension between preexisting social networks and emerging hierarchies. Meanwhile, processes of coevolution continued and deepened: cultural evolution persisted along with social evolution; the expanding footprint of agriculture came to influence climate; and the dispersion of humans among three great continental masses led to important parallels and distinctions. The variety, complexity, and diversity of social innovation expanded through these processes; in addition, historical documentation is more plentiful for the Holocene epoch – including material remains, genetic and linguistic evidence, and written records – that reveal the complexity of human social life.[1] To grapple with the complexity and diversity of Holocene social change, I introduce a simplified model that distinguishes society, and its clans and elites, to assist in describing the distinction between commoners and elites that arose in this era. I also rely on the work of Kent Flannery and Joyce Marcus, who trace, through the anthropological record, the repeated emergence of elites but also the elaborate devices that societies have implemented to limit the growth of elite strata and hierarchical organization.[2] The chapter concludes with analysis of Afro-Eurasian societies and institutional change from 1000 BCE to 1000 CE, a time in which these societies were the most elaborate within the Human System.

Holocene Dynamics: Environmental and Social

If the erratic climate of the Pleistocene helped build the resourcefulness and flexibility of the human species, the relative dependability of Holocene climate enabled humans to innovate and expand on an unprecedented scale. The social transformations of the Holocene derived clearly from the earliest steps in social evolution, yet they generated innovations of great intricacy and diversity. The interplay of the Human System and Gaia continued, but took a new turn.

Transformation into the Holocene. With the Holocene epoch, the previous epoch's extreme changes in climate eventually came to an end. The late Pleistocene included 8000 years of warming after the Last Glacial Maximum, then the cold snap of the Younger Dryas. The Holocene began with 3000 years of warming, up to 8000 years ago, followed by temperature stability until the Anthropocene. Geologists formally define the beginning of the Holocene epoch (and the end of the Pleistocene) at 11,500 years ago – to coincide with the end of the Younger Dryas event. Gaia, in the early Holocene, was generous. As the ferocious and erratic storms of the late Pleistocene gave way to stability in climate, rising temperature and humidity enabled plants and animals of every sort to thrive, though habitats were in perpetual adjustment. Humans found ways to take control of additional resources: certain plants were genetically modified through breeding, expanded in their populations, cultivated more widely, and made into food for a growing human population. Certain animals, bred in parallel fashion for food and hides, were then turned into sources of labor in tilling the soil, digging mines, pumping water, and transporting goods. Waste of various sorts accumulated in the Human System, and humans sought to expel the waste to other parts of Gaia, extruding human and animal excrement, wasted food, broken tools, shredded clothes, and remnants of former dwellings. In effect, these steps amounted to human colonization of Gaia. Of course, the natural biosphere prospered as well during the era of stable climate, so that no major conflicts had yet emerged between humanity and Gaia. Indeed, the very stability of the climate resulted in part from the expansion of agriculture, which generated greenhouse gases of carbon dioxide and methane that offset what would otherwise have been a decline in temperature after 8000 years ago.

Social Evolution, Expanded. I define the term *society* in much the same way as it is used in general discourse: as a generic term for social orders from localized agricultural, pastoral, and fishing groups to large-scale states. More specifically, I treat society as a social institution that formed in the Holocene and that has expanded in scale up to the present day. With the divergence in productive tasks and habitats of social groups, the specific form of *confederations* changed into the more general form of *societies*, some of which remained the same size while others grew with time. Many evolving societies emphasized decentralization, convening leadership groups only when the occasion required major decisions. The forms of society grew to quite different sizes according to their institutions: formation of states was a possibility but not the destiny for all. Within societies as they evolved, the preexisting structures of community and family were able to persist, though doubtless changing in interaction with the encompassing society.[3] Society, in sum, became a we-group network of individuals and communities and confederations intended to provide common identity and mutual support to members of the group. Its objective and dynamics were similar to those of the preceding groups it absorbed, but it also brought a broader sense of unity across multiple interests.

Viewed broadly, one may argue that the decisions to expand production of goods and services were also decisions to gain mastery over nature – to create resources, through human effort, that did not exist in nature. Taking it a step further, one might say that human groupings had implicitly declared their independence from Gaia: humanity would now make its own way rather than exist only as an aspect of the biosphere.[4] This is probably an overstatement, as those who sought to reshape nature did not reject the nature spirits commonly understood to govern the Earth, the waters, the skies, and the plants and animals. The vision of gaining control over nature developed slowly, as productive activity advanced in field after field of human exertion. Yet humans recognized increasingly the distinctive nature of their Human System, especially as it was reflected in the growing diversity of social orders and productive practices.

The size of social groups appears to be an important issue in analysis of social evolution, but it has not yet been possible to organize coherent data on this wide-ranging topic. The valuable insights of Aiello and Dunbar on human group size provide a baseline, but for now one can proceed only by hypothesis. So far I have assumed human family or household groups of 15 or more persons, speaking communities of at least 150 members, initial confederations of roughly 500 persons, and societies of about 1000 persons.[5] But for the society the limit on group size was no longer biological – group size depended on the social function of the groups and on the ability to coordinate relationships among larger numbers of people. As communications improved, the number of people speaking a single language could grow far beyond 150, while societies had the potential to grow substantially beyond a thousand members, so that the size of societies came to range widely.

Collective Intentionality: Networks and Hierarchies. Social networks, as I have emphasized, are groupings of nodes (or members) that exist in a spatial relationship, where the nodes can include individuals, I-groups, and we-groups. Society began as a we-group institution composed of the we-groups of communities. Within a society and beyond its borders, networked links of marriage, name groups, or food-exchange networks could take the form either of informal I-group networks or formalized we-group networks. Links among societies were principally I-group links, yet they were sufficient, in their entirety, to comprise the Human System.

A hierarchy is a logical extension of a network. The idea of hierarchy, while it expanded in the Holocene epoch, was not new. Hierarchy had arisen within the family and in the activity of workshops. A hierarchy, in parallel to a network, is a network of nodes, where the nodes are distinguished by rank and not just by space. But the nature of social hierarchy becomes both more complex and clearer when it is expressed in terms of collective intentionality. As with a network, a hierarchy can be an I-group hierarchy of individuals in a common situation who are different in rank; or it can be a we-group hierarchy in which the members, though of different ranks, agree to their place and their role in the hierarchy. As examples, debt slaves might find themselves in a hierarchy controlled by the holders of their debt. From the standpoint of the slave-owners, this is a we-group hierarchy. From the standpoint of the debt slaves, it is an I-group hierarchy – unless they become fully reconciled to their status, in which case it could be a we-group hierarchy for all. Similarly, though at a larger scale, an empire is a we-group hierarchy from the standpoint of the imperial elite; from the standpoint of the conquered, the empire is an I-group hierarchy. In response to this tension, the broad extension of citizenship in the Roman Empire expanded the number for whom Rome was a we-group hierarchy. In the analysis of cultural evolution, which does not allow for explicit agreements to form institutions, the hierarchy of tribal leadership was an I-group hierarchy.

The Problem of Social Fitness in Holocene Society. Holocene societies expanded impressively in their diversity and their complexity. Communities, confederations, and societies, with widely varying group sizes, brought variations in their residence patterns; their clan and lineage social organization; their occupations in artisanal, agricultural, pastoral, and foraging work; and in their cultural and religious expressions. Continuing migration opened new frontiers but also brought endless interchange among ways of life. Further, documentation of human experience is far more plentiful and more secure for the Holocene epoch than for earlier times: evidence from archaeology, genetics, and social anthropology becomes more common and, beginning as early as 5000 years ago, the written record becomes available for certain regions. Linking these sources makes it possible to view human society in the Holocene at an intimate, society-level scale, but also to pull back and view the Human System at continental and planetary scales.[6]

This diversity and complexity of society highlights a tension in the logic of social evolution. In the institution of society, with its expanded scales, the problems of selection and the standard of social fitness become complex and indeterminate. How can it be ensured that an institution provides the intended services to its beneficiaries? Do the beneficiaries have a way to regulate or to eliminate an institution that is not serving them well? Do elites, directing powerful institutions, focus on their own interests rather than those for whom the institution was formed?[7] On one hand, society included horizontal divisions: interest groups formed for foragers and farmers or varying clan groups that were remnants of the separate communities absorbed into confederations. On the other hand, vertical divisions of society could include the difference between an elite and commoners. How were these different interests to be served?

This is a complex issue, in which one could easily get lost in the details. To assist in the discussion, I propose a simplified model of Holocene society. Its purpose is to help in identifying the varying membership groups and interests associated with social institutions; to trace the processes of assessment and social selection, including regulation of institutions by beneficiaries; and to identify debates over social priorities. My model of Holocene-era society assumes (1) society as a social institution; (2) within it, clan identities, retained from the distinct communities that combined to form society; and (3) elites within society who hold special privileges and responsibilities. For the myriad institutions in society, one may argue that they were assessed and reproduced through this structure. The specific processes of an institution might give preferential benefit to society overall, to an elite, or to certain clans.[8] In response to an institutional focus on a certain beneficiary population, other groups within society could protest and call for regulation or even elimination of an institution. What claims might then arise for rejection or regulation of an existing institution by those who feel underserved? For a religion led by an elite, those who governed the institution would be in the best position to modify it, but society overall or clan groups would also be able to press for religious reform. Society was clearly involved in a balancing act, in which the creation of new institutions needed to be balanced against assessment of the fitness of existing institutions. For the numerous institutions within society, who was intended to benefit, who did benefit, and what were the mechanisms for resolving mismatches of benefits and expected beneficiaries?

For institution after institution, I turn now to their creation in the course of Holocene-era social change and to the recurring tension of their reproduction and fitness. Each institution developed under pressure from its inherent dynamics; more broadly, institutions interacted with each other through the expansion of networks and hierarchies in society, although expansion in hierarchy was often without the approval of many members of societies. I make occasional reference to the model of Holocene social tensions just introduced, but otherwise hope that readers will keep the logic of the model in mind as they explore the numerous institutions that populated and sustained the Human System.

Social Evolution, 12,000–6000 Years Ago

The institution of *agriculture* arose at first in the Levant but then with remarkable parallels in multiple regions, including the Sahara, Northeast Africa, Southeast Asia, New Guinea, western Africa, South America, and Mesoamerica. Jared Diamond's well-known account of the rise of agriculture emphasizes technology, the specifics of crops, and disease.[9] In addition to these factors, thinking of agriculture as a social institution draws attention to the commitment of men, women, and children to its processes, including storing, processing, and consumption as well as production. The dynamics of agriculture included the daily routine of tending fields and preparing meals as well as the annual routine of preparing fields, planting, weeding, harvesting, storing, and distributing. Agriculture imposed a stern discipline on its practitioners, a discipline that surely extended to other areas of the lives of farming populations. Fields could be moved from place to place, though on an annual basis. Control of the storage and distribution of harvests led to taxation, seizure, and war. Thus agriculture, while inevitably disciplined, could also become a hierarchical mode of life. Expansion in the scale of the social order expanded rather systematically with the rise of production; hierarchy within that order was not automatic.

The institution of *animal husbandry* arose in the same era as agriculture, sometimes earlier and sometimes later. Among the early domesticates were sheep, goats, cattle, chickens, donkeys, and pigs – while water buffalo, horses, camels, llamas, and elephants were domesticated later. Focusing again on the commitments of herding peoples, men, women, and children each played roles in daily feeding and control of the animals, in annual breeding, seasonal moves for grazing, slaughtering, and use of various animal parts as food and in artisanal production. Animal husbandry imposed a social discipline but quite a different discipline from that of agriculture – for instance, it could include personal relations between animals and their human caretakers. For societies that conducted agriculture and animal husbandry at once, the two types of logic had to be fit together.[10] Some pastoral populations became specialists who concentrated fully on care of their animals, but most were farmers as well and continued gathering and hunting. Their villages, if often smaller than those of farming specialists, sometimes included fenced-in areas to hold their herds. At times of seasonal migration, from summer to winter pasture lands, populations met and made the journey together, using the occasion for ceremonies and also for discussions. The improvised fires of the journey alternated with the built-up hearths of the home village. By 6000 years ago in the Eastern Mediterranean, female goats and cattle were being kept alive for longer terms, as the practice of milking began. Milking spread to most but not all of the pastoral populations. Dereivka and Krasnyi Yar, two major archaeological sites of the Eurasian grasslands, give insight into the community life of pastoral peoples.[11] They have received particular attention because they provide evidence on early stages of the domestication of horses. Dereivka, along the lower Dniepr River in Ukraine, has the remains of a settlement that includes horse burials. Dated to 6500–5500 years ago, the site indicates a mix of wild horses and tamed or domesticated horses. Krasnyi Yar, in Kazakhstan, is dated at a somewhat later time, and shows remains of fences or corrals that are presumed to have contained horses. The inhabitants of Dereivka probably spoke Indo-European languages, while the inhabitants of Krasnyi Yar are thought to have spoken Altaic languages (see www.cambridge.org/Humanity, in resource text 12.Eurasiatic).[12]

The *town* formed as a residential institution. The best-known early town – associated both with agriculture and animal husbandry – is Çatalhöyük, an archaeological site that lay at a boundary of grasslands and marshes of the Anatolian Plateau (Figure 6.1). It has been excavated and analyzed since the 1960s. The settlement has been traced back to 9400 years ago, and it was occupied until about 8400 years ago. Its population of 6000 and sometimes more people lived in rectangular mudbrick buildings packed against each other. There were no streets or yards: inhabitants entered and left their homes through the roof by wooden ladder. The town was not walled and there are no indications of warfare in the remains, though it is clear that fires sometimes got out of control. Over the thousand years in which this town thrived, new constructions were built on top of those that had become dilapidated, so that the town

Figure 6.1 Sketch of Catalhöyük archaeological site. [Copied and adapted with permission from Francis D. K. Ching et al., *A Global History of Architecture*, 3rd ed. (Hoboken, NJ: John Wiley & Sons, 2017), 13; © 2017 by John Wiley & Sons.]

became steadily more elevated on its mound. As archaeologists excavated the remains beginning in the 1960s, they found as many as 18 successive layers of construction before they reached the earliest and lowest level.[13] As excavation continues, the site has been protected with a roof to prevent erosion and by guards who watch out for looters. So far as is known at present, Çatalhöyük was a dense settlement of agricultural, herding, and hunting people. Remains of food in the homes show that the inhabitants produced and consumed wheat, barley, peas, almonds, pistachios, fruit, cattle, and sheep – and other foodstuffs in smaller quantities. The paintings and sculptures indicate appreciation – and perhaps reverence – for women, men, and animals both wild and domesticated. Yet so far there is no evidence of a political or religious hierarchy, providing worship or subservience to a central military or political leader or to a governing religious cult. Instead, the inhabitants of this village seem to have developed cooperative living to a new height.

Agricultural towns and societies provide the most abundant knowledge of societies in the early and mid-Holocene epoch. The available results of archaeological studies enable us to trace the expansion of town life, step by step, for the past 9000 years. The Nile Valley, eventually the location of magnificent monuments, supported the development of numerous agricultural villages and even large towns, especially in the Nubian areas of the middle Nile, several thousand years before the rise of Egypt. At the southern end of the Red Sea, archaeologists have shown that the highlands of Yemen included sites of towns that were continuously occupied from as early as 6000 years ago. In Mesoamerica, evidence of village life goes back to the beginnings of maize production, as early as 7000 years ago. Inhabitants of these villages began to make pottery by about 4400 years ago, and some of their settlements had grown to the size of towns by 3500 years ago.

The occupation of *fishing* expanded as an institution: populations probably experienced the greatest continuity in this era. It is certain that even before the rise of agriculture, some large fishing towns existed along waterways and at the ocean shore, in places where humans could harvest fish, shellfish, and other marine life. Locating and examining seaside villages and towns is difficult, however, because the sea level has risen a total of more than 100 meters as the planet has warmed with the end of the last great Ice Age. Yet techniques are emerging for investigating early Holocene fishing villages, for instance along the California coast. In Southern California, near modern Santa Barbara, shell middens were discovered to have been deposited on cliffs above the shore. The middens are heaps of discarded shells of abalone, clams, and other shellfish, after the flesh was consumed. The shells, dated by radiocarbon analysis, show that this population reached back to 9000 years ago.[14] These coastal fishing folk exchanged food with the inland foragers for fruits and nuts. Along coastal Peru, similarly, fishing populations exchanged their catches with inland farmers who grew gourds and sweet potatoes. For inland regions, archaeological records show that, in the era from 10,000 to 5000 years ago, substantial lakes existed in what is now the southern and eastern Sahara. Populations lived there by fishing and hunting. One site that has been excavated in northern Niger revealed 200 burials, including a young woman buried next to her two children, the children's arms stretched out toward her. In another burial from the same site, a woman wore a bracelet made of hippopotamus ivory.[15] As agriculture expanded, however, fishing and boat handling came to comprise a smaller portion of overall human activity. The advances in harvesting the resources of rivers, lakes, and oceans fell short of matching the advances in agriculture and pastoralism. As a result, the proportion of the human population living away from the waters grew. Farmers and herders relied on access to water, though in different ways.

Leadership. Expansion of these Holocene institutions took place through formation of networks reaching across space – both informal networks as in the exchange of goods and formal networks, as in links of families by marriage. While the social and technological change of the Holocene epoch broadened networks within and beyond societies, the same pressures brought continuing pressure for the creation of expanded structures of leadership to coordinate or benefit from the many areas of change. Flannery and Marcus identify two contrasting directions in which the pressures for diversification of leadership could lead: achievement-based organization of society and hereditary leadership.[16] I adopt their categories into my model of Holocene social evolution, treating them as institutions of societal leadership: as *achievement-based* and *hierarchy-based* leadership. The two approaches had different dynamics and ritual accompaniments. In achievement-based society, individuals were encouraged to strive to advance themselves – to benefit and to contribute in impressive fashions. At the end of their lives, they might benefit from a monument to honor their achievements, but they could not pass on their power or influence. The principle of hereditary leadership, in contrast, was that a leader who had proved himself through excellence should be able to pass his accumulated power and wealth to another in the family, often a son.

Within both achievement-based mode and hierarchy-based mode, disputes arose and required resolution. The institution of *the judiciary* took form for the handling of disputes and conflicts. Senior figures were consulted to resolve disputes, based on previous experience; law became a growing collection of principles, developed out of multiple decisions. In hierarchy-based mode, the leader was likely to be the judge as well. In achievement mode, the judiciary was more likely to rely on a council of elders. Monument-building in society can be seen as an indication of an achievement-based strategy. Other factors leading to recognition of achievement include formal steps to increase virtue by learning sacred lore, and successfully conducting distant trade. In extended examples, Flannery and Marcus explore the examples of the encouragement of ambition among the Tewa and Hopi peoples of the American southwest and the Mandan and Hidatsa peoples of the Missouri Valley: all created ways for talented people to rise in respect but avoid hereditary nobility. Achievement-oriented societies have tended to exhibit great stability.[17] As examples, along the Fraser River of British Columbia, ambitious individuals sought to acquire luxury items so that the resulting growth in their families brought expansion of their houses; with time, poor families joined the rich and became constrained by debt servitude. Angami Naga men of South Asia could rise up in status through successful headhunting or alternatively by becoming a holy man through study and a sequence of feasts, gaining the reward of a monument at the end. Both cases among the Naga may be seen as providing a social balance of personal ambition and public good.[18]

Did the expansion of agriculture provoke the establishment of patriarchy? This thesis is widely advanced. To put it slightly differently: did the emergence of societies – along with agriculture – bring about patriarchy and the subordination of women? There is little doubt that gender relations were renegotiated as societies emerged out of confederations and as agriculture arose – gender relations are restructured with any major social change. One of the great unsolved problems of world history is whether this era – the time of the agricultural revolution – brought the subordination of women to men. For literate societies, it has been almost uniformly the case that they were organized as *patriarchies*, with men largely monopolizing public positions of power and influence in families and in confederation. Males have dominated even in societies in which families were traced through the female line. The question is whether the rise of agriculture, pastoralism, and expanded artisanal work marked a clear stage in the subordination of women. It is clear that the overall degree of hierarchy in human society expanded as the range of social roles and responsibilities grew in the age of production. Substantial processes of conflict, oppression, and violence surely accompanied the creation of these new roles and the expanding importance of property. But we do not know how far such processes went. Whether the dominance of men over women expanded at a specific time in our history cannot yet be determined. The distinctions in dress and behavior of men and women are visible in the representations of humans from the earliest times. But there is no pattern so constant that it is repeated for every social order – in practice, it seems that there has been a good deal of choice in human affairs as to the roles and relationships of men and women.[19]

Nevertheless, as Flannery and Marcus argue, "the concept of the sacred ... would one day be manipulated to create a hereditary [society]."[20] Hereditary elites have arisen steadily and gained the upper hand in many cases; at times there were cycles of alternation between hereditary and achievement structures. One of the key paths to hereditary rank has been through taking in impoverished neighbors, loaning them resources and then gaining control of them through debt slavery. The institution of *slavery*, whether enslavement was through capture or through debt, grew to include the enslaved, the masters, and others who benefited from enslavement. The dynamics of slavery included capture, foreclosure, seasoning, renaming, service under hierarchy, and then the chance of self-purchase, escape, or rebellion.

Coevolutionary Factors

The process of social evolution, expanding through its own dynamics, encountered other influences. Three such influences are explored in this section, to consider their effect on social evolution: processes of cultural evolution, the warming of the Earth caused by farming, and the geographical and ecological differentiation of human life on the various continents.

Cultural Evolution. The process of cultural evolution, which had developed individual learning in the many millennia before syntactic language, arguably continued in the era of speech. Peter Turchin has articulated the latter-day implications of cultural evolution, arguing that it was the principal process for human social change during the Holocene. Here I summarize his argument concisely, in order to pose the question of whether the processes of cultural and social evolution were complementary, mutually exclusive, or in some other relationship. Turchin presents human communities as having developed slowly during the Pleistocene, as the gradual process of social learning developed the strength of cooperation among individuals, and as the most fully cooperative communities were able to displace and replace the less cooperative, raising the overall level of human cooperation, as described in the specific workings of the Price Equation.[21] Turchin's main historical argument centers on the transformation of society in early and mid-Holocene epochs. He argues that the expansion of agriculture, with concomitant population growth, led to increased violence because of struggles among communities for control of land and other resources. The logic of cultural evolution required that the adoption of agriculture be made all at once for the whole society, in order to establish a new set of norms and institutions, notably the idea of property. With that shift – and a lag of some 5000 years – the main competition among societies took the form of warfare, in which "war serves to weed out societies that 'go bad'." Though the costs of war to the health and social status of individuals were heavy, and though farmers were less well-nourished and less healthy than foragers, the bigger armies of unhealthy farmers could overpower the smaller forces of healthy foragers. That is, the rise of agriculture led, with a lag of several thousand years, to expanded intersocietal war and violence in a contest for control of resources. God-kings arose in consequence, benefiting from expanded warfare.[22]

Equally important in this analysis is the complex turnaround presumed to reverse this expanding wave of violence in later years. That is, the violence worsened with the expansion of horse-based warfare 4000 years ago, in which archers from the Eurasian steppes were borne first by horse-drawn chariots and later as mounted cavalry. These Eurasian steppe peoples posed a mortal threat to the agricultural states to their south. In response, the states produced effective innovations – Darius and the Achaemenids built an empire so large that horsemen from the steppes could not threaten it; then Ashoka of the similarly sized Mauryan empire, after years of conquest, turned to Buddhism to create a great, peaceful realm. While there were many fluctuations in later centuries, Turchin argues that the principal trend was a decline in the proportion of humanity that died violent deaths, as large states maintained peace in the imperial core, while the wars and casualties were pressed to the imperial fringes. The steady growth in cooperation and decline in violence is presumed to have continued to the present.[23]

The analysis, wide-ranging and engaging, is a prominent example of new studies in the framework of ultrasociety.[24] Yet it differs in several particulars from the analysis I propose for the same period, in social evolutionary perspective: could the two be combined? Turchin applies the logic of cultural evolution in its multilevel selection version, in which social groups become relatively homogeneous but with differences among groups in the proportion of their individuals with altruistic characteristics and the proportion with nonaltruistic characteristics. This analysis, while it yields a slow development of increasingly altruistic groups as projected by cultural evolutionary theory, nevertheless includes certain rather sudden changes, such as in the adoption of agriculture.[25] The model thus presents a sequence of the slow development of altruism, a sudden shift to agriculture, and the resulting expansion of violence. While the model is clear, and while violence is surely a significant issue in human history, I prefer to urge caution about centering analysis on the two poles of violence and altruism in human behavior, in that the ongoing analysis of emotions emphasizes diversity and interplay of several more emotions.

Agricultural Warming. Quite a different analysis of Holocene life focuses on human agency in climate change, with little relation either to social or to cultural evolution. William Ruddiman, a leading climatologist, has suggested that the expansion of agriculture and animal husbandry, which brought clearing of forest to create fields and grazing of animals, raised average temperatures.[26] Comparing the Last Glacial Maximum with eight preceding great ice ages, he finds that Northern Hemisphere temperature would have been expected to decline after 10,000 years ago, as declining solar radiation brought production of smaller amounts of carbon dioxide and methane, the greenhouse gases that cause atmospheric temperature to rise. His analysis shows that carbon dioxide, resulting especially from clearing of forest for agriculture, rose from its expected level beginning 7000 years ago. Then methane, resulting especially from the flatulence of growing herds of domestic animals, rose from its expected level beginning 5000 years ago. In this view, the first major influence of the Human System on Gaia began well before the Industrial Revolution's burning of coal and petroleum, with the Agricultural Revolution's clearing of forests to create sunny fields. Put another way, the remarkable stability of Holocene temperature was not only beneficial to farmers but was in significant measure also caused by them.[27]

Continental Comparisons. In a third view of Holocene coevolution and differentiation, I offer a global comparison of continental geography, population density, and human society in the era from 4000 BCE to 1000 BCE (6000 to 3000 years ago). I begin by comparing the geography of continents. Africa, humanity's home continent, has a total area of 30 million square kilometers (km^2). Eurasia, the other great portion of the Old World, has a total area of 55 million km^2. Thus, Africa's area is 55% of that of Eurasia. The Americas, North and South together, have a total area of 42 million km^2, halfway in size between Africa and Eurasia. The remaining significant area of human settlement is Oceania, with a total area of 9 million km^2.[28] Each of these continental regions has a very uneven landscape, able to support human population to uneven degrees. Eurasia and North America have great arctic regions that can sustain only sparse populations. Great deserts occupy the Sahara in Africa, the arid region from Arabia to the Gobi in Eurasia, most of Australia, and large deserts in North and South America. The result is that each of the three great continents has roughly similar regions that support agriculture and dense human population. In tropical lands, Africa has the largest nondesert tropical landmass; Eurasia has the second-largest tropical terrain, while the tropical lands of the

Americas are almost equal to those of Eurasia. In nondesert temperate lands, Eurasia has by far the largest quantity, followed by the Americas (especially North America), while Africa has quite small temperate regions at the north and south fringes of the continent. Oceania's tropical lands are habitable (including the northern fringe of Australia), while temperate Australia is mostly desert.

We do not have dependable estimates of human populations for times as far back as 1000 BCE, much less further back. Nevertheless, one can hazard guesses about population density in qualitative terms – distinguishing sparse, intermediate, and dense populations. For Africa, the great belt of savanna, south of the Sahara from the Atlantic to the Indian Ocean, was being farmed well before 1000 BCE and had an intermediate population density, with higher density in some regions. The narrow northern coast of Africa and the wider eastern coast of East Africa also had intermediate population density. Africa's equatorial forest and the deserts, while populated, were sparsely so. In Eurasia, the lands of South Asia and East Asia had become densely populated, while populations were of intermediate density for Europe and for West, Central, and Southeast Asia. The arctic regions and the deserts were sparsely populated. The Americas, because they were recently settled, were less densely populated than the Old World. Nonetheless, settlers in the Americas had no natural enemies, little in the way of disease to face, and they rapidly domesticated a significant variety of food crops. Maize was the most productive crop, but it only began to spread widely about 3000 BCE. I hypothesize that the Americas had achieved intermediate population density along the eastern Pacific coast, the area of initial settlement, and in nearby mountains. The eastern areas of North and South America had sparse populations. But all American populations grew at a high rate, tending to catch up to those of the Old World. In Oceania, population may have been densest in New Guinea, where agriculture had begun at an early stage. Most of the rest of Oceania was sparsely populated by foraging peoples until rice-growing Austronesian-speaking migrants began settling the region in about 2000 BCE. While the Malay Archipelago ultimately became quite densely populated through rice cultivation, most the region was sparsely populated up to 1000 BCE.

Were human communities actually able to maintain contact with one another? Within each of the continental regions, there were routes of connection among most populations, though the distances were often long. The Himalayas and the Asian deserts separated East Asia from South and West Asia, yet these obstacles are known to have been crossed in both early and recent times. In this section so far, I have assumed a clear frontier separating Africa from Eurasia. But that boundary is only conventional and can be viewed in quite different ways. So many people have crossed the boundary that one could simply neglect it, envisioning the entire Eastern Hemisphere as one great unit in geography and in human population. Alternatively, one can note the great ecological similarity of the Arabian Peninsula to Northeast Africa, and draw a northerly line that includes with Africa the territories of Iberia, Mediterranean islands, the mountains of Anatolia, the Persian Gulf, and Arabia. Yet again, one could do as is commonly done today and draw a line south of the Sahara, creating sub-Saharan Africa as a unit, while joining the Sahara and North Africa to Eurasia. This view has led to coining of the term "Afro-Eurasia," including all of Eurasia and northern Africa – a term I will use occasionally.[29] Finally, yet significantly, because all of these variations deal in one way or another with the long borderland between Africa and Eurasia – the Mediterranean, Red Sea, Iberia, northern Africa, Ethiopia, Arabia, and the Levant – one can treat this borderland as a region unto itself, developing its own special character out of its localities but also out of the contributions from both the African and Eurasian sides of the borderland. Each of these interpretations of Africa, Eurasia, and their shared borderland is valuable for studying aspects of the past.

In other comparisons of continental regions, it is striking that each of them – Africa, Eurasia, Afro-Eurasia, the Americas, and Oceania – was the site of a major development of agriculture in the early Holocene epoch. Further development and spread of crops took place within each region and, in the Eastern Hemisphere, domesticated crops and animals spread at remarkably early dates from one continent to the next. What was the degree of networking and hierarchy in the societies of each continental region? Networks of exchange of goods stretched within each continental region well before 1000 BCE. Hierarchy, especially as measured by the rise of states with urban settlements, developed most fully in Eurasia, though Eurasia had many societies that were not governed by states. Africa, in contrast, was striking in its resistance to developing hierarchical states except in northeast regions close to Asia. African technology in agriculture, ceramics, metallurgy, and commercial exchange was roughly on a par with that of Eurasia, but Africans had chosen small rather than large political systems and had chosen to rely more on network than on hierarchy. The Americas were an intermediate case, revealing a continuing emphasis on network rather than hierarchy, though with major exceptions in Mesoamerica and in the Andes. For Oceania, the expansion of Austronesian populations was to be the main source of expansion in hierarchy. Although we do not have dependable population estimates for the early and mid-Holocene, we may phrase the issue rhetorically by asking: at what point in time did Africa cease to hold more than half of the human population?

Further Institutional Evolution, 6000–3000 Years Ago

Mid-Holocene technology advanced with the use of plows for cultivating wheat and barley, the rise of wet rice, and the expansion of maize in the Americas. In social organization, the institution of the *state* arose, in large societies and small, in cases in which a formal governing structure recognized the leadership of a monarch, commonly hereditary. Some monarchs were sharply constrained by councils of elders, while others gained great individual power.[30] When viewing states in the context of social institutions broadly, as is the case here, it is appropriate to consider not only the structure and top-down governance of the state, but also the issue of the social fitness of the state in its societal context, including responses of those unsatisfied with the regime who called for reform, regulation, and even overthrow of the state. Further, states were not the only way ahead, as many Holocene-era societies took up the innovations of the age – agriculture, metallurgy, commerce – without taking up governance through the state. As empires began to rise in 1000 BCE, it may still have been the case that more than half of all humans lived in societies that were not governed by monarchs.

Workshops. The basic format of the workshop remained consistent but expanded widely in this era, with impressive results in the use of fire. The use of fire eventually became sophisticated enough to permit widespread production of ceramics – a type of production that generally required a workshop form of organization. Ceramics required the location and shaping of appropriate types of clay, and then firing in a kiln – an oven that achieved the desired temperature for firing the clay.[31] Additional workers gathered wood, stacked the pots, tended the fire, and prepared for firing the next set of ceramic works. The construction of wagons and carts, which began as much as 6000 years ago in lands north and west of the Black Sea, was another task that required a workshop. Richard Bulliet has shown that early copper mining in the Carpathian Mountains of today's Romania led to the creation of small wagons with fixed-axle wheels, used to carry ore from the underground mine face to the point at which it would be smelted. Larger wagons created later on enabled these populations to move their goods as they changed residences, in what Bulliet calls "vehicular nomadism."[32] By 4000 years ago, four-wheeled wagons had mostly been replaced by two-wheeled carts, and the solid oak wheels had been replaced by spokes so that they were lighter and sturdier.[33]

The organization of workshops remained valuable for artisanal production, creative works, for the expansion of abstract knowledge, and ultimately for the construction of public works. The leading figure of the workshop, who most commonly created the workshop and directed it as well as carrying out the work requiring the highest skill, relied on co-workers to complete the work. Workshops often relied on families and communities for their workers, but workshops could equally well function with workers drawn from a range of families, based on their interest, skill, or availability. Still other types of artisanal work included construction, the making of textiles, and the preparation of food. The construction of houses became more common but not universal in this era. Before the Last Glacial Maximum, people had lived next to rock shelters when possible, though also in caves and in shelters made of brush in open areas. With time, improved forms of home construction were developed, along with specialists for building. Houses were both round and rectangular. Some were made of planks or bamboo; others were made of adobe; tents made of skin or fabric, wattle, and daub (thin strips of wood attached to poles and then covered with mud); wooden houses on stilts; and eventually bricks. Roofs were commonly of thatch. The making of textiles advanced significantly in this era, though the physical remains are scarce. Spinners and weavers created cotton, linen, and silk textiles, along with clothing from quite different materials. Further, once ovens developed and pottery containers became widely available, cuisine could advance to include new types of roasts and stews.

As control over fire and ovens rose, and as temperatures of fires increased, the extraction and purification of metals became possible. Metals contained in mineral material melted once the temperature was high enough, and could be captured, purified, and worked into tools or decorations. Silver melts at 960°C and gold at 1060°C, and the fires produced in many parts of the world were able to purify these metals by about 7000 years ago. Gold and silver decorations appear in the remains of communities as far back as that time in eastern Europe, Central Asia, several regions of China, and along the Nile. At somewhat later times, the manufacture of gold and silver decorations developed in the Americas: in the Caribbean, Mexico, the Andes, and the Southwest of the United States. The institution of *metallurgy* took a clearer form once fires were able to sustain a temperature of 1080°C, at which copper can be melted out of ore. Copper ore, more plentiful than gold and silver, has the disadvantage of oxidizing readily, though Central African metal specialists long worked with it in pure form. But Eurasian metal workers were able to mix copper with tin to create an alloy, bronze, that melted at 900° C and was more workable. In addition, bronze was harder than any previously manufactured metal and was highly resistant to corrosion. From somewhat before 5000 years ago, metallurgy in bronze spread across much of Eurasia and came to be used more for tools than for decoration. Centers of bronze metallurgy included Mesopotamia,

Greece, the Indus Valley, North China, the highlands of the Yangzi Valley, and West Africa. Archaeologists have therefore used the term "Bronze Age" to refer to the period from 5000 to 1000 years ago in Eurasia and Mediterranean Africa. The melting and forging of iron was a more difficult matter: iron melted only at 1538°C. Iron production, therefore, had to wait until furnaces could reach this heat, so that the Iron Age is a story for later.

Workshops, in turn, enabled the undertaking of public works projects. As communities grew larger in this era of expanded production, they took on new projects in physical and social construction. As a result, expanding communities on every continent faced new problems in organization and leadership. It may be that the power to command and exploit emerged far back, in the earliest human days, but in this era arose new opportunities for leadership through command. The need for such leadership arose from disputes among families, decisions to create large projects, coordination of construction and maintenance, and defense against attack. In response to the heavens but perhaps also to some more basic motivations, communities of the mid-Holocene epoch constructed monuments – platforms and pyramids – of ceremonial importance. The most immense of these were the Ziggurat of Ur, in Mesopotamia, ca. 4100 years ago, and the great pyramid of Cheops along the Nile in Egypt, constructed several centuries earlier. But other platforms and pyramids were constructed in North and South America (in the Pativilca Valley of coastal Peru 4900–3200 years ago), and in North China.

The transformation of religious and spiritual beliefs and practices also drew on the workshop form of organization. Even the earliest of humans had spiritual beliefs, as indicated in the offerings recovered from their early burials. But in this era of expanded production, life had become more complex, knowledge had become more specialized, and problems of social organization had become more perilous. Religious beliefs and practices, therefore, became more complex as well: they reflected the new roles, tensions, and the growing hierarchy in human affairs. Included were the recognition of major gods; nature spirits who resided in and controlled such places as rivers, mountains, and forests; and celebration of ancestors – either of one's own family ancestors or the ancestors of elite social figures. Religion included healing practices through spiritual and pharmaceutical ministrations; it also included divination in attempt to learn the future and sometimes it included magic or sorcery to try to change the future.

Knowledge of Time. Observatories provided examples of enterprises that were at once collective laboring tasks and matters of specialized knowledge for an elite. In the dark nights of the world before recent times, people looked to the skies to puzzle at the mysteries of the heavens. The sun and the moon were inescapable; the stars and the planets became familiar to those who studied them. As early as 4000 BCE, societies in many other parts of the world sought to keep careful account of time through astronomical observations. Observations of both the sun and the moon were required. The phases of the moon conveyed the idea of months as a time unit; the movement of the sun from the high point of summer solstice to the low point of winter solstice conveyed the most accurate measure of a year, though many calendars relied on a mix of solar and lunar measurements. In addition to the detailed calendars of literate societies, there are the stone monuments of many other societies that give clear indications of the position of the sun at the summer or winter solstice, thus enabling an exact counting of the days in each year. The great circle of monoliths at Stonehenge in England, dated to between 2800 and 800 BCE, is now recognized as an observatory, for which the placement of the stones enables shafts of light to shine in recognized directions at the solstice. Of the numerous stone circles in Japan, the best known is Nonakado, 42 meters in diameter, which was constructed about 2000 BCE. Parallel discoveries of stones set in Ireland, Brittany, the Nile Valley, and Peru reveal that agricultural peoples throughout the world sought to understand the movements of the heavens with precision. The undated propping of stones in the lands of the Pueblo peoples of the American Southwest also created a shaft of light at the summer solstice. Still, it may be that time was handled differently in achievement-based and hierarchy-based societies, as written calendars arose in the latter.

In this book's narrative of human history, I now shift from one calendar to another: from "years ago" to "years BCE" and "years CE." In previous chapters – and in the first half of this chapter – the references to time have been expressed as "years ago" or "years before the present." This is a system of accounting for times long ago, used especially in geology, archaeology, and genetics but also useful, I argue, for long-term world history. Now I switch to an updated Gregorian calendar. This calendar, created in 1582, gained acceptance as the international civil calendar during the twentieth century, and has been relabeled in the twenty-first century to use secular rather than religious terminology, beginning with the year 1 of the "Common Era (CE)" as the approximate date of the birth of Jesus and moving ahead to the present; earlier times are labeled as years "Before the Common Era (BCE)."

I chose to set the moment of transition from one calendar to the other at 4000 BCE (6000 years ago) because calendars came into wide use soon after that moment, including the Sumerian, Babylonian, Egyptian,

Chinese, and Mesoamerican calendars – and these were not the only early calendars. Early calendars gave more attention to calculation of the length of the year in days – and to the subdivision of the year into months or weeks of varying length – than to the enumeration of the number of years since a starting point. Maya and other Mesoamerican calendars have been traced back to the fourteenth century BCE. The Chinese calendar (with the year 4720 in 2020) appears to begin at 2300 BCE; Sumerian and Egyptian calendars were somewhat earlier. The advance of these calendars was in literacy and the formulation of a distinctive type of thinking: not just the details of annual variations and changes across living memory but with labels for each year that could be preserved, perhaps for all time.

Cities. The mid-Holocene epoch brought population movements of two major types: urbanization in restricted areas of the world and migration in many parts of the world. The rise of cities – urban agglomerations of more than 5000 persons – brought populations from rural areas to live in compressed cities, either to walk to distant fields or to take up urban occupations. This was in one sense the logical next step after the emergence of towns in many parts of the world during the early Holocene. Yet cities developed primarily in regions where agriculture had led to dense populations, sometimes but by no means always under the leadership of a monarchy.[34] Uruk is the best documented early city, founded just after 4000 BCE. Ur and Eridu were other Sumerian cities, while Faiyum rose at much the same time west of the Nile in Egypt. Dynasties in monarchical cities displaced each other regularly until a new stage was reached with the conquest of most Mesopotamian cities by the Akkadian leader Sargon (2334–2279 BCE). Yet another stage opened with Hammurabi's construction of walls around Babylon in ca. 1800 BCE. Byblos, Aleppo, and Damascus took form before 2000 BCE. Quite different were the cities of Harappa and Mohenjo-daro in the Indus Valley, which persisted from 2500 to 1800 BCE without central authorities. The Chinese cities of Anyang and Zhengzhou arose in the lower Yellow River valley in roughly 1800 BCE; Carthage arose in 1200 BCE on the North African coast. Then cities began to grow in larger numbers and in additional regions: Varani on the Ganges and the earliest Maya cities arose shortly after 1000 BCE.

Agricultural Migrations. In the same era, rural migrations over long distances unfolded in a social process that was quite distinct from the contemporaneous development of urban societies. In most such cases, farming peoples spread to adjoining territories in waves of migration that started and restarted. New technologies arose in the course of centuries and millennia of migration, and varying social structures arose in different parts of each diaspora. Nevertheless, elements of common heritage persisted across each of these migratory streams: the heritage of related languages (which enables the reconstruction of migratory paths) and also common social structures, rituals, and religious practices.

At least four great migratory movements began in the mid and late Holocene epoch. Best known is that of the Austronesian-speaking peoples, who carried out maritime migrations that ultimately dominated maritime Southeast Asia and the South Pacific islands.[35] These migrants, starting at about 4000 BCE in the lower Yangzi valley – with their characteristic outrigger boats, houses built on stilts, and reliance on water buffalo in the cultivation of rice – settled in Taiwan and there differentiated into several subgroups. As of roughly 3000 BCE, certain Austronesian-speakers moved south to the Philippines; they steadily occupied more territory in this archipelago, though encountering the existing inhabitants (believed to have spoken Indo-Pacific languages).[36] By 1500 BCE, one group of Austronesian migrants moved southwest, settling the big islands including Java and Sumatra and maintaining rice cultivation. Another group settled in islands to the southeast, including the big island of New Guinea, where they encountered the taro-growing farmers speaking Indo-Pacific languages. In the combination of these communities, later generations spoke Austronesian languages, used Austronesian outrigger boats, and grew taro rather than rice. After 1000 BCE, their descendants were to settle the uninhabited islands of the Pacific (see www.cambridge.org/Humanity, in resource texts 6.Indo-Pacific and 9.Austric).[37]

Bantu-speaking populations began as farmers relying especially on yams and living in a savanna-forest fringe in what is now the Nigeria–Cameroon border. About 4000 BCE they began clearing lands in the forest to the south, sometimes moving larger distances along waterways by canoe. They encountered and largely absorbed the earlier inhabitants, including the Twa, whose descendants were able to stay on as hunters. After expanding throughout the Central African forest, Bantu-speaking farmers reached the highlands of East Africa. There they encountered inhabitants speaking Afroasiatic and Nilo-Saharan languages. Bantu speakers adopted the local economy, including farming of millet and care of cattle. From 1000 BCE they spread to the Indian Ocean coast and occupied most of southern Africa (see www.cambridge.org/Humanity, in resource text 4.Niger-Kordofanian). The volume *Bantu Africa* by Cymone Fourshey and colleagues provides a remarkably thorough and engaging tour through society and social change of this expanding group of migrants in the mid-Holocene epoch.[38] Rural agricultural migrations were mostly achievement-led, but they could change to hierarchy-led forms.[39]

Populations known as Proto-Indo-Aryan, speaking languages of an Indo-European subgroup, were based in the steppes of Central Asia where they grew wheat, cared for cattle, and raised horses. They shared in the steppe technology of horse-drawn chariots with spoke wheels. These chariots had sustained a culture of heroic warfare in their homeland for some time; eventually the war chariots ventured beyond their steppe homeland to challenge the large states. Starting about 2000 BCE, groups of these charioteers migrated in several directions, forming subgroups as they moved: west to Anatolia and the Levant, south to Iran and India, and east toward China. Those moving southward divided into the Indo-Iranians, who settled in Iran, and the Indo-Aryans, who settled in the Indus Valley and north India more broadly (see www.cambridge.org/Humanity, in resource text 12.Eurasiatic). In Iran and India, the language and religion of the settlers became dominant. In the Indus, the previous cities of Harappa and Mohenjo-Dara declined and disappeared by 1700 BCE.

Remarkably, shortly after Indo-Aryan migrants had settled into South Asia, a latter-day set of Austronesian migrants moved westward across the Indian Ocean. They doubtless stopped along the coast of South Asia (meeting speakers of both Dravidian and Indo-Aryan languages), then continued westward until they reached the eastern coast of Africa. There they left traces of settlement in Ethiopia, where their bananas and Asian yams were adopted by locals. With time they moved southward to the Swahili coast, where they met Bantu speakers who had arrived previously from the west, and where they left bananas, xylophones, and outrigger canoes as contributions to African life. By about the first century CE, these Austronesian migrants settled Madagascar and the Comoro Islands: their Austronesian language (traceable to Borneo) and rice cultivation took root in Madagascar, along with African population and traditions brought from the mainland.

One more group of this general type, though not as fully documented, is that of the Penutian-speaking peoples, whose homeland was among the oak trees and grasslands of northern California. This group differs from those just discussed in that they maintained a foraging economy rather than farming. Nonetheless, the pattern of their language distribution shows that Penutian speakers moved south and east, into what are now Arizona and Mexico (see www.cambridge.org/Humanity, in resource text 13.Amerind).[40] It may be that these settlers adopted agriculture and the cultivation of maize from those who already lived in the region. Penutian speakers moved eastward and became a major part of the population along the Caribbean Gulf coast. In further migration, Penutian speakers moved along the Gulf, ultimately settling in Yucatán and nearby lowlands. The numerous Maya languages of today are a subgroup of the Penutian languages. It is not known when Penutian languages reached Yucatán, but the construction of Maya cities has been documented as early as 1000 BCE.[41]

As indicated by the narrative of these migrations, especially the well-documented Austronesian, Bantu, and Indo-Aryan migrations, linguistic evidence has great potential for clarifying patterns of migration and cultural exchange, during the Holocene but even in earlier times.[42] Classification of languages can identify groupings and subgroups of language that reflect the historical process by which individual languages gave rise to daughter languages. It is of additional value that the inheritance of languages follows a different pattern from genetic inheritance: in language, each language has one parent, while in genetics each organism has two parents. The combination of linguistic and genetic analysis can therefore provide two types of observations on human inheritance.

This review of urbanization and migration in the mid-Holocene epoch provides a basis for asking: what was global in the world history of this time? Humans were widely dispersed across varied lands during the mid-Holocene, yet there remained at least four ways in which they sustained common traditions. First, their common heritage and common instincts provided an inherited legacy – the ancestral patterns of family life carried forth parallel behavior on every continent. Second, environmental changes affected all humans at once – so that the stability of mid-Holocene climate benefited communities in north and south, east and west. Third, interactions among communities spread ideas and practices widely: while global connections were not immediate, the gradual interconnections through exchange and migration eventually put people in contact over great distances. Fourth, human communities followed the shared logic of social evolution: innovative creation of institutions, learning their dynamics, building networks and leadership structures, and developing the means to replicate and regulate institutions.

The Afro-Eurasian Age, 1000 BCE–1000 CE

Two great changes took place, late in the second millennium BCE: one was a human collapse and the other was a human invention. The collapse was the near-simultaneous fall of most large states from the Eastern Mediterranean to the Pacific during the twelfth and eleventh centuries BCE. Natural causes, while central to the collapse of large-scale societies, did not include a great decline in temperature such as the Little Ice Age of the seventeenth century CE but instead a mix of earthquakes and droughts reinforced by shortages of tin and copper for bronze. In a set of events that remains mysterious, a "perfect storm" of factors unleashed the Sea Peoples, a confederacy of naval raiders who harried the coastal towns and cities of the Mediterranean region, ca. 1276–1178 BCE, concentrating their efforts especially on Egypt; not long thereafter, the Shang dynasty collapsed in the Yellow River Valley, later to be replaced by the rising Zhou.[43] The invention was the rise of iron smelting in several parts of Africa and Eurasia. In this second great change, furnaces reached the temperature of 1538°C, enabling the smelting of iron and the production of iron implements once the complex techniques for working with iron had been learned. This advance, achieved by about 1200 BCE in Anatolia, was matched at roughly parallel times at several locations in Africa and Asia, whether in great states or rural areas. As iron ore is much more plentiful than copper or tin, use of metals could now expand. Iron came to be used widely in warfare and for agricultural tools in Eurasia; in Africa, iron seems to have been used for agricultural tools but not so much for warfare. The Iron Age thus began seriously in the first millennium BCE in the centers of its invention and then in surrounding areas. By the beginning of the Common Era, iron production was known to most people in Africa and Eurasia, from southern Africa to the Baltic, and from the Atlantic to the Pacific.[44] Bronze production continued in Eurasia, but especially for decorative materials.

As of about 1000 BCE, societies in many parts of Afro-Eurasia experienced a remarkable era of social innovation and expansion. Sub-Saharan Africa relied on similar technologies in metallurgy and agriculture, yet with fewer domestic animals, as the tsetse fly limited the range of cattle and especially horses. A greater difference appeared in the social order, where African societies gave primary emphasis to what Flannery and Marcus call achievement societies, while Eurasian societies gave primacy to hereditary hierarchy.[45] The era from 1000 BCE to 1000 CE brought remarkable prominence for Eurasian societies in the global order: Afro-Eurasia from the Mediterranean to Japan stands out in this era for its leadership in many areas of accomplishment that had built effectively on hierarchical traditions.

A growing network of *commercial institutions* increasingly linked regions throughout Afro-Eurasia from 1000 BCE, supplementing the exchanges of goods that had been practiced since the distant past. Preexisting institutions for transportation were strengthened – for watercraft, wagons, pack animals, and by head.[46] So also with ports and caravanserai, the terminals and resting spots for merchants on the move. Merchants themselves created the trade diaspora as an institution making it possible to have trusted collaborators in distant markets.[47] Only in the sixth century BCE was the institution of *coinage* developed, with the first Lydian silver coins; other realms and especially Rome subsequently produced coins for use in trade. Cowrie shells, small and elegant shells, were most numerous in the Indian Ocean but circulated as well from the Etruscan region of North Italy to North China; by the fourth century CE they had become established as coinage in much of India and later in Yunnan and West Africa.[48] Bas van Bavel has emphasized the difference between the monetary market for goods, arising with the Lydians, and the systems for allocating the factors of production, especially land, labor and capital. He argues that markets for factors of production arose later and less widely than product markets, and that the institutions of factor markets became the core of capitalism.[49]

The institution of *literacy* took form initially in the fourth millennium BCE in Mesopotamia and Egypt, as hierarchical monarchies were taking form there. An associated institution arose as a companion to literacy, the institution of *schooling*, a grueling system of specialized learning that was necessary to train scribes for the complex cuneiform and hieroglyphic scripts. But by the first millennium BCE, a breakthrough had created *abjads* or alphabetic systems for closely related Semitic languages of the Levant. With only some 25 characters, each linked to a sound in the language, these systems could be learned rapidly and used widely. In the form of Phoenician, such script came to be used all across the Mediterranean. In the form of Aramaic, a similar script came to be used all the way to India. With time, these scripts came to be used for religious, literary, and scientific texts as well as for commercial purposes. Phoenician script inspired the creation of Greek, Latin, and Etruscan scripts; Aramaic led to the creation of Dravidian and many other scripts of South Asia, Central Asia, and Ethiopia. In China, meanwhile, a standard script, relying on thousands of characters, arose by the first millennium BCE and has remained in use since, with periodic reforms. Thus, the institution of literacy included all those who were able to communicate through a given script. The institution of schooling involved the masters and the students learning each language.

The institution of *military force*, formed well over a thousand years earlier, gained greatly in strength in the first millennium BCE. Initially the military was convened in time of war; eventually specialized military forces became permanent, for instance in the Akkadian conquests. The use of horses and chariots in warfare, invented in Central Asia before 2000 BCE, spread to many regions of Afro-Eurasia in the second millennium BCE. Initially, horses drew chariots with archers – their mobility defeated any opponent for a time. Later in the second millennium BCE, mounted cavalry, armed with bows and arrow, with sword or spear, made up a succession of armored units that replaced one another over the centuries. The raising and training of war horses became a central issue in states, and states grew in strength through the use of horses in war. During the first millennium BCE, the greater availability of iron implements made it possible to arm infantry so that they could face the challenge of cavalry.

In *institutions of water supply*, the sophisticated irrigation systems of the ancient Nile Valley and Mesopotamia provided the background for creating aqueducts. An early and impressive aqueduct was constructed in the seventh century BCE by the Assyrian state, bringing water to the city of Nineveh. But it was the Roman Empire, with its systematic construction and support of cities, that brought aqueducts to their peak in architecture and engineering. The aqueducts were created and maintained by a combination of factors: the large urban populations with need for water for drinking, cooking, washing, and bathing; the technical knowledge to construct aqueducts carrying water at a gentle downhill slope for long distances; and the state, able to recruit and direct workers and able to collect funds to pay the cost of workers, materials, and design. The city of Rome had 11 major aqueducts, nine of them built between 312 BCE and 52 CE; the longest was 80 km long. At the imperial city's height, according to one calculation, the distribution system was able to provide more than 1 cubic meter of water per day for each of more than 1 million inhabitants – a large amount by today's standards. In contrast to the large-scale work of aqueducts supplying cities, smaller scale *qanat* systems of water supply developed to supply water to fields and to communities. This system developed initially in the dry and mountainous Iranian plateau, where it brought water in brick-lined tunnels from underground aquifers in highlands to fields and reservoirs in lowland areas. The system spread all the way west to Spain and North Africa. As with aqueducts, *qanat*s moved water by gravity; their tunnels minimized the loss of water to evaporation.

The creation and exchange of knowledge expanded in the literate societies of the late Holocene. While the accumulation of literate knowledge was not yet supported by formal social institutions, the ample libraries of individuals and the copying of manuscripts for exchange helped to preserve written knowledge. Among the types of knowledge preserved in this fashion were the accounts of such travelers as Herodotus, the Buddhist monk Xuanzang, and the Persian diplomat Ibn Fadlan.[50] Quite a different factor, aside from philosophical curiosity, was ultimately to bring a much greater expansion in literate culture – the rise of large-scale religion and philosophy in response to widespread social discontent. The expansion of commerce and of states, along with the frequency of war, created discomfort and unhappiness for many people. The wealthy seemed to control access to the gods, so that religion was little source of comfort to those who were deprived and unhappy. The persistence of conditions that made ordinary people unhappy brought about recurring proclamations of systems of ideas that promised to bring order to the world, recognition to ordinary people, and an understanding of the world.

Here is a very brief and compressed narrative of the emergence and interaction of more than a dozen major religious visions, and the way that they changed each other and changed the world. While religion had existed and undergone transformations for thousands of years, I will refer to the prophets and philosophers who spoke so memorably from 600 BCE to 700 CE as the Founders of *institutions of large-scale religion.* In virtually every case, these new religious orders came to rely on scriptures or sacred texts that conveyed the core of their message.[51] Lord Vardaman, founder of the Jains, and Siddharth Guatama, founder of Buddhism, each lived in the Ganges River Valley in the sixth century BCE. Each felt the inequality and misery of the world, and suggested ways to address them: asceticism for the Jains and a "middle road" of prayer and meditation for the Buddhists. Each built up large communities of believers, and the Buddhists eventually converted the emperor of the Mauryan state, who proclaimed peace throughout the empire. The oral texts from the Founders were committed to written form after many years, and disputes broke out about the meaning of the texts. The competing religious tradition of the Vedas, dominated by the elite of South Asian society, sought to win back adherents among the common people by adopting ideas and practices of the Jains and Buddhists. Thus rose Hinduism, a new religion in practice, but one that was claimed to have existed in this form since the beginning. Map 6.1 shows the centers in which new systems of religion and somewhat parallel systems of philosophy arose from the sixth century BCE to the seventh century CE.

Philosophical traditions arose, in Greece and in China, with various objectives. In China, Confucius composed texts on how best to serve the monarch, and on how to serve the head of a family as an equivalent to the monarch. Texts drawn together and attributed to Lao Tze, in contrast, emphasized ways in which a person can accommodate to the existing world, though it may be unpleasant. When Buddhism came to China, centuries later, the interactions of the traditions led, on one side, to Confucianism and Daoism adopting the practices of a religion, and to Buddhism incorporating ideas from Confucianism (with its priority on orderly hierarchy) and Daoism (with its emphasis on patient survival) into

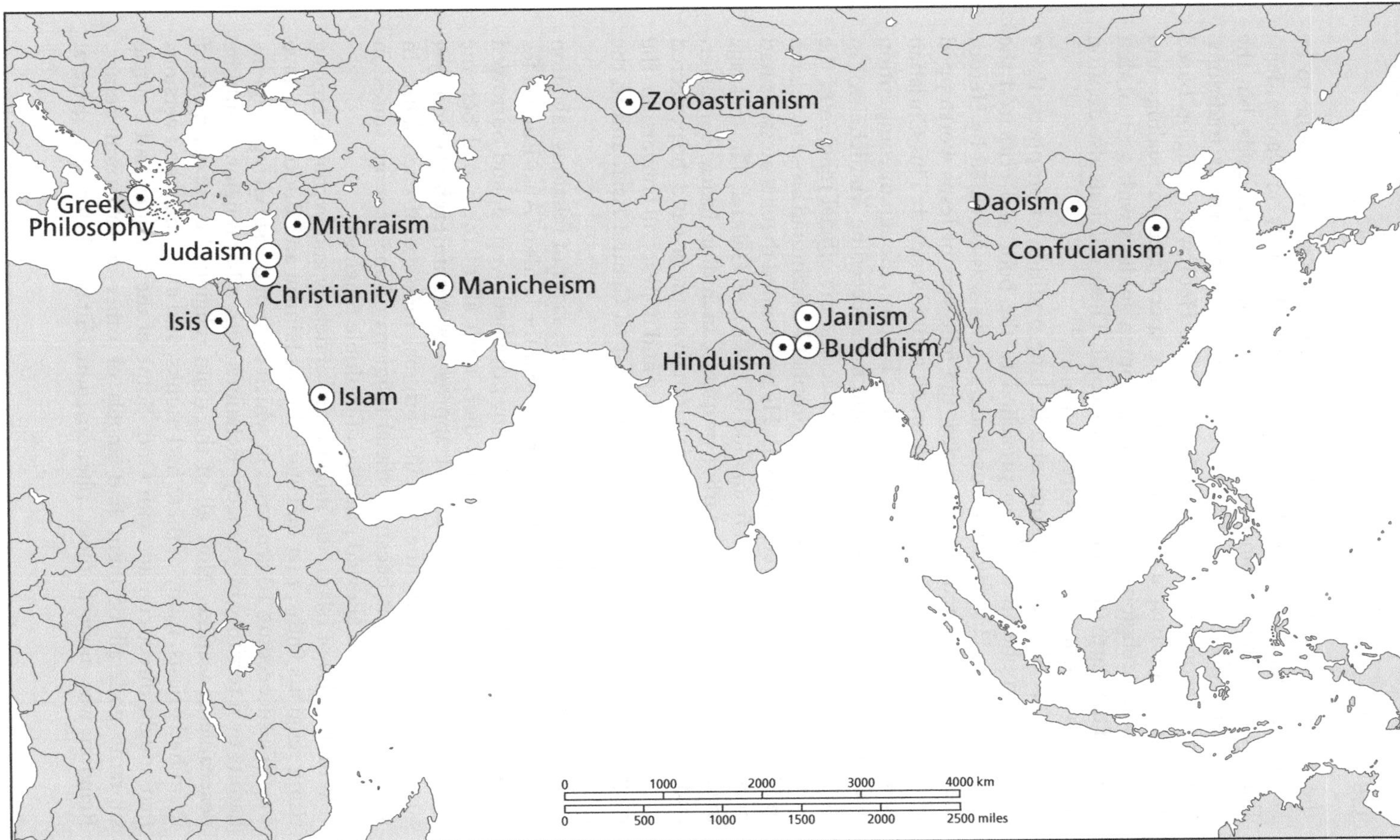

Map 6.1 Homelands of major systems of religion and philosophy. (Redrawn based on original image, courtesy of the University Library System, University of Pittsburgh; generated by Boris Michev, 2019.)

the Buddhist religious tradition. One result was Zen, as it came to be known in Japan, a form of Buddhism that relied heavily on Daoist ideas. Further west, the Roman Empire had its official imperial gods, but the empire also played host to a wide range of other religious traditions. Among these were the ethnic religion of the Hebrews; a religion based on worship of Isis, a female deity who specialized in healing; and Mithraism, a religion popular among Roman soldiers with a ritual that renewed the Earth through sacrifice of a bull. These religions also drew on the philosophical writings of Greek scholars.[52]

Within the Hebrew religion, with its long tradition of prophets, there arose the prophet Jesus, who preached to the poor and criticized the practices of the Hebrew religious leaders. Jesus was executed by Herod, the Roman-supported king of Judea; the followers of Jesus worshipped him as the son of God, through whom all could be saved. This Christian religion spread widely within the Roman Empire, though in competition with Mithraism, the religion of Isis, and the succeeding Manicheism. In just over three centuries, Christianity became the official religion of the Roman Empire. Centuries later, the prophet Muhammad arose to preach in Mecca in the seventh century CE. He proclaimed that his message had been passed directly to him from God by the Angel Gabriel. The message made explicit reference to earlier prophets, including Jesus, Zoroaster, and Hebrew prophets, but asserted that this was the last and the perfected message from God to humans. In warfare and in peaceful conversion, the religion of Islam spread west to Spain, east to Central Asia, and to the borders of the Byzantine state.

By the time of Muhammad, these new messages of religious salvation and ethical practices had spread throughout the lands of Afro-Eurasia for more than a millennium. The religions preached, warred, and borrowed institutions and texts from each other, until all of them had become comprehensive, with a variety of approaches within each religion, yet somehow maintaining the specific, original message of each faith. In some cases, religions became official religions of one or more states; in other cases, states allowed practice of multiple religions.

With the rise of Islam, the millennium of debate and exchange among rising religious faiths, that had started with Jainism and Buddhism, gave way to a succeeding 10 centuries in which each of the surviving religions had its own territory of primacy. Relations with other religions could be peaceable or warlike. Now all of the main religions had texts, most of them had priests, many of them had monks, and all of them had versions that sought direct contact with God. Some of the religions had a single supreme god; all of them had angels or other supernatural beings. Religion was now a mass phenomenon, intended for the general population as well as for the leaders of the society. Each religion had a formal theology, but that meant that debate was possible, so that religious reform and revolution arose periodically from within. As an example of changes that could now take place, the Tang dynasty in China, in a moment of fiscal stringency in the 840s, imposed a massive crackdown on Buddhist monasteries throughout the regime, seizing much of the wealth that was being accumulated in them.[53] In an example of the complexity of this transition, the Abbasid Caliphate, as it took power in 750 CE, rapidly built a prosperous economy and expanded scholarship through translation of works from eastern and western languages – and expanded the formalization of Sunni Islam. The Sunni tradition largely replaced the egalitarian values of early Islam with hierarchy in gender and social structure – and roughly matched the other religions in this regard.[54]

Empire arose at much the same time as large-scale religion – the two great institutions interacted in highly complex ways. Conquests had taken place for a long time, but there was a difference between military conquest of peoples sharing the same language or culture and the creation by conquest of a regime governing peoples of quite different language and tradition. I take as first the Akkadian empire of Sargon in Mesopotamia, succeeded by the Babylonian empire of Hammurabi and later by the Assyrian Empire which, in the tenth century BCE, extended its conquest from Mesopotamia to include the Levant and, for a time, Egypt.

As we address the expanding institution of *empire*, it is time to make explicit a further adjustment to the theory of social evolution because of the increasing complexity of social groups. As a society expanded to conquer other societies, one must now distinguish the homeland from the colonies: the range of elites and ethnicities grew in number with the colonial territories.[55] The imperial state might therefore have taken its beneficiary population to be that of its homeland territory or an elite within the homeland, rather than the population of the overall imperial realm. That is, imperial institutions were always under pressure in that it was difficult for them to satisfy the full range of contending interests. Imperial politics became a contest over which groups were able to set the policy of imperial institutions and even of local institutions within the empire. For religious and economic institutions on a large scale, the reality of hierarchy and diversity set up similar pressures on the social fitness of institutions.

Table 6.1 provides a compressed summary of 18 centuries of empires in eleven world regions. It distinguishes the most powerful empires in each period and region (in uppercase bold type), second-rank empires in each period and region (in uppercase regular type), smaller yet important

Table 6.1 *Empires, 500 BCE–1200 CE*

MAJOR EMPIRE – IMPORTANT STATE – Smaller State – *Invading Group*

	500 BCE–200 BCE	200 BCE–300 CE	300–600 CE	600–900 CE	900–1200 CE
Northern Europe		ROME (300 BCE–300 CE)	*Germanic invaders*	**CAROLINGIANS**	*Vikings* (800–1100)
Mediterranean	CARTHAGE (800–164 BCE) **ACHAEMENIDS** (550–330 BCE) **HELLENISTIC** (312–63 BCE)	**ROME** (300 BCE–476 CE)	*Huns* **BYZANTINES** (300–1450 CE)	BYZANTINES (300–1450 CE) **UMAYYADS** (634–750 CE) **ABBASIDS** (750–900 CE)	BYZANTINES (300–1450) **FATIMIDS** (909–1171) Almoravids (1061–1147) Almohads (1149–1269) Ayubbids (1169–1260)
West Asia	**ACHAEMENIDS** (550–330 BCE) **HELLENISTIC** (312–63 BCE)	**ROME** (300 BCE–476 CE) **PARTHIANS** (247 BCE–228 CE)	**SASANIDS** (224–642 CE)	**UMAYYADS** (634–750 CE) **ABBASIDS** (750–900 CE)	Seljuks (1087–1194)
South Asia	**MAURYANS** (322–185 BCE)		**GUPTA** (320–550 CE) *Huns*		Chola (850–1267)
Central Asia	Greco-Bactrian (225–130 BCE)	Kushans	*Xiongnu* Gok Turk (550–750 CE)	ABBASIDS (750–900 CE)	Ghaznavids (975–1187)
East Asia		**HAN** (206 BCE–220 CE)	*Xiongnu* WEI		**SONG** (960 – 1127) **SOUTHERN SONG** (1127–1279) Jurchen (1127–1220)

Southeast Asia			Funan (100–800 CE)	Srivijaya (650–1250 CE) Khmer (800–1300 CE)	Srivijaya (650–1250 CE) Khmer (800–1300 CE)
East Africa		Meroe (4th BCE–325 CE)	Aksum (50–550 CE)	Nubia (4th–13th CE)	Nubia (4th–13th CE) Ethiopia (from 1270 CE)
West Africa				Ghana (600–1076 CE)	Ghana (600–1076 CE) *Almoravids* (1050–1150 CE)
Mesoamerica	Olmec (to 400 BCE)	Monte Alban (1–500 CE) Teotihuacan (1–500 CE)	Maya (200–900 CE)	Maya (200–900 CE)	
South America					Chimu Tiahuanaco

empires (in lowercase type), and invading groups (in italics). (Note that Rome was a first-rank empire in the Mediterranean and West Asia from 200 BCE, but a second-rank empire in Northern Europe in that era.) It was with the rise of Darius and the Achaemenid state that the era of empires fully opened. In 550 BCE, Darius conquered the Assyrian state, the Levant, Egypt, and the Iranian plateau, combining them into the largest state yet created. Maintaining an empire required an army for conquest and an administration to preserve the state, both on an unprecedented scale. Two centuries later, Alexander of Macedon led armies of Greeks to defeat and occupy the Achaemenid state. The Mauryan empire, of equal size, arose in India immediately after the fall of the Achaeminids. From that time on, empires of roughly the same scale rose and fell, commonly with dynastic terms lasting some three centuries. The empires were mostly in Eurasia, but with time they began to appear in other areas. There were unusual empires. Rome spent 200 years conquering until, in 100 BCE, it had come to dominate the Mediterranean; then it maintained its rule for more than 400 years. The Byzantine empire, a relatively small successor to Rome, survived for a thousand years.[56] The Abbasid caliphate, founded in 750, remained at its full extent only until roughly 850. Within that time, however, the regime presided over a great flowering of scholarship and translation that became the basis of further advances in knowledge in succeeding centuries.[57]

While imperial dynasties rose and fell one after another, the literate communities associated with empires sometimes lasted longer and spread more widely. Literary scholar Sheldon Pollock coined the term "cosmopolis" to refer to the collectivity that wrote and read Sanskrit language, especially from 400 CE to 1400 CE.[58] Sanskrit speakers were not only Hindu and they were not only in India, but extended into Southeast Asia. Similarly, a Chinese-language cosmopolis spread to many parts of East Asia beginning in roughly 300 BCE. A Greek-language cosmopolis formed in 500 BCE; it continued after Roman conquest and lasted on a smaller scale into Byzantine times. A Latin cosmopolis formed with the conquests of Rome and survived, especially sustained by the Catholic Church, until vernacular languages displaced it in the sixteenth century. Arabic and Persian languages, as they became languages of scholarship in the eighth and ninth centuries, created two overlapping cosmopolis.

Throughout the Holocene, the flexibility of social institutions and of the societies they served was sufficient to support dramatic social and technological change. Under initial pressure from climate change, individual and group innovations created new institutions for production, governance, and cultural expression, linked by networks. Yet the inherent problems of institutional fitness and regulation – determining who is to be served and how institutions will be updated – grew along with society itself, allowing expansion in inequality and oppression.

In a well-known global overview, John and William McNeill greatly expanded the discussion of webs and networks in world history with *The Human Web*.[59] In it, they established the plausibility of interpreting the connections in history through a steadily expanding and thickening web. They traced the ways in which religious ritual promoted trade across distances, the importance of monsoon-powered shipping from the Indian Ocean to China, the role of the Silk Road in cross-hemispheric trade, and the long reach of imperial bureaucracy.[60] Their vision of webs at all levels, generally growing over time, held the narrative together, yet they depicted the webs at a metaphorical level, with little detail on how they functioned. My approach is to build on the image of webs by specifying particular forms of social webs – social institutions, social networks, and hierarchies – suggesting ways they were connected to each other and to processes of cultural and social evolution. This focus on social institutions, however, has tended to downplay the common emphasis on technology in world history. As an alternative, Edmund Burke III has documented the nature and the importance of "technological complexes" in the Islamic world and their broader influence. He identifies a total of nine technological complexes in human society and argues that they became standardized in the era 650–1700 CE. He presents in detail how the Islamic world brought about standardization of three of them: the writing/information complex, the water management complex, and the mathematics/cosmology complex.[61] This approach indeed demonstrates the importance of the technological complexes and their social links. If one were to restate these complexes in the framework of social institutions, there would be greater attention to the social agreements of the practitioners and the social dynamics of each type of work, in addition to the details of each technology.

While commercial networks, large-scale religions, and empires all arose at much the same time in the main Afro-Eurasian centers, these three types of human activity were quite different in their internal dynamics. Commercial networks commonly reached beyond the boundaries of empires. Since they were governed by the informal market rather than by a formal administration (though individual ports and marketplaces had administrators), one rarely has dates indicating their initiation or their termination. On the other hand, economic historians have traced qualitative and quantitative data to show the cycles of rise and fall in commercial transactions, the opening and closing of commercial routes, and the shifts in the identity of merchants and the character of the main commodities. Large-scale religions, once successfully created, tend to survive for a very long time – perhaps forever. They may be tied to certain states for a time, but they generally outlast the state. Religions, in their long histories, undergo periodic reform, administrative change, and even change in the guiding religious principles. Also, in their long histories, religions may find that their membership of believers changes place, through migration or conversion of believers. The number of believers may decline to a relatively small total, yet the religion generally finds a way to persist. Empires, on the other hand, have clear beginnings and clear ends. They are tied to political figures, armies, and certain territories. It may be that their administrative structures survive them, as the succeeding empire adopts their practice. Languages of administration may survive the empire that imposed them. In an unusual case, the long succession of dynasties in China has come to be seen as a continuous middle empire.

7 Collisions and Contraction

Collisions, at once natural and social, disrupted evolution of the Human System from 1000 to 1600 CE. The natural world of Gaia and the Human System each underwent independent shifts; these changes interacted with one another, bringing further disruption. From the standpoint of the Human System, the combination of these changes brought tumult and substantial transformation. The shifts in the patterns of Gaia began with a great warming and then a more severe cooling throughout the planet. Within the biological side of Gaia, exchange of biological species took place at all levels. At the tiniest level, invisible microorganisms moved to new regions, bringing disease to humans and to plant and animal species. At a larger scale, plants and animals spread from region to region, especially on the ships of humans. In sum, Earth's map of living creatures shifted significantly in this era. Within the Human System, the campaigns of ambitious warriors – fueled initially by global warming – brought migration of armies and refugees, leading to death, destruction, and replacement of regime after regime. Merchants, sharing in the ambitions of the age, pressed ahead to exchange goods over wider distances, on land and at sea, yielding interactions both peaceable and hostile. A third category of ambitions, in efforts to expand knowledge of the world, brought intensity of study in some centers along with travels and translations to link those centers.

Gaia and the Human System – one fluctuating, the other evolving – collided along multiple axes. Migration of every type expanded, in every region. The movement and mutation of microorganisms brought waves of disease. Infectious disease, communicated from one moving population to another, spread pathogens and brought illness and death long before the disease brought immunity. Rapid maritime connections enabled pathogens and other biota to survive voyages and take root on distant continents. The displacement of plants and animals changed the food chain on every continent and in the oceans. Thus, the path-breaking maritime voyages, although they were triumphs of technology and social organization, brought significant death, disorder, and oppression over a period of centuries. This planetary reshuffling of human and natural resources ended human growth for a time: it brought more decline and destruction than growth and construction, especially as seen in the declines of regional populations and overall human population after 1300.

These collisions, while imposing a heavy cost on humanity, nevertheless laid groundwork for a more integrated and hierarchical system of population, economy, society, and culture – a system that was to expand at an accelerating rate from 1600 to the present. Human ingenuity – and sheer good fortune for some – brought new uses for redistributed plants and domestic animals; expanding systems of social organization in slavery and colonial rule; new technology in war, watercraft, mining, agriculture, commerce; and new thinking in religion. The political map of the world changed repeatedly in the era from 1000 to 1600 CE, not only in Eurasia but also in Africa and the Americas. Ambitious elites launched efforts to subordinate common people through competition, conquest, enslavement, and ideological dispute. Cavalry and gunpowder, deployed intensively in the Afro-Eurasian empires, spread elsewhere in Iberian ships. That is, the momentum of sociocultural evolution was able to carry on, even in this difficult time.

Collisions with Gaia

Temperatures rose in the Northern Hemisphere in the period from about 900 to 1250 CE – the average annual Northern Hemisphere temperature increased by a total of 0.4°C over that period. The higher average temperature and humidity, though varying from one part of the world to another, led to expanded agricultural output and growing human population. This fluctuation, arising from the impact of shifting solar force on Gaia, provided the platform for a set of significant changes in human society. After 1250 came a further shift in temperatures, as Northern Hemisphere average temperatures had declined by a total of 0.8°C (with fluctuations) by 1650 CE, the low point of the Little Ice Age, after which temperatures again rose. The multicentury era of warming is best documented for England, where it is known as the Medieval Warm Period. Detailed observations of the natural world, combined with detailed documentation on English village-level society, confirm the growth in population and social welfare. Of course, the English experience cannot simply be projected to the rest of the world, but recent improvements in the precision of climatological observations show that the more we know about climate variations in various regions over time, the more we see that they are tightly correlated with one another. So I will proceed with the assumption of a general but fluctuating worldwide rise in temperature and humidity from 900 to 1250.[1]

Following the long warming period came the beginning of temperature decline, after 1250. Bruce Campbell has labeled this era – from the late thirteenth century to the late fifteenth century – as the Great Transition. Campbell's volume, centering on well-documented medieval Europe but including an impressive assembly of worldwide data, argues that a systemic framework reveals the broad patterns of change in that era.[2] Throughout the early second millennium, in times of rise and fall in temperature, human migration facilitated the migration of other biota. Disease vectors, other microorganisms, insects and arthropods, domestic and wild plants, domestic and wild mammals and reptiles – all of these spread by land and by sea. While the Black Death is the best known element of this shift to negative natural influences on human society, it was not the first. For instance, in the late twelfth century, epidemics of rinderpest and sheep scabs arrived from the East to afflict English cattle and sheep. As Campbell shows, from the eleventh to the late thirteenth centuries, medieval European societies were able to take their environmental and epidemiological conditions more or less for granted. But changes built up to a tipping point in the mid-fourteenth century that signaled irreversible change and the emergence of a new regime of interactions, reflected especially in a decline in temperature from 1250 to 1470, accompanying what Campbell as called "a massive implosion of Old World populations."[3] Then, after a half-century turn to modestly increasing temperature up to 1520, temperatures again continued to decline until the late seventeenth century, the low point of the Little Ice Age. Gaia's momentum shook the Human System.

The Black Death, in which the bacterium *Yersinia pestis* hitched rides on a range of insects and mammals, spread in East and Central Asia, devastated Europe, and took lives in every region of the Eastern Hemisphere, including much of Africa, for hundreds of years thereafter. The great outbreak of plague in the mid-fourteenth century may have started in China, where a devastating epidemic (for which the precise pathogen is yet to be documented) was reported in the aftermath of a 1331 famine in Hubei province. Another specific reference is for 1339, when burial records in eastern Central Asia show numerous deaths from bubonic plague.[4] By 1345, the Black Death had reached the lower Volga metropolis of Sarai and the Black Sea. In 1347 it reached Constantinople, Alexandria, and Genoa, and within another year it had engulfed both of the Mediterranean shores, then infected northern Europe by sea. From Cairo the Black Death moved up the Nile toward Nubia and along both sides of the Red Sea, reaching Mecca and Ethiopia.

By 1349 the pandemic had encompassed Palestine and Syria. Less well attested, but certainly important in its impact, was the 1346 spread of plague to Iran.

The pandemic reached Europe at the worst time, reinforcing deteriorating environmental conditions: it struck women especially, and its mortality rates were high even in sparsely populated rural areas, peaking in November and December of 1347. Roughly one third of the populations of Paris and of London died of the plague, and death rates in Scandinavia were similar. Death rates in Germany were lower; the pandemic had little effect in Bohemia or Russia. Social movements occasionally arose, reflecting feelings of fear and desperation, and a need to find a cause for the disaster. The earlier European practice of flagellism returned abruptly in the Rhine Valley, but also in Iberia, France, and Hungary: men stripped to the waist and flogged their backs with scourges, singing songs of religious penance.

In yet another human–natural interaction, the high level of human mortality is likely to have brought about climate change: mortality from the Black Death meant a decline in agricultural output and in animal husbandry. William Ruddiman has argued that, as forests grew up where fields had been and herds did not propagate, the production of carbon dioxide and methane resulting from farming declined. With lesser output of greenhouse gases, less solar radiation was retained within Earth and its atmosphere, so that temperature declined, roughly from 1400 to 1600.[5] Ruddiman traces an earlier instance of global temperature decline likely provoked by epidemics affecting humans, from 200 to 600 CE. The pandemic known as Justinian's Plague, in the fifth century CE, was part of this mortality peak. The recovery in human population and the expansion in domestic animal populations, after the seventh century, increased emissions of carbon dioxide and methane that contributed to the Medieval Warm Period, from 900 to 1300.[6]

For the era of the Black Plague, the effect of population change on temperature may have been larger than Ruddiman estimated. That is, while his calculations assumed population decline in Europe, China, and India, we now have indications that plague brought additional mortality in sub-Saharan Africa. Although no samples of pre-1900 *Yersinia pestis* have yet been discovered in sub-Saharan Africa, evidence is mounting in support of the hypothesis that the Black Death spread to that region, with some delay, and caused substantial losses in population. Gérard Chouin's archaeological work in the center of modern Ghana revealed earthworks that had been abandoned suddenly in the fourteenth century. Further study made clear that the well-known commercial city of Jenne-Jeno, on the middle Niger, had been abandoned at much the same time; the earthworks and the extraordinary bronze sculpture of Ife, in modern Nigeria, also seem to have been abandoned at the same time. A large-scale, multidisciplinary project is now pursuing those issues.[7] In parallel work, historian Monica Green has conducted innovative analysis of the full genome of *Y. pestis* as it is known today through its many existing strains. This has enabled her to reconstruct much of the genetic change in plague from the time of Justinian's Plague to the present.[8] An outstanding element of her analysis was that a new strain developed, probably in the Caucasus in about 1500, out of the principal strain that infected Central Asia, West Asia, and Europe. Further, she is able to trace this strain's movement through India on its way to East Africa, where it has apparently remained since. The growing impression is that, in the fourteenth and fifteenth centuries, plague caused great loss in life throughout Afro-Eurasia, especially in its initial spread, but also in the echoes of the epidemic for the following three centuries.

The other great spike in mortality was the loss of millions of lives in the Americas in the years following the opening of regular contact with Europe and Africa – and, eventually, Asia. Human migration was the initiating factor: migration brought warfare, conquest, formation of new families, and other social transformations. Equally important, migrating humans carried microorganisms within them and carried plants and animals of all biological orders on their ships. Indeed, the sixteenth-century movements of biota at all biological scales, in what Alfred W. Crosby called the Columbian Exchange, exceeded all previous exchanges.[9] While the Spanish-led social conquest of the Caribbean, Mexico, and Peru brought widespread death and disruption from 1490 to 1550, disease continued its devastating expansion through Amerindian populations of the entire hemisphere. The evidence in written documents for this population decline is clearest for Mexico, where it is argued that the principal sources of illness and death included smallpox, typhus, measles, influenza, mumps, pertussis – and bubonic plague.[10] These diseases, which had long since spread throughout Europe and Africa, gradually came to the Americas with migrants, and spread among unprotected populations. Similar processes spread disease and other biota throughout the Eastern Hemisphere as well. Thus, it has been shown that syphilis arrived in Japan by 1512: it reached Melaka in 1511 among Portuguese mariners who had contracted it directly or indirectly from the Americas. In Melaka, a trading vessel from the Ryukyus was in port at the time of the Portuguese conquest; as it reached Japan, some of the sailors had brought syphilis with them.[11] Through processes such as these, migratory collisions may have brought population declines in several parts of the Old World. To give one example for a slightly later time,

Siberia in the sixteenth and seventeenth centuries experienced epidemics and population decline as Russian merchants, missionaries, soldiers, and administrators spread out through the region. The eighteenth century brought parallel population declines to many Pacific lands.

The collapse in population of the Americas in the sixteenth and seventeenth centuries, as a result of new diseases imported from the Old World disease complex (Afro-Asian as well as European), similarly resulted in a decline in farming until the combined population of Amerindians and immigrants rebounded. The Amerindian population decline is confirmed in yet another fashion by genomic data, which suggests a population decline of at least 50% in the sixteenth and seventeenth centuries.[12] This same logic ought to be extended to the Old World: that is, researchers have tended to assume that only the New World suffered population decline from the global spread of disease. Genomic tests on Old World regions, parallel to the initial estimate for the Americas, may enable us to see whether Old World populations declined as part of the Columbian Exchange. As Ruddiman argues, the Little Ice Age, running broadly from the fourteenth to the eighteenth century, with its low point at about 1650, corresponds to the combined results of population loss from the Black Plague and the Columbian Exchange.[13]

What is the meaning of such pandemics in human history? In every age, people have divided their thoughts between fitting into the world as it is and trying to control its outcome. The Black Death and the American pandemics provided a powerful affirmation for people of each hemisphere that humans do not have full control over their destiny. This message had its relevance not only for the Mongol rulers who sought to control the whole world, but to the many others who sought to control small pieces of it. The disastrous spread of deadly bacteria may have resulted from the Mongol achievement of expanding human contact throughout the Old World, but the details of life and death were far more complex. Perhaps more clearly than before, the interactions and fluctuations in the Human System and in Gaia could be seen – thereby suggesting that the human project of remaking the natural world was to face unexpected limits.

Terrestrial Collisions: Ambitions and Warfare

How serious was warfare in this era of collisions from the eleventh through the sixteenth centuries? We lack any consistent measure of the degree of violence and warfare in human society – at least before the twentieth century, in which nations and national statistics are more specific but still dubious. Yet the analysis of violence and warfare remains an important issue in human history, and researchers have debated various approaches to documenting and interpreting it.

I join in the debate especially to explore the place of warfare in the process of social evolution – the ways in which warfare draws on individual motivations, the motivations of groups, and the interactions of social and commercial institutions with each other and their environments. I see the early second millennium as a war-torn era of the past and seek to affirm the level of carnage, the consequences, and something of the causes. While the regimes that fought each other were sometimes in hierarchies, rather rarely in networks, I will speak of them as individual regimes. Within the regimes, I center on the ambitions of warriors, merchants, and inquiring scholars – and the responses of subordinated commoners – asking how individual emotions resonated with the institutionalized ideologies and *zeitgeisten* of social structures. To this end, I find it of interest to explore Jeremy Black's concept of "bellicosity," which stems from "a model of transcultural variations" centering on "the notion of each state having a distinctive strategic culture."[14] I cannot now verify the hypothesis that social institutions such as armies and monarchies could channel the interactions among biological emotions, or the institutional dynamics of military discipline, or the expression of group emotions such as bellicosity. Yet I can at least draw on a narrative of medieval conflicts to illustrate the logic of positioning warfare at a crucial juncture in a multilevel social nexus, so that it appears as both cause and consequence in processes of social change.[15]

The years of the early second millennium CE were war torn. As I argue, the expanding vegetation of this warming era provided resources for ambitious warriors. Their campaigns encouraged each other and yet canceled each other out. Even China, in this era, serves as an example of the period's waves of destruction. In 960 CE the forces of the newly created Song dynasty conquered most of China and set up their capital at Bianjing (now Kaifeng), near the Yellow River in North China. After more than a century of battles along the northern frontier, the Song were expelled from North China by the Jin invaders; the Song moved their capital and its archives south to Hangzhou in 1127. Then the expanding Mongol army, after eliminating the Jin, challenged the Southern Song in 1253, attacking from all sides until the Song collapsed in 1279. Again, the capital and archives were moved, now northward to the new capital at Beijing. The Mongol-dominated Yuan dynasty lasted less than a century, and in 1368 a rebellion within China brought the Ming dynasty to power. After the fighting, the victors moved the capital and its reshuffled archives south to Nanjing.[16]

A parallel wave of conquests took place in North Africa, at that time a heartland of the Islamic world. The Fatimids (909–1187), Almoravids (1062–1147), and Almohads (1147–1269), each dynasty professing a different version of the Islamic religion, conquered each other in North Africa and Iberia.[17] In the same era, Seljuk Turks established an empire (1037–1194) from Iran to Anatolia. Crusader armies, venturing from Europe with the aim of conquering the Holy Land for Christianity, challenged the Seljuks episodically. Further wars included a Crusader sack of Constantinople, 1202–1204, and European reconquest of Jerusalem for much of 1229–1244. European Christians, expelled from the Levant, engaged Muslims in further Iberian and Mediterranean wars. The Christians of the Nile Valley lost in the fighting and lost again as Arab migrants moved into the middle Nile region from Egypt; in Ethiopia, however, the Abyssinian regime survived the Muslim threat from Yemen and grew in strength.[18]

Summarizing this picture, R. J. Barendse has argued for a "Eurasian process" of feudalism from the tenth to the thirteenth century. In it, warhorses were developed to unusual size and strength, while warfare came to be led by local lords who steadily subordinated independent peasants to become dependent producers of grain in a "class struggle" within the peasantry. The result, he argues, was an expansion in warfare and a subordination of common people in the immense space from Central and South Asia to Western Europe (if so, it included northern Africa as well). Stephen Morillo, in response, argued that local variety was such that "there was no warhorse revolution, no pattern of increasing warrior dominance socially, and no pattern of changing peasant production across Afro-Eurasia between 900 and 1200," and that there was no such thing as feudalism in this era.[19] This 2003 debate is not resolved, but recent data tend to add support to Barendse's vision of broad trends throughout Afro-Eurasia, especially in the Medieval Warm Period. Growth in taxation of commoners, enclosure of their lands, their enslavement especially in warfare, and the forced religious conversion that could take place – these processes are likely at once to have subordinated the commoners of wide regions and to have provoked uprisings. Thus, I reaffirm the argument that there was a continuing process of imperial competition and mutual destruction, accelerating that of the previous millennium. Empires never became stable; they failed to develop a social archive for reproducing their regimes beyond several generations.

The wars of the early thirteenth century created three major empires – the Delhi Sultanate, Mali in West Africa, and the Mongol Empire. The Delhi Sultanate of North India grew out of a 1206 invasion by Turkish- and Persian-speaking warriors based on the remnants of the collapsed

Ghaznavid state. The new sultanate rapidly imposed an Islamic state on the Indus and Ganges valleys, repelled an invasion by Chinggis Khan himself in 1223, and was later to undergo repeated but ultimately unsuccessful attacks by Mongol armies. At much the same time, the Mali empire arose in the West Africa savanna: in ca. 1230, the young Sundiata Keita gathered allies to challenge the regional hegemon of Sosso. Relying on cavalry, supplemented by archers and sword-bearing infantry, Sundiata's army vanquished Sosso within a year. The monarchy came to control a span of 1500 km of grasslands across the valley of the Niger River, an extensive commercial network and the major source of gold for the Old World.

The rise of Chinggis Khan and his unification of the peoples of the steppe led to extensive conquests by the time of his death in 1227. To the west, Chinggis confronted and destroyed the recently created realm of the Khwarazmshah in Central Asia; to the south, he defeated the Jurchen Jin. The Mongols were never more than 700,000 people in an empire that grew to rule perhaps 200 million subjects. The royal family as well as top administrators and generals – divided early into factions centered on Chinggis's four eldest sons – came from the Mongol elite that first organized a unified military machine out of competing tribes. Recent work led by Nicola di Cosmo shows the importance of environmental factors in the rise of the Mongols: a long period of cold and dry weather, limiting Mongol expansion, was followed by a period of rapid temperature rise supporting expanding pastures, making it possible to reproduce horses and expand the Mongol cavalry to conduct warfare across wide distances. The four contending Mongol states created new practices and exchanged old practices: they doubled the territory of Chinggis's conquests and stirred a continental pot of contact in religious, administrative, and cultural practices from Lithuania and Mesopotamia in the west to Java and Korea in the east.[20]

Warfare not only had its own internal dynamics; it also interacted both positively and negatively with commerce. The armies (discrete we-groups of institutionalized military force) and the commercial networks (overlapping I-groups of informal commercial links) did not fit precisely. Janet Abu-Lughod, in her classic study of Eurasian commerce, shows how commerce reopened between episodes of fighting but halted as war returned. She interpreted the late thirteenth century as a time of prosperity and commercial linkage throughout Afro-Eurasia, benefiting from the warm climate and imperially imposed peace.[21] Her interpretation, addressing the period 1250–1350 and focusing especially on the year 1300, provided a good description of optimistic and ambitious expansion at the end of the Medieval Warming. She traced the hemispheric flows of silk, pepper, spices, silver, gold, diamonds, pearls, emeralds, furs, textiles, ceramics, tea, sugar, wheat, horses, and cattle; cowries may have served as currency along with silver. Yet the first 30 years of her century-long study period were occupied by the ferocious Mongol war to destroy Song China. In the west, Fatimids and Ottomans limited the trading voyages eastward from Venice and Genoa, so that Mediterranean links to the east were only those of Venice through Alexandria.[22] Abu-Lughod argued, sensibly, that the thirteenth-century world system she described was different from the seventeenth-century world system described by Immanuel Wallerstein. She showed the rapidity with which commercial networks could be reopened across Afro-Eurasia, by land and sea, in relatively short periods after the end of hostilities.[23] But disruptions – Mongol conquest, wars of Mongol succession, battles at the imperial fringe, and the great pandemic – combine to show that the success of the thirteenth-century world system was episodic and relatively brief. Warriors, with their conquests and taxes, disrupted the marketplaces of merchants.

Other interpretations of war, commerce, and social change combine to show that no single interpretation will suffice for the Era of Collisions. For the lives of commoners, Bruce Campbell's analysis of climate and disease, at a European level, shows long-term and sometimes sudden fluctuations. But multiple analytical perspectives are inevitable because of the variations in topics, scales, and time frames of this complex topic. At the level of citizens in towns on three continents – active in governance, commerce, charity, or militias – Maarten Prak emphasizes the stability of these institutions even in difficult times. For Italian towns, Bas van Bavel argues that popular rebellions succeeded in bringing equality that permitted economic and institutional growth, though with later consequences of inequality and social sclerosis.[24] These and other examples indicate that, while there is significant evidence of intense warfare in the early second millennium CE, it is not easy to link that warfare to the social welfare of commoners.[25]

Further up the social scale, scholars and bureaucrats – elite specialists who might benefit from post-conquest prosperity – could lose everything with the next war. The institution of the *university* had taken form in Buddhist theological institutions of the first millennium CE, though Buddhist universities did not survive Islamic conquest. Subsequent Islamic universities, such as al-Qayrawiyin in Fez, focused on theology and law; Christian universities, beginning with Bologna, later added medicine to theology and law. In these fragile institutions and, more commonly, through individual study and informal exchange, learned individuals explored many overlapping issues. Occasionally there were efforts to institutionalize knowledge, especially for astronomy. As the Mongol leader Hulegu destroyed Baghdad and the Abbasid Caliphate, creating the Il-Khan regime in 1258, he moved immediately to support construction of an astronomical observatory at Maragha in northwest Iran, under the noted astronomer al-Tusi. The observatory gathered a large staff and library, and widely distributed manuscripts on its observations – it survived for roughly a century.[26]

In subsequent fortunes of war, two of the four Mongol states – in China and Persia – collapsed in roughly 1350. The other two Mongol regimes of the Chagatai and the Golden Horde (in Central and Northwest Asia) survived, but only because of substantial reorganization. Was it then the Black Death that brought the collapse of the two most populous Mongol regimes? Or had the Mongols already created enough enemies and divisions among themselves to bring down the empire? Did the cultural exchange of the Mongol era come to an end after 1350? Recent research gives a clear answer to the last of these questions: while there was never again a unified Eurasian realm, Mongol patterns of governance, commerce, and intellectual exchange reappeared for centuries in successor states across the Eurasian land mass. The Mongols themselves persisted as elite families in larger societies, and they maintained their homeland in the eastern steppes. Turkish peoples, closely associated with the Mongol regime, remained very significant in Iran, in the realm of the Golden Horde from the steppes of the Volga region to the Black Sea, and especially in the Chagatai lands of Central Asia. Territories at the fringe of the Mongol realm had been restructured by their tributary relationship, as with the Lithuanians, Muscovy, and Korea. Horses and equestrian life continued to be central in government and warfare: horse populations grew in all the regions that could support their use in warfare and regal display.[27]

Two major successor states arose in ca. 1370. In China, the leader of a peasant rebellion proclaimed the Ming or "brilliant" state, vowing to renounce all Mongol practices. Later monarchs, however, adopted many of the administrative structures and practices of their Mongol predecessors. For a time, the dynasty encouraged maritime commerce and navigation, but then pulled back to build a great wall in the north and fight the steppe peoples.[28] In the western Chagatai state, the amir Timur seized military power and extended a new realm from his capital in Samarqand to all of Persia and lands to the west. He relied expertly on a mix of artillery, cavalry, infantry, and archers, and focused more on destroying his enemies than on seizing their lands. He seized Persia, occupied Moscow, splintered the Golden Horde regime by destroying the great commercial center of Sarai in 1396, and later destroyed Aleppo,

Damascus, Baghdad, Delhi, and Ankara. After Timur died in 1405, his Timurid successors ruled smaller territories until 1550.

Let us pause at this point to consider the broader question of war in society in Afro-Eurasia from the eleventh through the fourteenth centuries CE. In this book as a whole, I seek to identify specifics of a missing link of biology and social structure. That is, I argue that individual emotions, generated out of biological ancestry, can be reinforced and transformed in the group atmosphere of institutions and networks. I ask whether the warm conditions of the early second millennium reinforced emotions of ambition among individuals, which contributed in turn to ambition and bellicosity among warriors and their leaders, so that a broader zeitgeist of ambition for control spread to social institutions throughout the world, encouraging warfare and hierarchical government, accompanied by individual and group motivations of ambition, greed, and desires for domination. Jeremy Black's identification of a culture of "bellicosity," varying among societies, is valuable in considering how readily groups go to war and how energetically they fight.[29] That is, bellicosity can be seen as the Afro-Eurasian zeitgeist for this time period. As such, the concept may help us peer into biological and social-evolutionary dimensions of warfare. As I argued in Chapter 6, the role of a zeitgeist in history is that the character and dynamics of leading institutions – for instance, agriculture and artisanal work in that earlier time – privileged certain motivations.[30] Out of the full range of biologically supported emotions and motivations, certain outlooks gained reinforcement from social institutions. Feedback, linking institutional behavior reinforced with the motivations and outlooks of individuals, could lead to a prevailing outlook or ideology – a zeitgeist. Yet such a prevailing outlook was sure to lead to disagreement in some quarters and might also lead to results that were harmful to the interests of society.[31] The era from 1000 to 1600 CE opened with ambitious expansion within many human communities: expanded migration on land and by sea, within continents and among continents.[32] In short, global interconnection brought dreams of power, wealth, and knowledge, but also brought conflict and disaster within society and difficult interaction with Gaia.

Further, I argue, ambition at the group level in the early second millennium aimed not only at displacing competing dynasties but also at subordinating common people to the will of elites. High taxation, recruitment for war, land grants to nobles, enslavement, gender exploitation, and suppressing selected ethnic groups were among the techniques of subordination. To some degree, commoners accommodated to the expanded pressures. They also fought back in one way or another, such as with peasant revolts, slave revolts, and court cases.[33] Institutions of large-scale religion, in their early days, had occasionally held sufficient power to restrain elites and provide support for common people. But by the second millennium CE, religions had come to be closely associated with states and the power of governing elites. Thus, in this era of collisions, religion did more to encourage warfare than to restrict it.

For additional observations on issues of ambition, war, and zeitgeist in the early second millennium, we may turn to the writings of Ibn Khaldun (1332–1406), perhaps the most outstanding observer of society in his era and a deep believer in a Sufi-influenced approach to the Islamic religion.[34] In his North African homeland, the preceding four centuries had seen recurring rise, decline, and mutual destruction of empires which, though powerful in their time, left little in their wake. Ibn Khaldun challenged hereditary realms, identifying the cycle by which early monarchs, exhibiting discipline and reliance on law, gave way to reigns of luxury and greed, followed by collapse. He debated the institutional structure, not just the character of the rulers.[35] Near the end of his life, Ibn Khaldun experienced imperial crisis at a personal level. Residing in Cairo under Mamluk rule, he was sent to Damascus in 1400–1 as Timur was besieging the city. Timur, knowing of the scholar's reputation, sought out conversations with Ibn Khaldun – the memoirs of the latter show that the discussions were wide ranging, but that Ibn Khaldun was discreet in expressing his opinions about imperial politics, offered praise to Timur, and was at last allowed to return to Cairo.[36] Timur completed the destruction of Damascus as he had destroyed Delhi, Baghdad, Sarai, and Ankara. His realm, however, declined during the century after his death and then underwent conquest by the Safavids. Ibn Khaldun's analysis and his account of debates with Timur, meanwhile, survive as a critique of flawed political order.

John Darwin, tracing imperial history after the lapse of the Mongol and Timurid states, effectively documents the continuity of bellicosity. Widespread imperial expansion – of the Ottoman, Ming, Iberian, Safavid, Mughal, and Songhai states, in the fifteenth and sixteenth centuries – encompassed great territories but failed to establish lasting dominion.[37] The Delhi Sultanate, after a century of consolidation, briefly extended its control to South India but was driven back by both Muslim and Hindu armies. Throughout this era, Turkish and African men, in slavery and freedom, were brought to South Asia as warriors in these and other struggles. Meanwhile, far to the west, the states and religious institutions of Europe experienced the expansion that would bring the rise of national monarchies; nevertheless, continuing environmental calamities, even in Italy during the sixteenth-century Renaissance, confirm the recurring difficulties of the age.[38] In Africa, beyond the reach of Eurasian empires but in contact with commerce of the Mediterranean and Indian Ocean, states arose in the Great Lakes, Malawi, Zimbabwe, Kongo, and along the West African coast from Yoruba to Akan. In the Americas, impossibly far away from the Mongols, lived peoples who nevertheless experienced substantial expansion in states during the fourteenth and fifteenth centuries.[39]

Oceanic Collisions: Encounters in the Seven Seas

As armies of men charged across plains, as the besieged of both sexes held out behind walls, and as peasant women and men tilled their fields while artisans crafted their wares, so also did the crews of ships, mostly male, sail the seas. Maritime life, which sustained a tradition going back to the earliest days of human transportation, expanded and changed dramatically in the early second millennium. To address maritime changes, I draw especially on the recent work of Lincoln Paine, who has provided a remarkably thorough and consistent narrative of maritime life in human history.[40] Paine has emphasized the links and commonalities across the seas, making good progress in overcoming the isolated regional and temporal segments within which maritime history has traditionally been written. Mariners moved neither as armies nor as individuals, but as complements on ships, usually small, each under leadership of a captain. Captains were generally backed up by their crews – through a mixture of loyalty and fear – so that the encounters of maritime voyagers with other ships or with people in port, in reality, were encounters of small groups more than of individuals.[41]

Human existence has entailed an ongoing balance in its landward and seaward dimensions. Initial human occupation of the continents beyond Africa depended heavily on movement by water, and human skill in fishing developed early, although agriculture and domestic animals later brought terrestrial predominance. Several independent improvements in watercraft arose from the third to the first millennia BCE, so that the islands and waters' edges again became thickly populated. The Mediterranean Sea and the Indian Ocean became the best traveled and most cosmopolitan of waters; their long tradition of being criss-crossed by vessels from near and far expanded into steadily denser trade. It may be that terrestrial history conveys the cycles of politics and trade, while maritime history highlights the longer-term advances in technology and exchange.[42] In reality, both sides of the littoral are linked essentially in sustaining the human experience.

While the great Iberian voyages are central to this maritime history, it is important to give attention to their maritime antecedents from the eleventh to the fifteenth centuries. The Indian Ocean, linked to waters both east and west, sustained a great commercial network. Srivijaya and Chola entered into competition and warfare in the eleventh century; both declined in the thirteenth century.[43] Parallel commerce took place among China, Japan, Korea, and the Ryukyu Islands and in the waters of the East Indies and the South Pacific, especially the spice islands but with possible ties to the great voyages of Polynesians. In the Mediterranean, Roger de Lauria, admiral for Aragon, gained prominence for his victories in galley warfare, 1283–1305. In the same century, the massive maritime ventures of the Mongols included two unsuccessful attacks on Japan, defeat of the Song navy, seizure of Annam and Champa and, briefly, of Java. Thereafter, Majapahit arose in eastern Java to dominate the commerce of Southeast Asian trade to the fifteenth century. Venice and Genoa persisted as great maritime powers of the Mediterranean and the Black Sea, occasionally reaching into the Atlantic and the Red Sea; the first commercial fleet from Genoa reached Bruges on the North Sea in 1277. The Hanseatic League led in Baltic commerce of the thirteenth and fourteenth centuries; fishing voyages from the western coast of Europe sailed to Iceland and Greenland, and later to the Grand Banks. This era also saw the growth of merchant guilds, institutions that represented groups of visiting long-distance merchants in the commercial centers of Europe (and also the Mediterranean and the East) – until their decline in the seventeenth and eighteenth centuries.[44]

Maritime voyages spread the trade in commodities through purchase with silver and other currencies. But they also expanded the factors of production: building a market for labor especially through enslavement but also with wage labor; gaining access to land by purchase but also by seizure; and concentrating capital for financing voyages.[45] The commodities of greatest value in long-distance maritime trade were textiles, slaves, precious metals, and spices. Textiles were the core commodity that pulled the world economy together. Every region produced its own textiles, but the most prestigious textiles flowed from region to region: cottons, especially from India; silks, especially from China; linens, especially from West Asia; and woolens, from Europe and South Asia.[46] These fabrics – spun and woven by hand, dyed and painted and otherwise decorated – were the most valuable items of trade. The next key element in the newly forming global economy was slavery, which grew steadily from the fourteenth century. The institution of slavery, which had long been strongest in the Mediterranean and West Asia, deepened in its home region as the expanding Ottoman Empire seized great numbers of captives in regions north and west of the Black Sea.[47] Enslavement also expanded in regions of the Indian Ocean, in Northeast Asia and, later, in the Atlantic world. A third key element of the global economy was precious metals – gold and silver. The available quantities of silver tended to exceed the amount of gold by roughly 10 to 1, and the price of gold long averaged about 16 times the price of the same weight in silver. But from the tenth to the sixteenth centuries, the supply of precious metals in the Old World relied particularly on gold from sub-Saharan Africa. Merchants also sought spices from the Asian and African tropics, along with ceramics from producers throughout Asia.

The first great maritime leader of the fifteenth century was the Ming admiral Zheng He, commissioned by the Yongle Emperor to lead great fleets of ships into the Indian Ocean. Zheng He led seven voyages between 1403 and 1435: they took his fleets as far as India, Arabia, and East Africa. These unique parades of maritime power included as many as 350 ships, with crews totaling thousands. The routes Zheng He followed had been charted by merchants of earlier centuries in a web that corresponded in large measure to the extent of the Islamic religion. Yet the Ming imperial voyages ceased in 1435, when a new emperor chose to give higher priority to China's Inner Asian frontier. Still, private Chinese traders continued to ply the waters. The most noted navigator of the era from the Arabian Peninsula was Ahmad Ibn Majid (ca. 1432–ca. 1500), born in Oman. He was a navigator more than a captain: he wrote 40 works of varying length, including a long poem on maritime life, lamenting that the role of navigator required leaving one's family. In the 1490s he completed his book *Kitab al-Fawa'id*, a compilation of principles and lore on navigation of the Indian Ocean and the Red Sea – how and when to use the monsoons, other winds, ports, and harbors.[48]

At the western end of Afro-Eurasia, and for significantly different reasons, the Portuguese monarchy began giving support to oceanic voyages in the mid-fifteenth century. The Christian monarchies of the Iberian Peninsula, increasingly successful in their wars against Islamic states, extended their crusade for land and commerce beyond the limits of the peninsula. In a remarkable technical advance, Portuguese mariners and Jewish mathematicians collaborated to develop improved methods for measuring latitude and longitude while at sea: tables based on trigonometric calculations were linked to mariners' sightings of the sun and moon with a modified astrolabe. Portuguese and Genoese mariners sailed the difficult south Atlantic and produced accurate maps of Brazil, the African littoral, and the Indian Ocean.[49] With this level of collaboration in oceanic travel, one can say that the first institution of an *oceanic network* had been created.[50]

Christopher Columbus, born in Genoa, gained his experience by sailing north to Ireland and south with Portuguese vessels to Elmina, the newly opened center of West Africa's gold trade. He read widely on geography and navigation and concluded that it was feasible to sail west across the Atlantic to Asia. It was understood among navigators that the Earth was spherical but estimates of its size varied. Columbus's estimate of the size of the Earth was far too small, making it appear that Asia could be reached easily by sailing west from Europe. Whether this was wishful thinking, willful distortion, or simple error is still not known. In any case, on his first voyage west, Columbus sailed south to the Canaries and then straight west, where he encountered and visited several Caribbean islands.[51] Bartolomeu Diaz reached the Indian Ocean in 1487, having navigated all the way along the African coast to its southern tip, and then turned back for home. Then in 1497–98 Vasco da Gama sailed for India: he went far to the west of the Atlantic (for more favorable winds), then south, and headed east on a latitude to the Cape of Good Hope and into the Indian Ocean; from Malindi on the Kenya coast he sailed northeast to Calicut.[52]

Oceanic travel showed that each ocean has its character – its winds and currents, its navigational aids, its stormy seasons, its archipelagos, its good ports and forbidding shores. When Ferdinand Magellan's ships entered the southeast Pacific in 1519, he declared it to be a great, peaceful expanse of water. Yet in fact this huge ocean, covering more than a third of the earth's surface, was far from pacific; its reaches below the equator undergo treacherous variations in weather and great storms. Yet it was inhabited: virtually all of its islands had been settled by 1000 CE. In its rougher seas, mariners developed two solutions to ensure sturdiness of their vessels: outriggers for the proas of South Pacific islands, and watertight compartments for the junks of China. The South China Sea connects directly to the Indian Ocean, through connections either north or south of Sumatra. Relative calm prevails in the Indian Ocean except during the stormy seasons between monsoons. The Atlantic has consistently rough seas: Europeans, the only population on Atlantic shores to sail systematically into the high seas, had to learn to make sturdy vessels.

The new oceanic routes added a maritime "keystone," linking the robust web of existing Old World routes. Map 7.1 emphasizes that, as of 1580 CE, most of the world's trade routes, on land and sea, had existed for some time.[53] The preexisting routes would continue to carry the great majority of the freight, but the new routes, pioneered by Iberian vessels and shown with dotted lines, completed the global network of maritime trade and enabled each region of the world to have contact with every other region. The expanded maritime voyages resulted in encounters

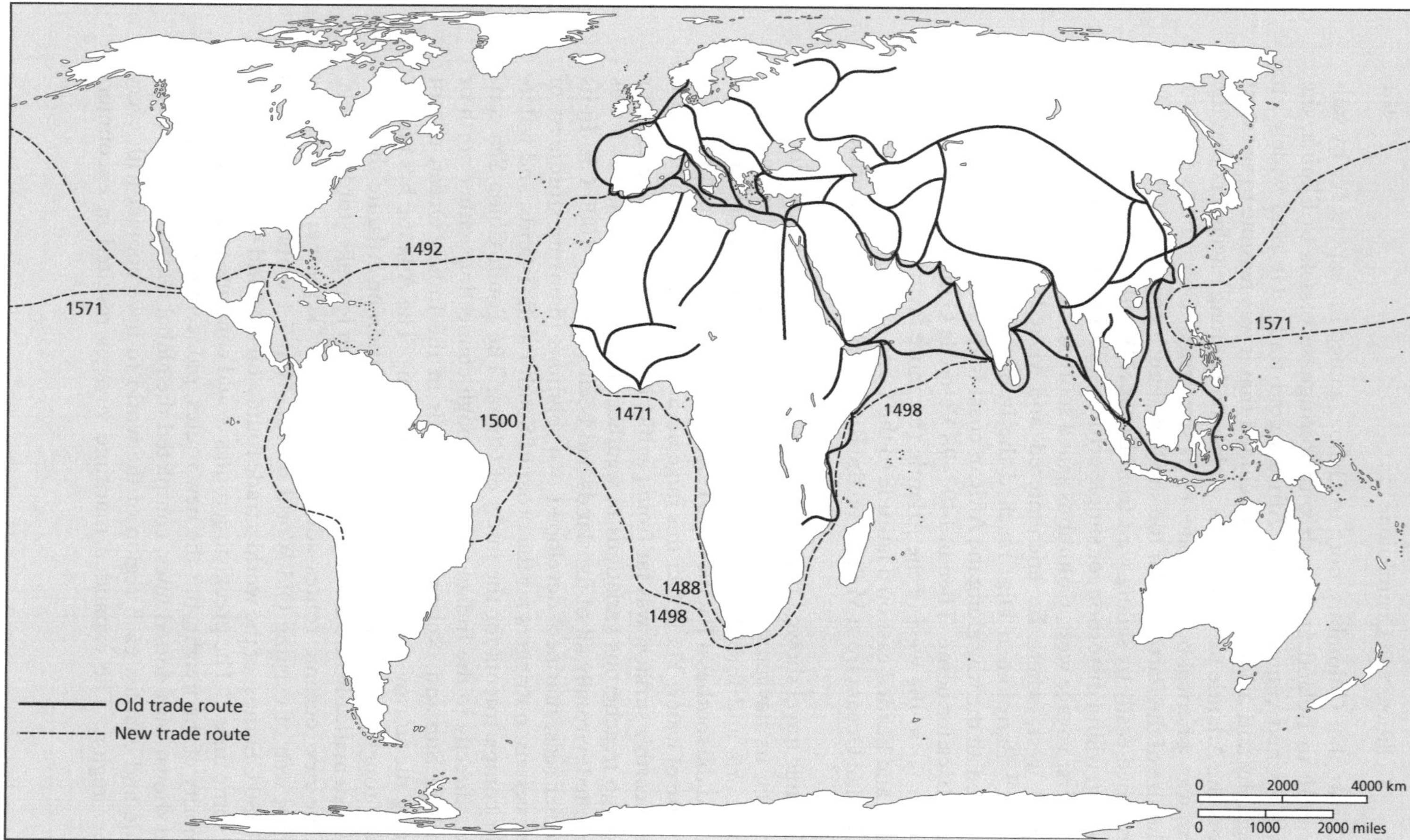

Map 7.1 Global and regional trade routes, old and new, ca. 1580. Dotted lines are new routes. (Redrawn based on original image, courtesy of the University Library System, University of Pittsburgh; generated by Boris Michev, 2019.)

among peoples who had no previous experience with each other, encounters that were renewed with each generation. In the Indian Ocean and the East Indies, European vessels visited most ports, in competition with regional shippers; few European vessels ventured into the South China Sea. Europeans monopolized shipping in the Atlantic and in the long route across the Pacific, benefiting increasingly from these routes.

The labor of maritime activity was most basically that of the sailors and of those who built the watercraft and loaded or unloaded the cargoes. Increasingly, however, the seizure, transport, and exploitation of enslaved workers became central to maritime trade. All along the major routes of trade, where laborious tasks needed to be done – or where sudden wealth permitted conspicuous consumption – young men and women were seized by adventurers and sold to merchants who delivered them to regions where labor was in short supply. In the fifteenth and sixteenth centuries, the regions of high labor demand included Atlantic and Caribbean islands, the Iberian Peninsula, Italy, Yemen, Gujarat, Bengal, the Melaka Strait, and Java. The origins of the captives taken to these destinations included West and Central Africa, North Africa, East Africa, South India, Bengal, and Indonesian islands. The merchants and planters who purchased captives put both men and women to work in agricultural and artisanal tasks; but they also displayed their wealth by using children and adults as domestic servants and by selecting women as concubines. For instance, the sixteenth-century wars of the Christian Kingdom of Ethiopia and the Islamic Sultanate of Adal led to large numbers of captures, mostly sent to the Punjab and Gujarat regions of South Asia. As these wars came to an end, the shipments of captives to South Asia declined.[54]

Connections to global trade opened up along the shores of western Africa and all of the Americas. Contacts of China and India expanded; new contacts opened among Europe, West Africa, and the Americas. The Pacific continued to be traveled busily by large vessels along its western shore and small vessels in its equatorial archipelagos, and also came to be linked to the Americas by the slender but significant thread of Spanish vessels. The numbers of people involved in these voyages rarely exceeded the several hundred people who could sail in a fleet of ships, yet the cumulative influence of maritime migrations over three centuries did much to change the world. After two or three centuries, the shocks brought by the global maritime connections died down, and a new system of intercontinental linkages entered a period of expansion.

Century after century, the Founders of empires pursued visions of creating a new world order, in which they would use the available technology and social innovations in pursuit of their expansionist goals. Most often, the emperors were soon frustrated and their regimes replaced. Technology did change for all: by the sixteenth century, the expanded use of gunpowder increased the use of infantry in place of cavalry, and the rise of artillery made city walls nearly obsolete; the construction of navies and especially the advances in naval gunnery made sea power increasingly important. Horses and elephants, nevertheless, retained their importance in warfare and imperial statecraft.[55]

Changes in the social order centered on the expansion of colonies, especially overseas colonies. In the Mongol realm, almost all outlying areas could have been considered as colonies under direct administration, until Khubilai tried to shift the imperial homeland to Han China. In the fourteenth-century Mongol successor states of the Timurids, Golden Horde, Ming, and Songhai, the regimes had arguably laid the groundwork for expanded trade with their modest colonies. Empires founded in the sixteenth century – the Ottomans, Safavids, Mughals, Russia, Spain, and Portugal – arguably extended Mongol models on land and by sea, controlling a larger proportion of the world's surface than Mongols had conquered three centuries earlier. That is, never before had so many people been governed from distant capitals by authorities who took tax moneys and commercial profits out of their region. Map 7.2, showing empires in ca. 1580, conveys both the continuity of Eurasian empires and the changes of maritime empires.

Representation of the World

As I have argued, emerging global interconnection brought as much trauma to human society as it brought excitement and opportunity. One must ask, then, about the spiritual and temporal effects of global collisions. From many perspectives, people encountered challenges to their established views of the relations between man and God, between humanity and destiny, between individual will and global structures, between authoritative knowledge and their own experience. Linkage of the Americas and the Old World brought a spectacular reorganization in representation of the world, yet I argue that the voyages of Columbus and the accompanying Columbian Exchange are not the whole story. They should be seen in addition as a metaphor for a wider set of shifts in understanding of the world.[56] Numerous additional factors disrupted old views. These included relentless warfare; expanding commerce; expansion of literate culture and elite power; subordination of common people; pandemic disease; diffusion of unfamiliar plants and animals; new visions of spirituality; and new knowledge on geography, cosmology, human populations, and practices.

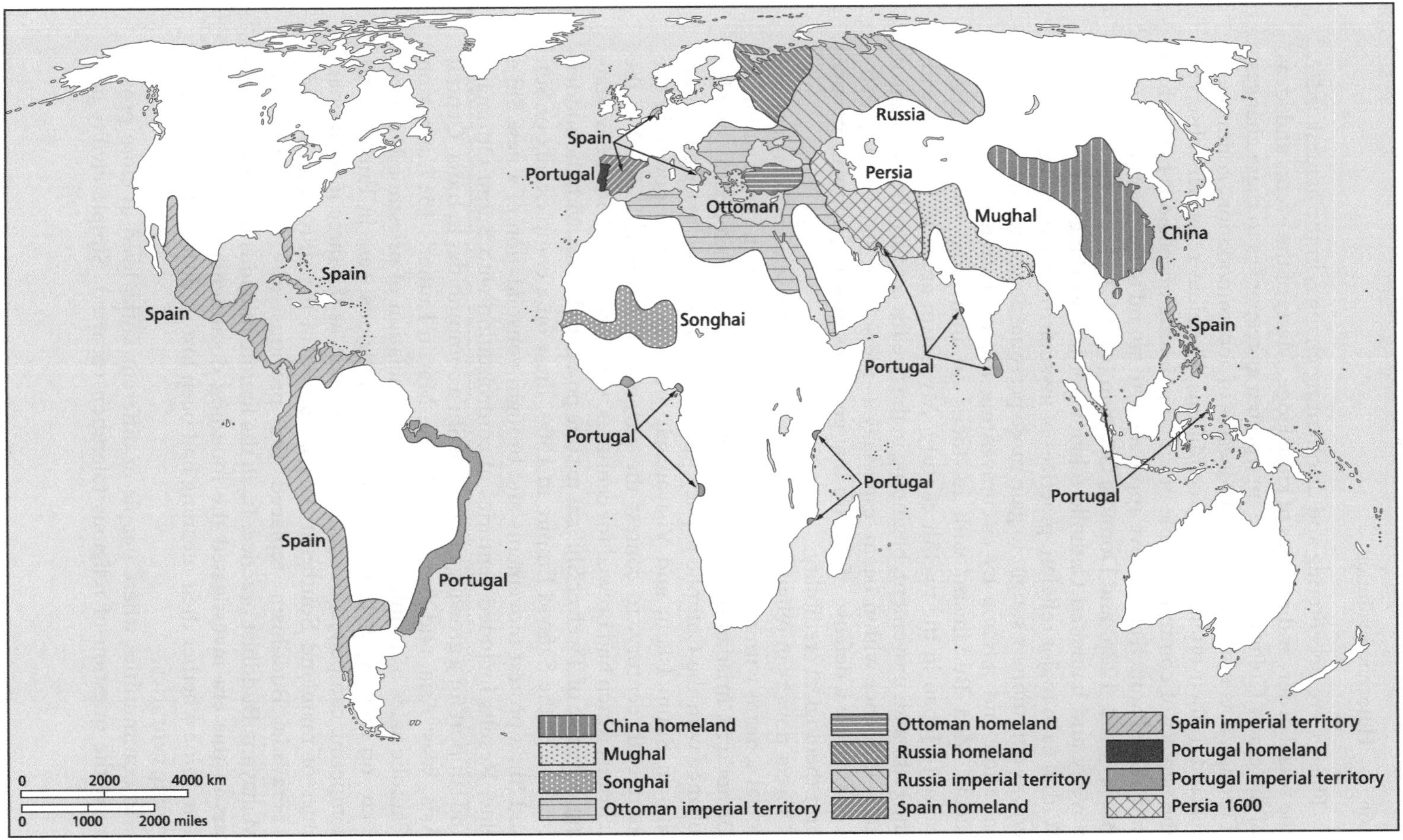

Map 7.2 Empires, ca. 1580. (Redrawn based on original image, courtesy of the University Library System, University of Pittsburgh; generated by Boris Michev, 2019.)

These were decentering and disorienting changes in the natural order, the social order, and regimes of knowledge. Many of the events and ideas of the era of Collisions can be interpreted as responses to these changes. Through acts of rebellion and processes of accommodation, commoners reacted to their often-unhappy conditions. Religious reformers observed the neglect of common people in the functioning of religious hierarchies and helped launch alternative approaches to attend to the poor. Francis of Assisi (1182–1226) and Dominic Guzman (ca. 1170–1221) took vows of poverty and launched Catholic religious orders working with the poor. Within each major religion, practices of mysticism gained strength in this era, as commoners sought to gain more personal access to God or sought salvation and *nirvana* by circumventing the mediation of priests and ulema. Zen Buddhism, with its focus on individual contemplation, came to Japan in the twelfth century. Mysticism arose in Judaism with the Kabbalah movement, beginning in the twelfth century. Sufism gained great influence within Islam, especially as the scholar al-Ghazali (1058–1111) wrote a defense of Sufism that gave this growing popular practice firm theological credibility.

Great powers in religion arose in this era, yet there were doctrinal and social schisms within most major religions. The Byzantine state sustained Orthodox Christianity; the Holy Roman Empire developed a complex alliance with the Catholic papacy; the Ottoman Empire asserted leadership in Sunni Islam; and Vijayanagar arose as a Hindu power in the fourteenth century, to contest the advance of Islam in India. Yet the schisms, which had brewed for centuries, were drastic and often devastating. The rise of Twelver Shi'ism to state power in Persia, with the sudden creation of the Safavid Empire in 1500, drew on a complex mix of Sufi and Shi'a protest movements and brought consolidation of a new Shi'a elite. Results included centuries of confrontation with Sunni Ottomans and support for a growing network of Shi'a communities in India, Central Asia, West Asia, and eventually Africa. Martin Luther's 1517 attack on Catholic sales of indulgences led to renunciation of priesthood, a great focus on literacy, well over a century of religious wars in Europe, and competing campaigns of evangelism. Further, as the three main monarchies of mainland Southeast Asia collapsed in the thirteenth century, Theravada Buddhism replaced the previous mix of Hindu and Mahayana Buddhist practice. In all the literate religious traditions, scholars in this era investigated the founding documents of their faith, to determine whether their meaning had been lost or distorted in the intervening centuries.

In communities where people of different faiths lived in close proximity, the question of religious toleration recurred. Simple civility and toleration were one approach. On the other hand, such toleration not uncommonly led to violating the basic principles of one's own religion. As a result, the questions of religious toleration on every continent were almost never resolved with simple peace-making. Instead, they provoked desperate struggles about the principles that made each religion most valuable. The Mughal sultan Akbar (1542–1605) pursued conciliation in another fashion. He successfully pursued a vigorous policy of conquest early in his reign, then turned from warfare to peace and piety, echoing Ashoka, 18 centuries earlier. Yet where Ashoka had proclaimed Buddhism throughout, Akbar presided over discourse among representatives of several established religions, emphasizing that a tolerant Islam could coordinate and prevail. Toleration did not last long after the reign of Akbar. At a hemispheric scale, one may contrast east and west by noting the greater religious multiplicity and overlap in the east and stronger tendency, in the west, for dominance of each region by a single religion.

Perhaps surprisingly, there were few major efforts to found new religions in this era of collisions. The main exception was Sikhism, which arose through preaching by Guru Nanak, beginning in 1500 in North India. His original synthesis of Islam and Hinduism gained adherents and survived, despite persecution, but did not expand to the degree that Manicheism (which drew on the preaching of Buddha, Zoroaster, and Jesus) had expanded 13 centuries earlier. In other cases, communities converted from one religious affiliation to another, hoping to find greater consistency between their religious practices and the changes in their worldly experience. For communal religions on every continent, we have little direct documentation. Yet we learn something about those communities through the efforts at converting them, such as with the spread of Islam in Africa and Southeast Asia, Orthodox Christianity in Siberia, and Catholicism in the Americas and Africa.

Conceptualization of the world as a globe gradually became established firmly. Magellan conceived of the voyage that would confirm the dimensions of the sphere. He departed in 1519 and did indeed circumnavigate the world: although he died in battle in Cebu, Philippines, he had sailed east of there to the Maluku Islands in 1511.[57] In another affirmation of knowledge that Earth is a globe, Jesuit scholars in Beijing created a terrestrial globe of 70 cm in diameter and Chinese-language labels: they presented it to the Ming emperor in 1623, explicitly to make the case for a spherical Earth.[58] This gift provided an interesting parallel to the globe presented some 350 years earlier, in 1267, to the Mongol emperor Mongke by a Persian scholar, Jamal al-Din. One big difference was that the Americas were included on the Jesuit globe, while the Persian globe necessarily lacked the Americas and otherwise relied heavily on guesswork.[59] Two Florentine intellectuals, Machiavelli and the historian Guicciardini, diverged on whether to look to the classical time period or to an expanded globe to gain inspiration for understanding the world.[60]

As for the new knowledge that arose in this era in what we now call the natural sciences, previous interpretations are undergoing some change. European scholars long viewed the natural sciences as having appeared rather suddenly in a "Scientific Revolution" centering on scholars from Copernicus in the sixteenth century to Isaac Newton in the seventeenth century. Recent scholarship in the history of science, however, is seeking more wide-ranging and interactive approaches to scientific knowledge in the era from 1000 to 1600 CE.[61] It may be that periodic fluctuations in learning led in the longer term to a slow but accelerating rate of change in the quantity, quality, and geographical spread of scientific knowledge.[62] That is, rather than a sharp inflection in the accumulation of knowledge in the seventeenth century, there may have been no single inflection point at which knowledge began to grow, as knowledge accumulation persisted both in earlier and later centuries. On the other hand, posing this hypothesis of continuity focuses attention on the related question of how and why centers of scientific knowledge, associated with campaigns of translation, have shifted geographically with time. Thus, the ninth-century Abbasid campaign of translation into Arabic – drawing on works in Greek, Latin, Sanskrit, and Syriac – made Arabic the key language of scholarship for a time.[63] The Mongol era brought thirteenth- to fourteenth-century translation into Persian and from Persian into Chinese, and expanded translation from Arabic to Hebrew. Then a growing movement in the Western Mediterranean, from the thirteenth to fifteenth centuries, brought translation among Arabic, Latin, and Hebrew, with additional translation from Greek to Latin as the Byzantine empire met its eclipse. At this time, vernacular languages grew as vehicles for scientific analysis in European languages but also in vernacular Ottoman, Persian, Chinese, and Japanese. Today, scientific publications are global though especially in English; they are based everywhere but are strongest in a few academic centers. What is it that explains the periodic shift of scientific discourse from one center to another, from one ethnic group or scientific community to another? What is the relationship of such shifts to the overall growth in scientific knowledge?

I have placed the sixteenth century at the end of a six-century period from 1000 to 1600, rather than at the start of a new period of five centuries leading from 1500 to 2000. From 1000 CE, humans exercised their agency with such force as to open new projects that not only transformed the human order but also shook the natural environment in equal measure. Warming and prosperity from the ninth century on had encouraged displays of ambition, especially among those in elite positions. Desire for imperial dominance brought assault and conquest. Most empires arose on the ruins of previous empires; some seized new areas. All brought destruction as well as accumulation. Europe was distinct and perhaps fortunate in that no one regime ever gained control of the whole subcontinent. Members of civil society, recruited or enslaved to serve in war, lost their fields and towns. Escape from war brought more migration and new cycles of disruption, dispersing families and cultural practices throughout the planet, yielding wealth to some, misfortune for most, and decline in population.

This same outburst of human energies provoked changes in Gaia. Human travel on land and sea – as in search of gold, silver, jade, and pearls – fueled ambitions, bringing exchange of biota as well as minerals. Plants, animals, and microbes arrived in distant continents. Rampant disease struck not only humans but also herds and crops; newly arrived species crowded out the old. With population decline, whether from war or disease, fields and herds shrunk in size, so that declining output of greenhouse gases reduced temperatures. The negative feedback continued from 1250 into the seventeenth century; only in the 1660s did average temperature in the Northern Hemisphere get past its low point and begin to rise. But commercial networks, fluctuating with the course of war and disease, somehow benefited from new technology and greater investment. Merchants continued their expanding focus on textiles, slavery, and the search for precious metals and moneys. Trade routes, shifting with the political and climatic order, expanded even as population declined, reaching out to become a fully global nexus of commerce soon after 1500. New and old forms of knowledge expanded, in a most unequal fashion but at a greater scale than before. The intensity of collisions – in war, disease, migration, commerce, and knowledge – brought reconceptualization of the world. Although religious and philosophical conflicts had arisen earlier, the schisms broke into the open in this era. Toleration could not keep up with hostility; independent belief hid from authority. In sum, the world seemed a smaller place in 1600 than in 1000 – more easily navigable but more dangerous.

My emphasis on the continuity of the sixteenth century with the preceding centuries, however, can only go so far: this century brought unique and extraordinary changes as did any other. Three such changes were at once culmination and initiation. The expansion of worldwide maritime commerce was the first of these. Second was the large-scale use of gunpowder and artillery in warfare, especially in the rise of "gunpowder empires," as the Ottomans, Safavids, and Mughals created great empires by deploying muskets and cannons to great effect, ending the era of city walls and reducing the importance of cavalry. Third and perhaps most significant was the wave of major changes in religious beliefs – the splits of Catholic and Protestant Christianity, the rise of Shi'ism to contest Sunni Islam, and the divisions within Orthodox Christianity and between Theravada and Mahayana Buddhism, combined with the missionary expansion of Catholic Christianity and Sufi Islam – much of it within the sixteenth century.

8 From Global Networks to Capitalism

By the seventeenth century, the era of collision and contraction had faded, so that the Human System – now more tightly bound – returned to growth in population and economic output. In this chapter I explore the economic, social, and cultural changes in the world's regions from 1600 to 1800 CE, focusing on two successive waves of economic change. The first wave of change was led by economic initiatives and innovations in many parts of the world during the seventeenth century, building upon the global economic network that had come into being from the thirteenth through the sixteenth century. Each region, in recovering from the various collisions of the preceding age, drew on its local resources and interregional connections to expand its position in global commerce. The second wave of change brought the rise of capitalism as a set of institutions dominated by owners of wealth who managed to achieve significant influence over major states. Their alliances of wealthy proprietors with each other and with state authorities, sustained through war and peace, enabled the economic leaders of northwestern Europe to gain expanding influence throughout the global economy. Confirmation of this two-step hypothesis is not yet fully feasible, yet the steady assembly of new data may soon give it a clear test. The chapter concludes by identifying cultural and political implications of these two stages of economic transformation.

Seventeenth-Century Regional Changes

Mercantile and productive organizations carried out the exchanges of the global economic network. Trade diasporas had long existed and continued to thrive: families and ethnic groups posted members in separate commercial centers in order to facilitate trade between them.[1] State trading ventures, including African monarchs trading slaves for textiles and Indian states marketing textiles, competed with smaller-scale private merchants in their home markets.[2] Chinese overseas merchant firms conducted much of the commerce of Batavia and the Pacific emporium of Manila. Workshops, some of large scale, conducted spinning, weaving, and decoration of cottons in India; while slave plantations, especially in the Americas but also in Southeast Asia, India, and Africa, required large-scale management and attention to possible worker rebellion and escape. Each form of enterprise required disciplined labor to complete its activities.[3] Peasant workers, not surprisingly, were largest in number, conducting agricultural and domestic labor. Artisans, working at many specific tasks, conveyed the range of economic activities.[4] Having survived centuries of fluctuations brought by warfare and climate, the economic regions of the seventeenth century were in many ways parallel to those identified by Janet Abu-Lughod for the thirteenth century – with additional regions in western Africa, the Americas, and Siberian forests as participants in the global network.

There was, however, one more natural crisis to survive: the low point of the Little Ice Age during the 1650s and 1660s. This was a cold snap that, concluding centuries of cooling, was more severe than any in the preceding 3000 years.[5] Geoffrey Parker, in an impressive and comprehensive study, has documented the resulting disruption, disease, and population decline in Europe and worldwide, especially at mid-century. Parker argues that the political and commercial policies constructed in response to this climatic crisis set Europe's emerging capitalist system on its way to growth and eventual global hegemony.[6] As I argue, however, a longer-term perspective sets the Little Ice Age in the context of preceding collisions and suggests that the seventeenth-century crisis was one of many overlapping factors in transforming the socioeconomic order of the world's regions.[7]

The port cities of Fujian province – Fuzhou, Quanzhou, and Xiamen – thrived in the late Ming era, and revived after the Qing regime became consolidated. From the great center of porcelain manufacture at Jingdezhen, inland from Fujian, came elegant ceramics, including the blue and white porcelain purchased and held in quantity by Safavid and Ottoman rulers.[8] Similar porcelain was carried by the English East India Company both to Europe and to Japan. Japan's Tokugawa shoguns, in the early seventeenth century, provided red seals to vessels that conducted trade with the nearby Ryukyu kingdom and with the Spice Islands, the Southeast Asian mainland, and Manila, but only indirectly with Chinese ports. Sampans from Fuzhou to Manila facilitated exchange of silk textiles for silver.

Perhaps unique in this commercial system was Potosí, the great silver-mining city of the Andean highlands, with a population of 150,000 in 1600. Miners and artisans were brought from distances in the Americas, Europe, and Africa to extract and purify the silver of Potosí and,

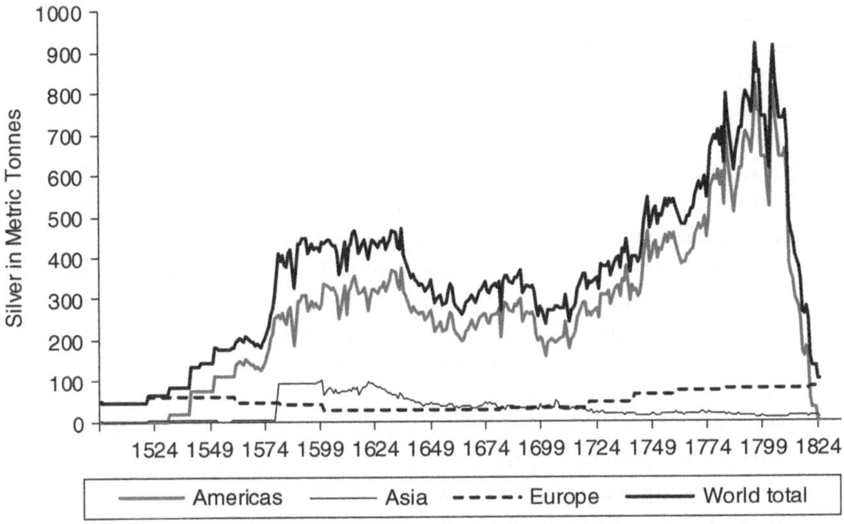

Figure 8.1 World silver production per region, 1500–1825. (Courtesy of Pim de Zwart and Jan Luiten van Zanden, *The Origins of Globalization: World Trade in the Making of the Global Economy, 1500–1800*, 72.)

increasingly, Zacatecas in New Spain. In the mid-sixteenth century, the Spanish conquerors of Peru and Mexico recognized the incredible wealth of the silver deposits in Potosí and in Zacatecas and began organizing their exploitation. This required a work force and a technology for extracting silver, transforming it into ingots, and shipping it.[9] The Spanish, in addition to sending most silver through Havana to Seville for distribution, created the ports of Acapulco and Manila and initiated in 1571 the system of sending annual galleons across the Pacific, carrying silver to the Philippines for sale to Chinese merchants, and purchasing silks and other commodities to cross the Pacific to Mexico.[10] Thus was completed a circuit of trade going literally around the world, as the silver of Peru and Mexico traveled east across the Atlantic and west across the Pacific, arriving in each case in India and China.

The growth in worldwide silver production is shown in Figure 8.1. The expanding supply of silver appears to have been distributed relatively widely throughout the global economic system, although research has yet to specify the details of regional flows. Almost the full amount of silver produced in Japan's silver boom (from 1570 to 1640) went to China, at the time when China's demand for silver was most intense.[11] Meanwhile, Portuguese merchants became central in another monetary exchange, as they began shipping cowrie shells from the Indian Ocean to West Africa, laying the groundwork for slave trade – while pearls and cloth also served as money in interregional trade.[12]

The silver mined and distributed around the world accumulated steadily, building the size of the money supply – especially in China, but also in India, Europe, and the Americas. Figure 8.2 gives estimates of cumulative stock of silver held throughout the world from the fifteenth to the nineteenth century. These figures suggest that from 1700 to 1800 (and on to 1900), global holdings of silver grew at 0.7% per year, nearly twice the rate of world population growth: this is consistent with steady monetization and with expansion of the volume of commerce, suggesting that the global market economy grew through that period on a per capita basis.[13]

Constantinople, along with Beijing, was a massive city and the capital of a great empire. It was also a commercial hub for the Black Sea and eastern Mediterranean, expanding its labor force with slaves brought from the north.[14] The city was decorated with tulip blossoms each spring – they had arrived from Kazakh lands in the sixteenth-century era of Sultan Suleyman. Dutch merchants, linking the Baltic to the Mediterranean, encountered tulips in Constantinople and brought bulbs to Amsterdam in 1593, so that the tulip gardens of Amsterdam eventually became as prominent as those of the Ottoman capital. Coffee, once a locally produced stimulant in Ethiopia, came to be shipped from Yemen to the coffee houses of Constantinople.[15] North of the Ottoman state, and a potential rival, was the duchy of Moscow. The capital grew in significance as its duke, Ivan IV, took the title of tsar of Russia in 1547 and thereafter completed conquest of the Volga valley and lands beyond the Urals to the Yenisei. Russia formalized a growing fur trade by requiring Siberian subjects to pay a tax on furs from 1600. Sables and other furs went west through Moscow to Leipzig for European trade, or went south to Bukhara for sale in Persia and China. In efforts at import substitution in this interregional trade, silks came to be produced not only in China but also in India, Iran, the Ottoman Empire, and Italy. In India, for instance English and Dutch merchants hired Bengali laborers as silk-reelers.[16]

Regions distant from imperial capitals, more often than before, became central to imperial economic and political systems. How they were governed – how they participated in economics, culture, and law – now mattered unmistakably.[17] Colonies varied widely, but I categorize them into four types that persisted across the centuries. A first sort of colony consisted of lands with sizeable conquered populations, ruled by small imperial forces, as in the European and Arab lands of the Ottoman Empire. In India, as the Mughal regime expanded to the east and south, it had to develop techniques to absorb and pacify the conquered populations; the fertile valley of Mesopotamia fell under alternate domination by Ottoman and Safavid forces, so that the region was always under foreign rule; the Spanish maintained their colonial rule over the dense but declining populations of New Spain and the Andean lands. Second, for settlement colonies, migrants from the imperial homeland landed, made initial alliances with local groups, and then expelled or outnumbered them through continued migration. Spanish and Portuguese settlements began in the sixteenth century; the Qing dynasty encouraged eighteenth-century movement and settlement by Han Chinese into the western provinces of Shanxi and Yunnan. Third, trading and supply posts were essential for long-distance trade, especially on the oceans: the Portuguese trading post at Melaka and the supply stop at Mozambique provided fine examples from an earlier century; Russian fur traders created trading posts at points across the Siberian forests; ethnic Chinese merchants predominated in Southeast Asian trading posts. Fourth, plantation colonies, demographically dominated by enslaved workforces, were commonly small in area but highly valuable in their production and concentration of wealth: French plantation colonies on Guadeloupe and Martinique, and later in Réunion, produced great wealth in sugar.

Elmina, on West Africa's Gold Coast, became a center for Portuguese gold purchase in 1482; a Dutch expedition seized Elmina castle in 1637 and held it for more than two centuries. In the seventeenth century, gold exports through Elmina exceeded the value of slave exports from all of Africa. Neighboring Gold Coast regions experienced an economic boom based on village-level artisanal production and exchange but, in a great turnabout, the rising Akwamu kingdom launched waves of enslavement late in the century, relying on firearms in their wars. By 1700, the price and volume of slave exports had risen so far that Gold Coast ceased exporting gold – instead, the region imported gold from Brazilian mines in exchange for slaves.[18] It was at this late-seventeenth-century juncture that notions of "race" took on new meaning and new power, gaining fuller codification in Christian Europe through slave codes such as the *Code Noir* of France.[19] This racialized view of slavery and the racialized stigmatization of Africans spread to the Islamic world through interaction with Atlantic slavery. In particular, free black people in Morocco were enslaved in the late seventeenth century, to make them into an elite slave army. This enslavement of free Muslims broke a fundamental religious rule. Yet the success of the campaign – ironically imposed by Mulay Isma'il, a king whose mother, a skillful courtier, was long a slave – helped spread a racial ideology elsewhere in the Islamic world to rationalize the practical enslavement of blacks among others.[20]

European long-distance shipping is known to have expanded during the sixteenth and seventeenth centuries in the Atlantic, the Mediterranean, the Indian Ocean, and the Pacific. What has now become clear is that non-European merchants, especially in the Indian Ocean and the western Pacific, expanded their shipping and commerce as part of the same process.[21] The arrival of European vessels in eastern waters was not a zero-sum game. While European warrior-merchants destroyed Asian vessels, such warfare did not preclude the overall growth of commerce. Records describing shipping for Asian and East African merchants, while far more sparse than those for European shipping, do tend strongly to indicate that the volume of trade in the Indian Ocean and western Pacific was still dominated by merchants based in the region in the late eighteenth century.[22] Further, we can now begin to make comparisons of the volume of shipping in the Atlantic and the Indian Oceans in the seventeenth and eighteenth centuries, especially by assembling estimates of the tonnage or shipping capacity of merchant vessels. In Table 8.1, I compare the recent estimates of eastern oceanic shipping by

Table 8.1 *Estimated tonnage of shipping in Atlantic and Indian Oceans*

Decade	Atlantic Carriers	Atlantic Tonnage	Indian Ocean (IO) Carriers	Indian Ocean Tonnage
1600s	Dutch	240,000	Asian IO shipping	325,000
	English	80,000	European IO shipping	25,000
	French	80,000	Shipping to/from Atlantic	9,000
	Other	100,000		
	Total	500,000	**Total**	359,000
1710s	Dutch	500,000	Asian IO shipping	400,000
	English	300,000	European IO shipping	70,000
	French	100,000	Shipping to/from Atlantic	40,000
	Other	100,000		
	Total	1,000,000	**Total**	510,000
1790s	Dutch	398,000	Asian IO shipping	300,000
	English	880,000	European IO shipping	290,000
	French	730,000	Shipping to/from Atlantic	70,000
	Other	400,000		
	Total	2,408,000	**Total**	660,000

Matthias van Rossum with the 1992 estimates of Atlantic shipping by Richard Unger.[23] Van Rossum's analysis, relying especially on Dutch documentation, estimates tonnage for European shippers in eastern waters, then for Asian vessels in the China Sea, Southeast Asia, the Bay of Bengal, and the Arabian Sea. Unger drew on a wide range of documents to conclude that shipping tonnage for the combined Mediterranean, North Sea, and Atlantic totaled 1 million tons in 1600, 1.5 million tons for the 1670s and, based on a 1786 French survey of European shipping, 3.5 million tons in the 1780s. Each of the estimates is preliminary; scholars believe that available Indian Ocean data still substantially underestimate coastal shipping.[24]

These estimates are certain to change as more complete collections of data become available. Unger argues that Atlantic shipping grew rapidly in the fifteenth and sixteenth centuries, regaining its earlier medieval peak by 1600. Comparing the two oceans for the seventeenth and eighteenth centuries, one may argue, at present, that Asian shipping in the Indian Ocean was roughly equal in volume or perhaps in excess of the volume of European shipping in the Atlantic in 1600 (this at a time when population of the Indian Ocean substantially exceeded that of the Atlantic world). Intra-Asian shipping greatly exceeded the volume of shipping between the Indian Ocean and Europe. With time, the volume of European shipping grew at an accelerating rate in the Atlantic and also in voyages to the Indian Ocean, while the volume of shipping in Asian vessels remained relatively constant or may even have declined somewhat. In sum, no single power gained dominance in global commerce of the seventeenth century. In commercial exchange and competition, revised colonial systems took shape for both European and Asian mercantile powers, relying heavily on slavery, textiles, and luxuries. European powers controlled the Atlantic and trans-Pacific routes; Asian shippers dominated the Indian Ocean; African and Asian powers also conducted trade on land – for instance, transporting captives both by land and sea.

Capitalism and Accumulation

Capitalism is commonly understood to be an economic system that has arisen in the last few centuries. The term "capitalism" itself was not coined until the late nineteenth century, as industrialization spread from Western Europe and North America to other regions and as social conflict between workers and capitalists took center stage. Subsequent debate has caused the term "capitalism" to be extended in its application back to the eighteenth and seventeenth centuries, and even beyond. But capitalism, whenever it began, drew on preexisting institutions. That is, many underlying elements of capitalism had come into existence long before, even thousands of years ago: these included markets for goods and services, networks of local and long-distance commerce, circulation of money, institutions of banking, state economic regulation, and labor systems including wage labor and slavery.[25]

If capitalism began in the last few centuries, the essence of its innovation must lie in a different sort of change. It must be a set of innovative institutions that transformed an existing and mature commercial network into a different sort of network.

Throughout this study, I have explored social evolution through attention to the creation and transformation of social institutions. Within that perspective, I seek to identify "capitalism" as a set of interacting social institutions. To identify an institution, I have emphasized the formation of a group of persons, the purpose that unites them, dynamics of the institution arising from its specific activities, and reproduction of the institution over time. A capitalist institution, like any other, would need to face the evolutionary test of fitness: to satisfy the needs of an appropriate beneficiary population.

I offer a four-step model for the rise of capitalistic institutions, with the key step as its expansion to the transnational level at the opening of the eighteenth century. (1) The basic capitalistic institution was the *enterprise* or firm – a private, profit-making enterprise. Such enterprises arose everywhere in the global economic network, as I have suggested in the preceding section, but gained relative strength in Europe. (2) Merchant guilds arose as the global economic network took form: they facilitated long-distance trade in commercial towns of Europe, Africa, and Asia from the eleventh century to the eighteenth century.[26] (3) A step essential to the expansion of capitalism was the *national proprietors' association* – a gathering of individual enterprise owners for their mutual benefit, initially in the Netherlands and in England during the seventeenth century, and focusing especially on influencing the state to follow pro-capitalist policies.[27] (4) From 1689, with William of Orange leading both Netherlands and England, it became feasible to establish a transnational capitalist alliance consisting of an Anglo-Dutch *transnational proprietors' association* and its efforts in ensuring pro-capitalist policies in both countries. From this base, capitalism spread further. In the remainder of this section, I trace these four steps in the crystallization of capitalism and their consequences during the eighteenth century. In Chapter 9, I argue that a global capitalist system took form in the mid-nineteenth century through the alliance of selected states with pro-capitalist policies and strong national proprietors' associations.

Enterprises of the second millennium CE are clearly described in the historical literature, though there has been some tendency to exaggerate the distinctiveness of European enterprises. Jürgen Kocka, in his short history of capitalism, effectively traces enterprises and their development in Asia and Europe over the centuries – mostly mercantile enterprises but also banks.[28] While an enterprise centers on a group of owners or managers building their holdings of capital, it includes employees who work to support its objectives, but who do not have governing authority. But enterprises of significance needed to satisfy additional criteria. Entrepreneurs needed to be knowledgeable in types of administration, such as accounting of profit and loss, management of capital, managing the production of the goods and services, distributing and marketing their products, and managing a system of security (to respond to thieves, recalcitrant customers, and unruly laborers). Enterprises needed to handle such inputs as labor (from family, wage labor, slaves, etc.), raw materials (both agricultural and mineral), and such intermediate goods as buildings, mechanical equipment, and shipping services. Two more criteria became increasingly important: the need to make a profit in multiple markets – both national and international – and the need to gain influence over the state, both at home and abroad. Interactions with states included taxes, regulations, law, public works, finance, and war.

Merchant guilds united visiting merchants in commercial centers to negotiate with local authorities for contracts. Two criteria just listed for success of an enterprise – participation in multiple markets and gaining influence in relations with states at home and abroad – called merchant guilds into existence. The guilds negotiated with local authorities and gradually ameliorated the conditions for contracts, security, and equity for visiting merchants: by the seventeenth and eighteenth century, there was little more need for their local-level intervention.[29]

Yet the old problems of commercial links with multiple markets and multiple states arose at another level – the level of emerging national states. National associations had the promise of gathering financial resources to fight wars of interest to entrepreneurs, and in pressuring the state to give high priority, in its various responsibilities, to the financial and military interests of entrepreneurs. In addition, the diverse interests of various enterprises needed to be coordinated, and proprietors' associations were often able to provide such coordination. For instance, while merchants might find themselves negotiating with officials of several states, other proprietors (landowners, manufacturers, and bankers) were more likely limited to the realm of a single state. In combining interests of the various proprietors, it made sense for merchants to join with others and focus on lobbying a single state. In this way, national interest came to be privileged among proprietors overall even though the merchants among them had multinational interests. Capitalists and their associations had to account for the dynamics of their institution and the global economy more broadly: business cycles, currency fluctuations, labor unrest, shifting consumer preferences, and changing laws and regulations. The dynamics of individual enterprises required devoting energy to proprietors' associations; the work of the proprietors' associations required devoting energy to developing common positions and finding ways to impose those positions on the state. By this reasoning, with the seventeenth-century resolution of hostilities between Dutch and English capitalist camps, a next step in the expansion of the capitalist system ought to have been the formation of proprietors' associations within France and the ability of those associations to persuade the French state to become pro-capitalist.

In this analysis of institutions, one must be clear on the term's definition. Which is the key element of an institution: the norms that it supports or the members and organization of the institution? The view of North and colleagues distinguishes institutions from organizations: it defines institutions as "the rules of the game" and "the patterns of interaction that govern and constrain the relationships of individuals." Organizations, in contrast, are "specific groups of individuals pursuing a mix of common and individual goals through partially coordinated behavior."[30] My definition of institutions, relying especially on Tuomela, combines North's organizations and institutions and adds more. In my view, institutions form through the collective intentionality of groups of individuals, in which members make group decisions that are irreducible to individual decisions. The institutions articulate norms and rules of behavior, but also exhibit inherent dynamics arising from their subject matter.[31] North's view, while treating institutional norms as central to social change, treats their change as exogenous – changes arise through the choice of elites rather than emerging through social evolution of institutions. In my analysis, the rules and norms of institutions can be explained endogenously through a combination of common choices by members of institutions and the influence of institutional dynamics.[32] In a recent update of institutional analysis, Eric Alston and colleagues have added some insights to the analysis.[33] In practical terms, North's focus on property and contracts fits more closely with the concerns of merchant guilds than with the concerns of warfare and national financial policy that were addressed by the Dutch and English national proprietors' associations.[34]

I believe that the history of the Netherlands and England, from the sixteenth to the eighteenth centuries, confirms the institutional changes proposed here. The Netherlands, thriving in northern European trade, had fallen under Spanish rule in the 1520s but became restive under imperial political and religious restrictions.[35] After the Netherlands provinces in general rose unsuccessfully in a 1568 revolt, the northern Netherlands again revolted against Spain in 1581, forming the United Provinces: this republican state gained practical independence by 1600 and formal recognition in 1648. During this era, as Pepijn Brandon succinctly put it, "Privatised warfare was the hallmark of capitalism in its infancy."[36] In 1602, leading Dutch merchants formed the Dutch East India Company (VOC), a *joint-stock* trading company: its trading ventures to the Indian Ocean focused on military assaults on Portuguese trading posts, as Portugal was under Spanish rule. The VOC seized such ports as Hugli and Melaka in the Indian Ocean and achieved the fiery destruction of a Portuguese fleet in Makassar, in the Spice Islands. In 1621 a somewhat similar Dutch West Indies Company (WIC) formed and attacked Portuguese and Spanish holdings in the Atlantic, successfully seizing Elmina and Gorée, in West Africa, and Pernambuco for a time. After the WIC failed to hold Portuguese Angola, the VOC created a post, further south, at the Cape of Good Hope in 1652. Meanwhile, small settlement colonies were established with state support: French settlers began colonies in Acadia (Nova Scotia) in 1604 and New France (Quebec) in 1608. English settlers formed colonies in 1607 (in Virginia) and 1620 (in Massachusetts Bay).[37] Dutch settlers created the prosperous colony of New Netherlands in 1614, in the Delaware and Hudson valleys of North America.

The creation within the Netherlands of national proprietors' associations and their substantial influence within the state arose in the course of the long war for independence from Spain. The republican form of government, adopted in 1588 with the reconfirmation of the Articles of Utrecht, affirmed the autonomy of the provinces in regional government but the leadership of the States General in warfare. Merchants and towns had overall leadership, but small-scale proprietors held substantial influence as a "historic bloc" of smaller interests linked to the great merchants. The ensemble, in the words of Pepijn Brandon, was a "federal-brokerage state" – it was not the first choice of the participants, but was adopted after potential monarchs declined the task or failed in accepting it.[38] Merchants of Amsterdam, while important in the republic, never ruled it – they had to haggle and collaborate with merchants from other provinces and with aristocratic, manufacturing, farming, and financial interests. The most remarkable effectiveness of the federal-brokerage state was in gathering funding and manpower for wars against much larger states in Europe. This Dutch system, more oligarchy than democracy, continued through the eighteenth century. As Brandon argues, the Dutch system of decentralization persisted when it might have been better for it to centralize.[39]

English merchants, in contrast, had to wait much longer to gain substantial influence within the state. They found themselves restricted under the autocratic Stuart monarchy of the early seventeenth century.

As a result, merchants, landed nobility, and the monarchy fought it out, with civil war during the 1640s and again in 1688. In this era, Dutch domination of markets within England led to the three Anglo-Dutch wars from 1652 to 1674. (It was in this era that Oliver Cromwell sent an ambassador to the Dutch Great Assembly of 1651 to propose "a closer union" of England and the Netherlands; the Dutch response was negative.)[40] The Dutch won two wars against England but lost New Netherlands to England in the third, in 1674. During the wars, English commerce and English naval warfare grew in strength. The English had focused on building specialized warships as well as innovative positioning of ships in battle; both sides built their skills in naval gunnery, which required elaborate coordination of cannon, powder, and timing in rolling seas.[41]

In a remarkable transition, the opportunity arose to form a transnational proprietors' association, along with a pro-capitalist alliance of states, with England's Glorious Revolution of 1688–89. With the overthrow of King James II and his pro-Catholic faction, England established a new domestic coalition. The new monarchs taking office in 1689 were Mary, Protestant daughter of James II, and her husband William of Orange, the Stadtholder of several Dutch provinces. William became the dominant figure in both countries until his death in 1702. The result confirmed that parallel policies would prevail in the two countries. In England, the nobility in alliance with the merchants gained parity with the monarchs, resulting in booming external commerce. In the Netherlands, the stadtholder was at most a first among equals in the federal-brokerage state. The Dutch still needed to repel repeated invasions from France, 1680–1714, and did so with their own financing and recruitment and with support of England and other powers.[42] Yet while both English and Dutch fleets improved greatly, neither power had yet established dominance within the Indian Ocean. The Maratha Confederacy and the independent Siddi mariners of western India maintained naval dominance over the English East India Company (EIC) in the seventeenth and early eighteenth centuries, using smaller ships with cannon at the bow for fighting close to shore. In an important struggle, the Maratha confederacy and its Siddi navy gained submission of the English at Bombay in Child's War of 1686–90.[43]

In sum, capitalist firms developed in the Netherlands, England and, somewhat more slowly, in France and elsewhere in Europe, with uneven support from states.[44] For capitalist firms to expand their scope and increase their profits, it was necessary to achieve dependable support from powerful states. The key institution, therefore, was that of the clubs of proprietors and their achievement of stable power within the state. At the crucial turning point of 1689, the English and Dutch had rough parity in commerce, in naval power, and in the influence of proprietors over the state. A brief Anglo-Dutch alliance in the era of William and Mary was beneficial not only for each national capitalist system, but also for the launching of transnational capitalism. The system was to expand, thereafter, because of the economic expansion of both English and Dutch capitalists, and also because capitalists from additional countries were gradually able to draw their states into the same system. Dutch merchants, though second to the English in wealth in the eighteenth century, nevertheless profited from the capitalist alliance. By this reasoning, a next step in the expansion of the capitalist system might logically have been the formation of proprietors' associations within France (as well as Denmark, German states, and other states) and the eventual ability of those associations to persuade their state to become pro-capitalist.

Other authors have affirmed a similar chronology but with different emphases. Immanuel Wallerstein coined the term "Modern World-System" in his major 1974 effort to redefine capitalism.[45] Wallerstein's most valuable point, I think, was regarding empire. As he argued, the sixteenth-century failure of Ottoman, Habsburg, and French empires to dominate Europe left a situation in which merchant interests had considerable liberty to design their own institutions. Merchants were able to build systems of investment and law that emphasized maximum profit, expanded tax revenue to support wars for trade, and expanded enslavement and other forms of labor control. The results encouraged commerce, profitability, success in warfare, and control of colonies. At times, truces or coalitions could be achieved among monarchs, aristocrats, and proprietors, perhaps reducing alienation within the general population. In such cases, the common trust brought by the coalition permitted the establishment of tax rates high enough to build a strong military force and support public works, thus enabling merchants based in this state to expand their external markets.[46] The commercial class was able to gain significant political and social power in many parts of Europe by benefiting from the competition of states that were powerful yet fragmented. Merchant classes in other regions failed to gain the political influence of those in Europe. But Wallerstein's world-system addressed the whole world only implicitly, not explicitly. In a key example, the World-System framework excluded from analysis the Dutch VOC and other European ventures trading beyond the Atlantic. The VOC, as Wallerstein saw it, by retrieving high-value goods from Asia, was conducting luxury trade with external regions but not contributing to the Modern World-System.[47] In my view, this modular approach, excluding the VOC and its profits from analysis of Amsterdam and the Atlantic economy, undermines the analysis of each region by neglecting interactions among them.[48] Wallerstein advanced the argument that collaboration among proprietors was essential to the rise of capitalism in Europe, and that it was possible in the absence of a continental empire. But he conducted his analysis in Europe and the Atlantic, not the world as a whole, thus treating European capitalism as unique. In my view, the argument benefited little by leaving out the rest of the world.

Jürgen Kocka, in his recent, major work on capitalism, has summarized the definitions of capitalism by major writers – Marx, Weber, Schumpeter, Keynes, Polanyi, Braudel, Landes, and Wallerstein.[49] Kocka then identified his own list of the key elements of capitalism: decentralization, commodification, accumulation, property rights, markets for allocation, and capital to invest in future gains. These convey a focus on the norms of capitalism more than on the social structure of its institutions. Still, Kocka refers briefly to "representatives of the financially potent population groups politically enfranchised in the Netherlands and England" and their role in consolidating public debt.[50] In spatial terms, Kocka appears to retain the approach of Wallerstein in downplaying enterprises beyond Europe and treating European capitalism as the single instance, rather than comparing or linking Europe with other nodes in the global economic network. The details of Kocka's later chapters include the full global context – including the intrusion of merchant capital into protoindustrial manufacturing on the European continent; but his theoretical section is based heavily in Western Europe.[51]

Nuala Zahedieh's analysis of London merchants is especially clear in emphasizing the nature and functioning of merchant associations, trust networks, and political networks – and also their accumulation of wealth.[52] As she shows, these proprietors' associations were relatively small and highly variegated, yet they were able to articulate the collective interest of entrepreneurs. In the tumultuous political and economic life of seventeenth-century England, they gained steadily greater influence so that, by the end of the century, the English state could be fairly described as pro-capitalist. Zahedieh has also developed very useful details on accumulation of capital centering on 1686 data regarding commercial imports to London. Out of a list of some 1500 merchants recorded as having imported colonial goods in 1686, she developed a sample of 159 merchants whose imports exceeded £ 1000 in that year. For the 59 of those merchants with the largest values of 1686 imports, she traced the wealth at death of as many as possible, ending up with information on 25. Of those, she found that nine died with estates of more than £ 30,000, eight had estates between £ 10,000 and £ 30,000, and the rest had lesser amounts, including three who died bankrupt in debt.[53] This research does give an idea of the substantial accumulation of wealth by certain London merchants in the late seventeenth and early eighteenth centuries.

Pepijn Brandon provides a parallel analysis of merchant associations and political networks for the Netherlands. Brandon's analysis, focusing on the state and warfare, is less specific on the details of proprietary associations than that of Zahedieh. Nevertheless, Brandon gives a clear idea of how Dutch proprietors carried out their negotiations through his detailed discussion of the VOC and WIC and their relations with the Dutch state.[54] In his critique of Giovanni Arrighi's analysis, Brandon accepts Arrighi's broad interpretation but argues that he made the common mistake of interpreting the Netherlands simply through Amsterdam rather than as a complex network of provinces and regions; instead, Brandon treats the Dutch state as "a mediating structure within a complex of conflicting social forces."[55] The work of Jonathan Israel laid important groundwork for this type of analysis.[56]

In my view, therefore, the rise of capitalism resulted not simply from the development of capitalistic business firms. It depended as well on the emergence of pro-capitalist social institutions: proprietors' associations, their successful inculcation of pro-capitalist policies in monarchical and republican states, the eventual construction of transnational networks giving priority to capitalist goals – and the contestation by laborers and communities that had to give way to this economic accumulation. These were intricate, multiparty relationships, distinguishing manufacturing, shipping, and finance, class, and community. A fuller story would give substantial attention to the labor forces recruited (often forcibly) to work within the system and by the communities of both metropole and colonies that sought to preserve their lands and traditions. Marcel van der Linden's global analysis of laboring perspectives addresses this need.[57]

When viewed in the European and oceanic context, these conflicting trends reveal a shift in the overall order. By 1700, the alliance meant that, in two national units at least, capitalism could accumulate, innovate, and expand through war and peace, somehow enabling finance, workshops, and public works. The arising system of capitalism survived the hegemonic shift from Dutch to English predominance. My argument on the institutional structure of capitalism can be linked, I believe, to the results of new studies in economic history. Such studies are beginning to confirm that capitalism did make a measurable change in global trade before the nineteenth century – a difference in the level of profitability as well as in the volume of trade.[58] This issue leads us into discussion of the relations between capitalism (as an institutional system) and globalization (as a description of dynamics in economic change). Pim de Zwart and Jan

Luiten van Zanden have pulled together much of the recent research and debate on economic globalization, emphasizing regional comparison and opening the door to global analyses in quantitative terms for the seventeenth and eighteenth centuries.[59] De Zwart's monographic research confirms a degree of market integration for Dutch shipping between Asia and Europe in the seventeenth and eighteenth centuries. For commodities such as cinnamon, mace, and nutmeg, produced under Dutch colonial monopoly, mark-ups in price decreased by negligible amounts.[60] But for commodities traded in competitive markets of Asia, Europe, or both - tea from Canton, saltpeter from Bengal, Indian textiles - mark-ups declined clearly during the eighteenth century. It was competition rather that decline in transport costs that brought price convergence.[61] These results clearly extend and improve the results developed earlier by Kevin O'Rourke and Jeffrey Williamson for the nineteenth century. O'Rourke and Williamson had been unable to find price convergence in eighteenth-century data, while de Zwart used data of larger quantity and higher precision to confirm aspects of globalization.[62]

To return to the narrative of expansion in capitalism and capitalist empire, the British displaced the Dutch as the leading slave-trading power in the early eighteenth century – for instance, obtaining control of the Asiento or contract for selling slaves to the Spanish empire in 1700.[63] Once the War of Spanish Succession ended in 1714, establishing a Bourbon monarchy in Spain, the level of warfare in Europe declined for a time. Yet a showdown could have erupted out of any number of local conflicts. As it happened, the Seven Years' War was declared in Europe in 1756 (fighting had begun in 1754 in North America). Britain and France were the principal combatants: the war resulted from mutually reinforcing tensions in North America, the Caribbean, Central Europe, and the Indian Ocean. Britain and Prussia led one camp; France, the Holy Roman Empire, Sweden, and Russia led the other. The Netherlands, Spain, and Portugal – all of them economically and militarily weakened – remained neutral. The Dutch Caribbean island of St. Eustatius, declared a free port in 1756, prospered greatly, providing a measure of international recognition of capitalism. War on the European continent led to its usual stalemate, except that Prussia gained in power. The British navy was able to achieve nearly all of its objectives, seizing French and even Spanish territories. The English East India Company (EIC) and its Indian allies won a key battle at Plassey in 1757, making Bengal the base of EIC power in India.

British industrial production, which had been expanding gradually along with that of other European states, expanded more rapidly beginning about 1760. There have been long debates about the relative importance of textiles, iron and steel, steam engines, coal mining, other industries, and agriculture in British economic growth. Recent work of economic historians, however, emphasizes a broad range of advances in productivity in many fields rather than a single sector that led in overall economic growth. In particular, steam engines, powered by coal from Newcastle, seem not to have expanded greatly in productivity and profitability until the nineteenth century. For the eighteenth century, the greatest increases in productivity appear to have been in cotton textiles and other industries.[64] Growing British imports of raw cotton and growing exports of finished cotton goods, produced in and near Manchester and traded through Liverpool, became the mainstay of the British industrial economy. This industrial expansion, relying on workers displaced from rural areas into factory towns, benefited from Britain's increasingly dominant position in global commerce and naval warfare.

Britain sustained its maritime domination after 1763: Britain was the leading supplier in the slave trade and began experimenting with cotton production in the Caribbean and North America. The English East India Company now began a long-term campaign to dominate India, contesting European competitors, the growing Maratha confederacy, and other Indian states. The EIC strengthened its hold on Bengal and India's great resource in saltpeter – the main constituent of gunpowder.[65] It gradually displaced Dutch and French competitors but met defeat in the first Maratha war, 1775–82 – at much the same time as Britain was defeated in the successful revolt of the United States, 1776–81. In another venture, the British sent the First Fleet of convicts to Australia in 1788, where they settled in Sydney Harbor.

The economy of China remained the world's largest during the eighteenth century. Recent scholarship has documented the relatively high level of Chinese production, the growth in population, and the effectiveness of its market economy.[66] Britain's McCartney expedition to China, in 1793, was an unsuccessful effort to expand trade with China on terms favorable to Britain. On the other hand, new work is suggesting that China's GDP per capita underwent decline throughout the eighteenth century.[67]

The dramatic changes in the European balance of power, along with the expansion of the Atlantic economy, had still not brought worldwide transformation. Thus, the extent of empires in 1790, as shown in Map 8.1, was rather similar to that of two centuries earlier (see Map 7.1). New dynasties had seized the imperial realms of Persia, North India, and China; the European powers and China had expanded their terrain somewhat. The Qing state had expanded in the 1720s, occupying Tibet, Mongolia, and the western lands of the Uighur, and had encouraged Han Chinese settlers to

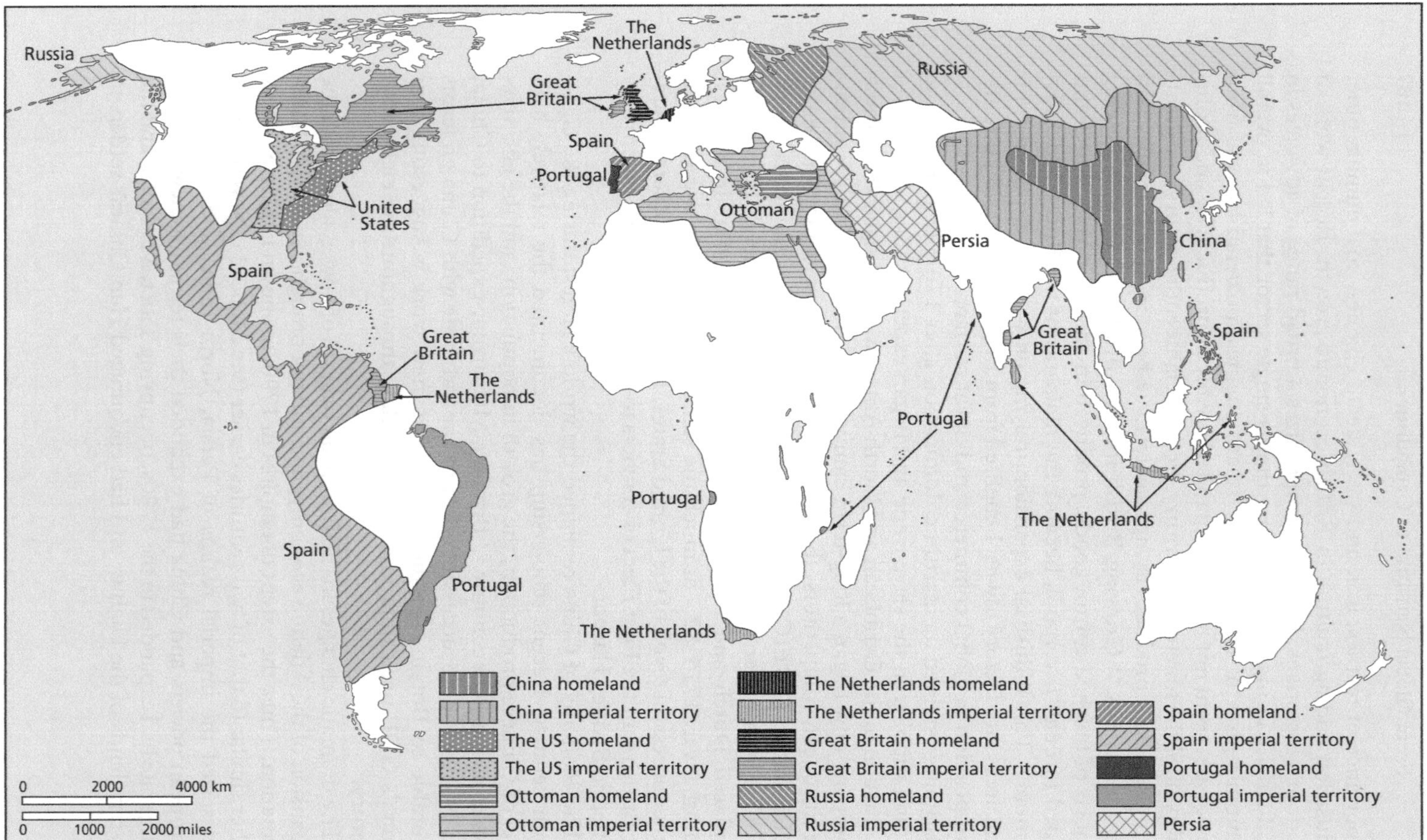

Map 8.1 Empires, ca. 1790. (Redrawn based on original image, courtesy of the University Library System, University of Pittsburgh; generated by Boris Michev, 2019.)

move further west. Iran prospered briefly under Nader Shah 1736–47 but weakened thereafter; the Qajar dynasty formed and took control of Iran in 1796. The Ottoman Empire maintained most of its territory, but lost lands along the Black Sea to Russia. Russia expanded to the east and the south in the eighteenth century, especially in the reigns of Tsar Peter and Tsarina Catherine. In India, the Mughal state had been nearly replaced by the Maratha Confederacy in the 1750s, and the Marathas retained the upper hand in wars with the British until 1800.

The end of the eighteenth century brought another set of world-shaking conflicts: the French Revolution and Napoleonic Wars. War began in 1792, with fighting in most years to 1815. British troops, not committed on the continent, were able to fight in numerous overseas engagements, including an unsuccessful campaign to defeat the rebels who had overthrown slavery in Haiti. The British conquered further lands in India and seized Dutch colonies. The United States, as a neutral power, expanded shipping in every ocean, playing a role that the Dutch had filled earlier. Atlantic slave trade reached a peak in the 1780s and 1790s, reinforcing growth in the population of the African diaspora, as shown in Map 8.2; the trade in African captives was to continue for more than another half-century. After a long internal dispute, Britain declared slave trade illegal in 1807, followed by the United States in 1808. When Napoleon invaded Spain in 1808, Spanish colonies in the Americas soon declared independence, with British support. Then Napoleon, defeated by Russia, met his final defeat at Waterloo in 1815. Britain and Austria-Hungary set up the Concert of Europe after the Congress of Vienna in 1814–15.

Over the longer term, Ottoman, Habsburg, and French attempts to conquer Europe in the sixteenth and seventeenth centuries had failed. The Dutch skillfully survived wars with Spain, France, and England, but ultimately lost their leadership to English economic power. The wars of the English and the French ultimately left the English as the dominant force in the capitalist economy. Britain's military victory of the 1760s facilitated industrial production and control of labor at home, so that by the early nineteenth century English cotton exports began crowding out domestic cottons in India. During this long era of conflict, Britain had avoided war at home, so that investments in continuing overseas commerce and new productive techniques enabled the domestic economy to grow steadily. As of 1815, Britain had just achieved its dominance of India, and turned next to establishing primacy in Latin America, China, and parts of West Asia.

In sum, I would say that the Dutch capitalist class gained hegemony at home by 1600; the English capitalist class achieved hegemony at home by 1688. In the Netherlands it was remarkable that the proprietors' class

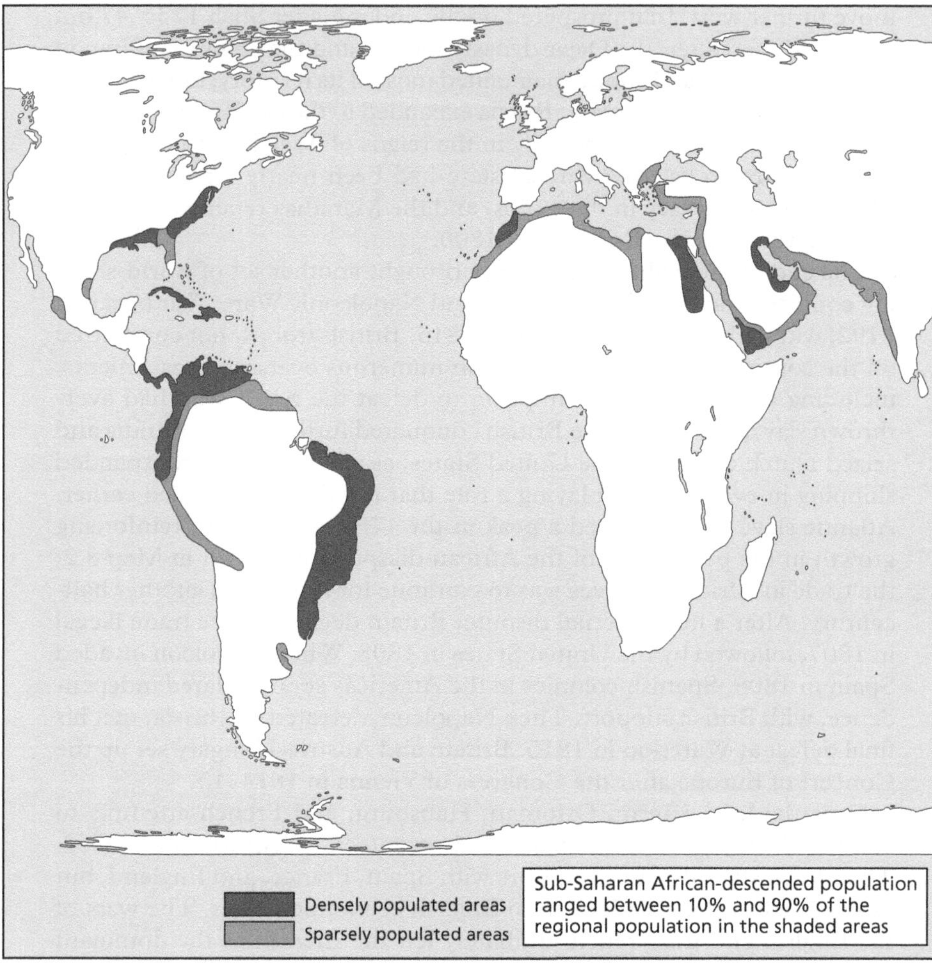

Map 8.2 The African diaspora, ca. 1800. (Redrawn based on original image, courtesy of the University Library System, University of Pittsburgh; generated by Boris Michev, 2019.)

was able to form; in England it was remarkable that the class could expand despite the Civil War, Restoration, and religious dispute. The Dutch and English proprietors worked out a modus operandi during the course of a century. As the system grew, contending national interests fought for dominance within it. The Dutch lost economic hegemony within the emerging capitalist system by 1700 but did not lose the basis for earning ample profit. Proprietors based elsewhere, as they formed into groups, then understood what steps they needed to take – in France and later the German states, the United States, and Japan. Indeed, one may argue that there were additional efforts to form capitalist classes in Mysore, the Ottoman homeland, Brazil, and in Egypt. My analysis of this transformation, while it is not greatly different from others, does give additional attention to the social institutions of the rising capitalist class. To confirm my hypothesis that there arose collective capitalist institutions for lobbying states, it would be necessary, first, to document associations and their influence within the Netherlands, England, and other countries to create and maintain a pro-capitalist state at home and, second, to document cross-national versions of such institutions that would make each pro-capitalist state provide support for transnational capitalist links even as they sought in war and peace to advance their own national position.

What portion of the global economic network was under capitalist control by 1800? We have no estimates yet, though it seems fair to argue that the capitalist portion of the global economy remained a minority in that its scope was able to continue expanding through the nineteenth and into the twentieth century. One further limitation of this eighteenth-century summary is that it is expressed in terms of national hegemony more than in terms of social institutions. Most studies have focused on the economic experience of nations and of enterprises (especially such trading companies as VOC and EIC). I suggest that more attention be given to proprietors' associations, their lobbying of the state, and cross-national alliances of these associations. Such research might confirm that the success of capitalist proprietors depended on their banding together into associations, their development of close association with institutions of the state (be it national or monarchical), and their growing ability to manipulate military and tax-collection power. Indeed, it may be that the confusion of commercial and national interest served the needs of proprietors by labeling capitalist influence in terms of a national interest, rather than the class interest of proprietors. Such obfuscation may have brought additional support to the campaign for capitalist expansion. Another confusion that may have served a similar purpose was the common tendency to confuse the private East India companies with the state of European powers – this confusion facilitated capitalist development but relabeled it as national expansion.

Cultural Exchange

Where does cultural change fit into the story of social evolution? Felipe Fernández-Armesto, in a recent and highly accomplished historical analysis of culture and nature, made the case that cultural change is responsible for most of the great variety among human societies but that neither biological nor cultural evolution has much to do with the changes in culture. He reviewed the literatures in biological and cultural evolution, evaluated seven distinct evolutionary explanations of cultural change, and found them wanting: none explained how and why culture changes; none provided "descriptions of the range and reach of cultural divergence."[68] Fernández-Armesto's argument provided a clear reminder that cultural practices and social organization are not the same thing. As he argued, culture arises especially from imagination, in eclectic fashion and not always with productive results: "Imagination is the motor of culture" and, again, "Ideas drive cultural change."[69] This hypothesis, while denying that cultural change is directed by evolutionary processes – be they biological, cultural, or social – does not necessarily imply that group-level culture is entirely cut off from evolution, for two main reasons. First, in the origin of culture, Fernández-Armesto argues that the remarkable strength of human imagination arose from long experience with hunting. That is, evolution of the skills that facilitated hunting – especially the ability to anticipate the next move one's prey might next make – led to selection for greater imagination. Second and at least equally important, human imagination and representation create the innovations that bring about social institutions.

Fernández-Armesto's argument presents a challenge but not an obstacle to my analysis. My purpose is to explain the evolution of the Human System: in so doing, I focus especially on social organization, as it evolved through participation of people in social institutions. In comparing culture and social institutions, one may argue that social institutions provided the platform on which people created cultural exchange and group-level culture. Cultural production, arising via the imagination of individuals and their interaction with evolving social structures, developed a relatively autonomous existence. At the same time, it also provides essential feedback to the process of social evolution. Thus, my narrative of human evolution traces institutional changes in detail and explores changing technology, mostly through the people and institutions supporting the technology, but does not attempt to explain cultural change. I treat community-level cultural change as exogenous – I rely on its changes to explain aspects of evolutionary trajectory, but I do not attempt to explain the trajectory of culture. I do argue that cultural imagination was a major source of the innovations that enabled human communities to invent social institutions once they had the benefits of spoken language and the ability to form explicit groups.

Having just reviewed how the global economic network of 1600 had changed dramatically by 1800, I turn now to the question of whether culture changed in parallel fashion. To do so, one must consider culture in the various scales of its possible global interaction. Was there a global cultural network in 1600? What changes in global cultural interactions, at large or small scales, took place in the seventeenth and eighteenth centuries? In one parallel to the global economy, we can affirm that, while the seventeenth-century world experienced interregional cultural interactions, these were not the first such phenomena. Before the times of globe-circling voyages, the Islamic institution of the *hajj* stands out as an institution of cultural linkage. This pilgrimage, strongly encouraged at least once in life for all Muslims, brought many thousands of people together each year. The pilgrimage to Mecca – understood to be a home of the early prophet Ibrahim (or Abraham) – had linked people in Arabia well before the rise of Islam. Then, beginning in the year 629 CE, seven years after Muhammad led his followers from Mecca to Medina, the *hajj* became established as a strictly Islamic pilgrimage of a week's ceremonies, to which people came from all parts of the growing Islamic community. The ideas and personal connections nurtured during this religious celebration not only unified the community but helped to spread innovations throughout a vast region. From the sixteenth century, the Ottoman Empire invested in expanding and supporting the *hajj*.

Here, for the years after 1600, I focus in turn on material culture, expressive culture, and reflective culture.[70] Foodways, a very basic element of material culture, changed rather dramatically in the seventeenth and eighteenth centuries. The world trading system brought a global agricultural transformation in taste, in processes for producing food, and in the commerce of food. Portuguese migrants introduced tempura to Japan, where it remains popular. The Chinese welcomed such new foods as maize and peanuts. Africans made new use of those foods, as well as of pineapples and squash. The coffee plant, native to Kaffa in today's Ethiopia, spread to Java, the Americas, and later to other African regions. Tomatoes, native to the Americas, became widely popular as an addition to stews, sauces, and salads. Each one of these foodstuffs – the tomato, for instance – caused a small revolution in agriculture, economics, and taste. While seeds and seedlings were doubtless carried from place to place by the earliest visitors, it appears to have taken about two centuries for the maritime discoveries to give rise to really significant changes in the crops grown in various parts of the world.[71] Indeed, the lag time may have been even longer: that is, the exchange of such foodstuffs as carrots and lemons between the eastern and western edges of Asia had been launched as a result of Mongol experiments, with gradual adoption thereafter.

The slowness in spread of crops from one part of the world to another was more than a simple reluctance of farmers to try something new. It took time to develop breeds of each crop that would thrive in different soils, temperatures, and precipitation patterns. In addition, it took time for farmers to learn the best timing for planting and techniques for tending, weeding, and harvesting. Potatoes, for instance, had to undergo significant experimentation before they could be transferred from the Andean highlands to the soils of northern and eastern Europe. But once potatoes were successfully adapted to Europe, they brought an extraordinary expansion in productivity and in population growth. Maize came to be widely adopted in both western and eastern Africa. The great westward movement of population in eighteenth-century Qing China relied significantly on expanding fields of maize. On a more geographically restricted scale, manioc, a root crop that was the principal source of nourishment in Brazil and Venezuela, crossed the Atlantic and gradually became a major crop in the forest and grassland areas of Central Africa.

Domestic animals moved more rapidly and were adopted more readily than crops. Horses, cattle, sheep, and goats, which had already spread to all the regions of the Old World where they could survive, spread in the seventeenth century throughout the Americas and islands of the Atlantic. In the eighteenth and nineteenth centuries, they spread throughout the Pacific. Domesticated guinea pigs and wild agouti spread from South America to the Old World tropics. On the Australian continent, there were no placental mammals except humans and dogs at the time of Captain Cook's visit in the 1770s. Of the many mammals introduced to Australia, the rabbit turned out to have the most devastating impact, changing the ecology and reducing the populations of many other species.

Expressive culture begins with the languages we speak. The era of the seventeenth and eighteenth century brought the continuing spread of Arabic as a second or commercial language in the Indian Ocean and in Africa; of Chinese as a commercial language in southeast Asia; of Spanish and Portuguese as languages of colonial government in the Americas; of Portuguese as a trade language on Asian and African coasts; of pidgin and creole languages in slave and free populations of the Caribbean and Indian Ocean coasts; and of English, Dutch, and French in North American and Caribbean settlements. Through every language, expressive culture underwent new pressures for shifts in formulas of greeting and politesse, social roles by age and gender, standards of dress, family structure, inheritance patterns, and the treatment of the ancestors and the gods.

Exchanges in expressive culture included the transfer of expressive media and also of the values expressed in those media. Traditions in music, literature, and dance, for instance, traveled and influenced each other around the world. Guitars provide a valuable reflection of global connections in expressive culture. Stringed instruments had earlier developed all across Afro-Eurasia, where they are known as lyre, oud, mandolin, vihuela, and balalaika. Yet guitars, while they developed as elite instruments in Iberia, became popular instruments as they entered almost every port, brought first by Iberian mariners. They were sturdy and compact instruments; they could be readily constructed and strung repeatedly. Variety and diversity developed in traveling guitars, as they were drawn into many local traditions: they became especially prominent in Iberian America but were played all along the coasts of Africa, India, Japan, China, and the Philippines. The ukulele, a small form of guitar, has remained prominent in the music of Brazil and Hawaii, where it spread in each case through Portuguese migration.[72]

African drums spread too, though not as far nor as fast. Drums as material culture could be reproduced in any region, though people in every area of the world had their own drums. The African rhythms, however, were quite distinctive. Rapid, complex polyrhythms were deeply expressive, and could be adjusted to meet the needs of many social situations. In Africa, drums and drumming were employed in every area of life, accompanying all sorts of social activities. Wherever Africans moved and settled, but most commonly through slavery, they brought and created drums and dances to accompany them. By the nineteenth century, drums and drumming from West and Central Africa had arrived in most areas of the Americas. At the same time, movements of Africans across the Indian Ocean brought drums and drumming from East and Northeast Africa to West Asia and South Asia. At a slower pace, the African polyrhythms became attractive to communities beyond those of African descent. In the nineteenth and especially the twentieth century, the rhythms and the drums as well became widely and cross-culturally adopted in a pattern that reflected the earlier spread of guitars.

Expressive culture extends beyond speech and song to the subtle communication of literature. Literary scholar Ning Ma argues for a remarkably unified shift in literary culture that came with shifts in global travels. She shows that the literary form of the novel arose simultaneously in East and West, in the context of the shifting social conditions brought by expanded trade and by silver in particular. She documents her case with five early novels, written in fairly quick succession: *The Plum in the Golden Vase*, an anonymous Chinese-language manuscript circulated in the 1590s, portraying the mundane world of money and sex; *Life of an Amorous Man*, published in Japanese by Ibaru Saikaku in 1682, tracing the lives and sentiments of merchant townsmen and prostitutes; *Don Quixote*, published in Spanish by Cervantes in 1605, with an additional focus on silver; and *Robinson Crusoe*, published in English by Daniel

Defoe in 1719, for which a sequel took Crusoe as a merchant to the Far East, where he complained of the English economy's lack of centrality in the world market.[73] These examples suggest that further study may encounter additional examples of connections in expressive culture that were synchronically mediated by the global economy.[74]

At a broader, macrocultural level, one may ask: was an overall global culture in existence or in formation by the end of the eighteenth century? The initial instinct may be to say "no," and to divide the world up into separate, civilizational groups, each with a distinctive cultural tradition. On the other hand, elements of globally shared culture appeared as early as the seventeenth century and perhaps earlier. The argument for distinctive cultural traditions commonly centers on the European tradition of reflective culture. Reflective culture, most fully recorded in its literate forms, includes philosophical, scientific, and religious culture. For the seventeenth and eighteenth centuries, studies of philosophical and scientific culture have focused on the contributions of European Enlightenment thinkers, including the institution of the *republic of letters*.[75] The outstanding work of Vico on language, Newton on mechanics and calculus, Linnaeus on biological classification, Diderot on technology, Voltaire across the humanities, and Kant's crystallization of the Enlightenment are but the upper crust of a deep tradition in thought. From a more global perspective, Sebastian Conrad, seeking to contest a Eurocentric bias in Enlightenment studies, has argued forcefully that scholars in Asia addressed key issues in social and humanistic affairs during the eighteenth and nineteenth century, though they were generally neglected by European scholars.[76] The debate continues on how broadly "Enlightenment" may be defined, but the ground has shifted toward breadth with Conrad's intervention.

For the natural sciences, however, we do not yet have an argument that worldwide discourse continued during the seventeenth and eighteenth centuries. In that era, the expanding world of science, global in its conceptualization, appeared overwhelmingly European in its practice – in contrast to the sharing of material culture from region to region of the world. In both the global economy and in worldwide scientific knowledge, Europeans had achieved remarkable global predominance by the early nineteenth century. Yet the paths to European hegemony in these two arenas appear, on present evidence, to have been different. For the economy, current scholarship is making unmistakable the incremental and globally distributed expansion in global commerce and the incremental advances in productive systems. Europeans gained economic predominance not by the sudden imposition of a new system, but through steady struggle, innovation, and exchange with people of other regions. For scientific knowledge, in contrast, we see an apparent disjuncture. For the two preceding millennia, the literate societies of the world had developed elements of scientific analysis, exchanged and translated them, and shared much in scientific knowledge and technique. This process was not smooth: instead, there were periodic campaigns to translate works from previously leading scientific communities, followed by sudden explosions of new knowledge in the wake of these translations.[77] From the thirteenth through sixteenth centuries, a wave of translations brought worldwide knowledge into Latin and European vernaculars from Arabic, Greek, Hebrew, and other languages. Thereafter, scientific analysis seemed suddenly to be concentrated, at least for a time, uniquely in Europe and among Europeans. The apparent intellectual hegemony of Europeans became an important factor in itself as the Anthropocene epoch opened. This apparent disjuncture in scientific knowledge requires further analysis.

Stirrings in Political Outlook

At the end of the eighteenth century, capitalism and its proponents, while gaining rapidly in influence, had not yet made a claim to dominate the world. Expansion of capitalism and its institutions brought growing problems within the Human System: the question of whether the social fitness of institutions was to be measured by the welfare of a society as a whole or of privileged groups within society; the question of what sorts of regulation would serve effectively to improve the functioning of institutions; and the overall expansion of hierarchy in social institutions. Relatively new questions arose in initiatives supporting emancipatory, democratic, and nationalistic impulses, each of them contesting hierarchy in one way or another.

Elements of what became the institutions of the nation-state and democratic form of government stirred in this era; the two institutions, reaching the world stage separately, came to be tightly bound to each other in the nineteenth and especially the twentieth century. Here I seek to show that, while Eurocentric histories of both nationhood and democracy will surely retain their present significance, both issues also have deep and global roots that can be documented in discussion of the world of the late eighteenth century. James Kloppenberg – in an analysis of thought in the United States and Europe – argued that both the achievements and the tragedy of democracy were evident from the early days.[78] He identified three principles in the nature and achievement of democracy: popular sovereignty, autonomy, and equality. More fully defined, these are the will of the people as the source of authority (though with a debate as to representation or participation in implementation); the autonomy of self-rule; and equality of opportunity and of result in political processes. Further, Kloppenberg argued, democracy relies on three additional ethical principles: deliberation (in which provisional truths arise from free inquiry); pluralism (recognition that there is no unitary vision of the good life); and reciprocity (agreement that all should be treated with respect, and their visions weighed in the discourse). The principles of democracy thus provide a process for unifying or balancing contrasting value judgments.

Most centrally, Kloppenberg advanced the caution that "the tragic irony of democracy – for it is nothing less than that – has been the recurrent creation of social and political arrangements that, although often initially appearing to mirror popular desires, ended up either freeing previously repressed impulses that undermined democracy or generating other pressures that produced new and unanticipated forms of dependency and hierarchy."[79] The hopes for satisfying the will of the people commonly go awry, so that an initially inclusive outlook ends up imposing exclusive limits on some of those who expected to be within the community. Kloppenberg nevertheless went on to argue, optimistically, that "The flexibility and open-endedness of democracy make it particularly suitable for cultures increasingly attuned to a scientific sensibility, committed to testing multiple hypotheses through trial and error and examining the consequences through open inquiry."[80]

As for nationhood and nationalism, Benedict Anderson's reflections, which have sustained near-iconic status since their first publication in 1983, focus on print capitalism as an initial basis for the imagined communities of nationalism. Anderson opened his detailed discussion with Latin America as a region in which these ideas gained broad application in the eighteenth and nineteenth centuries, yet turned to Europe for his principal narrative on the nineteenth century.[81] Anderson's observations gave attention to the contradictory tendencies within nationalist thought and action: to achieve inclusion but also to carry out social and political exclusion.

How did the issues of nationhood and democracy gain worldwide relevance? Did the spread of democracy and nationalism take place through diffusion of specific sets of ideas from their points of origin along well-worn paths of discourse? In contrast to the common notion of diffusion, I propose the process of "crystallization," in which conditions that were already shared throughout the world provoked responses and innovations of great similarity, even though there was no linear conveyance of the specific institutional or discursive responses from one place to another. For nationhood most evidently, it became clear during the eighteenth century that, in the increasingly connected world market and in the significance of polities, war, and taxation in gaining wealth and recognition, polities with clear political and cultural unity were better able to advance their interests in the increasingly competitive global array of states. In a striking instance, the Cherokee nation – having experienced both peaceful trade and assaults on their independence in interactions with English settlers and other Native American groups – opted in the early eighteenth century for a policy of diplomacy with Great Britain and, later, with the United States. The Cherokee established a national government, a press, and an energetic campaign to maintain their independence; nevertheless, it was rejected by the United States, which denounced and dispersed the nation, so that most walked the Trail of Tears to Oklahoma. In the same era, the polities established by Tipu Sultan in Mysore and by the khedive Muhammad Ali in Egypt, while in no sense democratic, nonetheless carried out campaigns of internal unification and economic ambition that fit well with national projects.[82]

For democracy, the breadth of the underlying idea is far greater. Village-level tradition throughout the world has included many types of institutions enabling participation and representation in decision-making, allowing for resolution of conflicts and appeal of decisions to higher authorities. Only in a few places did philosophical discourse on political democracy develop in detail. Nevertheless, whenever a new polity arose, as with Mali in the 1230s, the Safavids in 1500, or the Marathas in the seventeenth century, the founding political leaders drew on the local government institutions with which they had grown up as a basis for designing their larger regime. The result naturally included an expansion in hierarchy, but also included councils, regulations, spaces for discourse, and other restraints on central power. In these ways, people throughout the world had experiences that prepared them to join in the political debates and struggles of the Anthropocene.

Anthropocene Evolution

9 Systemic Threats

We humans have lived since 1800 within the Anthropocene, the era in which humans have become the principal agent for change on Earth. The exact date for Anthropocene initiation cannot be known, but I have chosen a start date of 1800 because of changes both in human society and the natural world.[1] The Human System, viewed in terms of its growth, has been thriving during this past two centuries. Yet the Anthropocene epoch has brought instability and crisis both to Gaia and to the Human System. The ancient dynamics of Gaia are now shaken by the impositions of the human order and by the new patterns resulting from intense interaction of the two systems. Habitats are disrupted, species are extinguished, waters are diverted, and climate turns back toward the fluctuations of the Pleistocene.[2] The Anthropocene has brought parallel instability to the Human System itself: to its economic, social, and cultural components. The threatening patterns include political conflict, relentless growth that reinforces socioeconomic inequality, social protest and revolution, wars of unprecedented scale, and reorganization of relations with the environment. When did this instability begin? Perhaps it arose as repercussions of expanding capitalism. At the same time, the tensions of the system may be seen as a culmination of the innovations of early civilizations. Still further, difficulties in the Human System might also be traced back to the founding elements of the Northeast African social order.

In addition to the twin crises of environment and social order in the Anthropocene age, we must acknowledge the existence of a third. That is, humans have inadequate knowledge on how to face the twin instabilities of the Human System and of Gaia. We have a hint on where to look for solutions: for better or for worse, it seems that human agency is at the core of each issue, though there are those who deny the existence of any threat. Yet the third crisis presents the greatest challenge. Natural scientists know enough to be certain that the Human System is destabilizing Gaia; social scientists know enough to be certain that human social practice causes the inequality, hierarchy, and antagonism in society. But none know how to solve either the crisis of human or natural instability – much less how solve the two together, as we must.

Given the immense accumulation of knowledge, especially in the past century, it is ironic to find our societies unable to develop effective responses to these systemic threats. For myself, as must be the case for many other people, I find it frustrating to be unable to respond to the challenge. But I do suggest that the perspectives developed in this book indicate some of the materials from which solutions might be developed. The Human System, as a framework for analysis, is comprehensive yet also attentive to subsystems and to the intricacies of interactions. The perspective of social evolution traces the processes by which groups of humans create and update social institutions with which to construct their destiny. These analytical tools focus explicit attention on human group behavior, individual behavior, and interaction of the two. The latter point requires more theoretical work on group behavior, to match the effort that has already gone into analyzing individual-level behavior. To explore these issues for the Anthropocene epoch, this chapter begins with three descriptive sections that confirm the magnitude of Anthropocene transformations, then turns to three sections of analysis. The analysis centers on balancing the assumptions and the ideologies of group and individual behavior.

Economic Change: Capitalism and Divergence, 1800–1900

The economic institutions of capitalism came to dominate the world economy during the nineteenth century, led especially by British economic and imperial expansion, under the flag of an ideology emphasizing both civilizational superiority and individualism.[3] The expansion of British capitalism – traced in the previous chapter up to 1815 – accelerated to the late nineteenth century. The results brought global divergence in economy, politics, and ideology. From 1815, British enterprises and policymakers worked to extract the wealth of India through cotton, tea, and opium; they exported opium to China as a device to gain free access to the Chinese domestic economy, they carried out antislavery campaigns to limit the influence of Spanish, Portuguese, and Ottoman competitors, and invested in railways and other resources in Latin America and the United States. The Bank of England reaffirmed the gold standard in 1818 to strengthen the British pound and encourage exports. In the same era, British industrialization expanded, especially as measured by the construction of canals and railways in England.

Slavery reached its worldwide peak in 1850, as shown in Map 9.1, as a result of contrasting trends. While Britain abolished its slave trade in

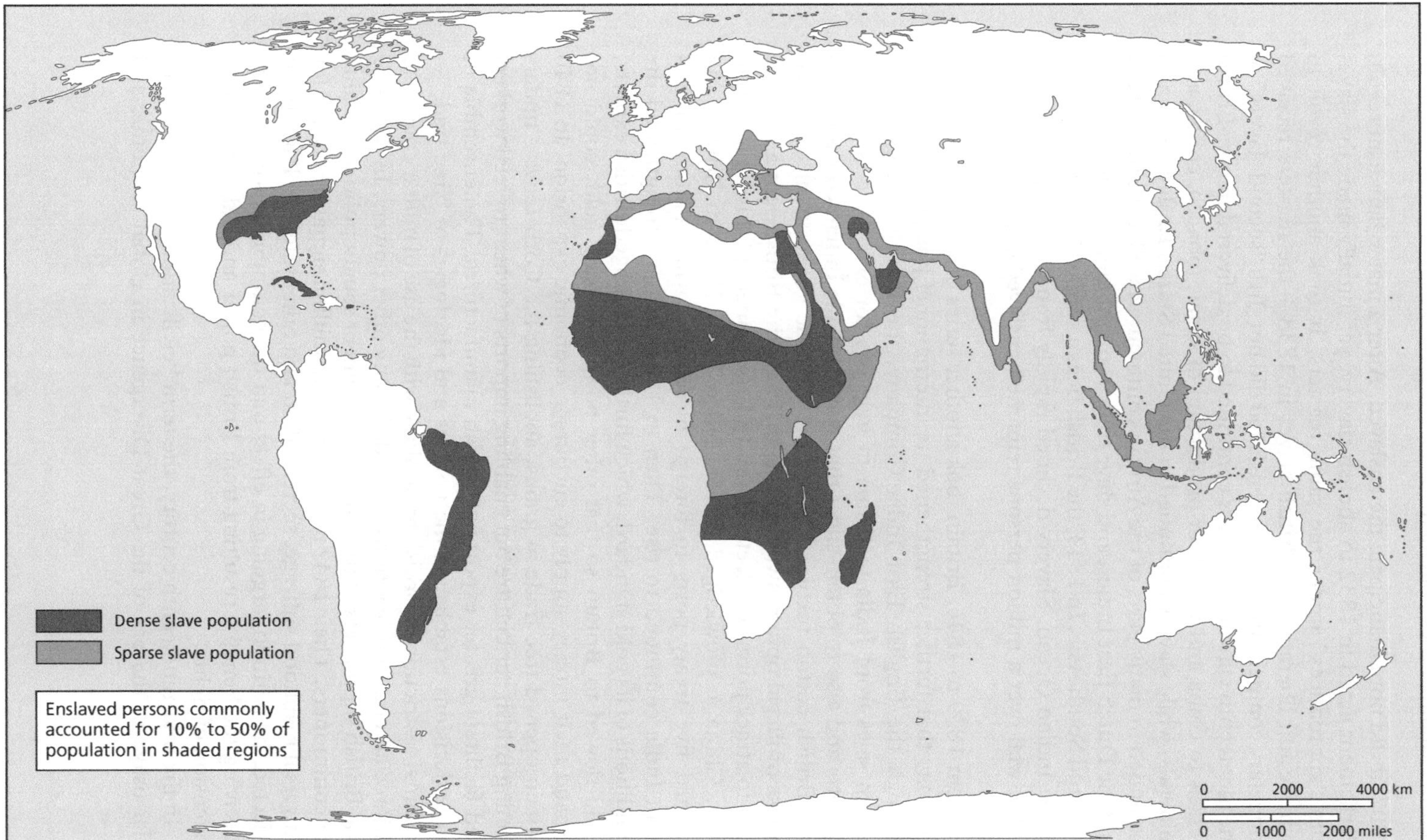

Map 9.1 Slavery at its peak, ca. 1850. (Redrawn based on original image, courtesy of the University Library System, University of Pittsburgh; generated by Boris Michev, 2019.)

1807, it did not emancipate the enslaved. After a major Jamaican revolt, Parliament acted in 1832 to abolish slavery, yet initially made the slaves into "apprentices" under the same masters. It took another revolt to achieve full liberation of British slaves in 1838 – the masters received monetary compensation but the enslaved did not. In a "second slavery" of the nineteenth century, captives continued to be delivered in non-British vessels to Cuba and Brazil for production of sugar, tobacco, coffee, and minerals, while slavery continued in the United States. Enslavement also expanded in many parts of the African continent as well as Burma, Siam, and the Dutch East Indies, and slave trade brought more captives to West Asia and South Asia. In 1843, the British began suppression of slave trade in the Indian Ocean. Slavery declined rapidly from 1850 but, as late as 1900, well over a million persons remained enslaved each in Africa and Asia.

From 1839 to 1846, Britain took another series of steps in policy and ideology that further strengthened its international position. Indirectly, through the English East India Company, the British had expanded opium sales from India to China to help pay for expensive British tea imports and also to obtain silver that would enable Indian consumers to buy British cotton textiles.[4] When the Chinese government destroyed stocks of illegal opium awaiting sale in 1839, the British sent a punitive fleet to attack ports and fleets. In the 1842 settlement of this First Opium War, China acknowledged defeat, ceded Hong Kong to Britain, and opened five treaty ports to foreign merchants, while opium imports from India continued to rise. Then in 1844, Britain strengthened the conditions of its gold standard, which further encouraged British exports. Gold flowed to Britain so that other powers, lacking gold, could not expand their money supply to encourage economic expansion. In 1846 Britain opened free trade in grains, abolishing the Corn Laws – undermining British agriculture but enabling imports of wheat to feed workers and facilitate greater exports of British manufactures. The ensemble of these decisions established the policies and ideologies of "free trade" – minimal restriction in British trade along with the use of military power to enforce an end to restrictions on trade by any other powers. In the same era, British supporters of the Chartist movement sought political equality for commoners. Their 1842 petition with 3.3 million signatures called for universal manhood suffrage, secret ballot, and other reforms. Parliament rejected the petition, arguing in effect that the commoners were already receiving the benefits resulting from British power and did not need an extension of rights.

In the mid-nineteenth century, the term "civilization" came into popular usage in the era of the Great Divergence in global economic and political power. "Civilization," a racially based term classifying regions and peoples, served as a triumphal celebration of European hegemony and a rejection of emancipatory social movements. Perhaps the most forceful popularization of "civilization" was that published by Arthur de Gobineau, in his *An Essay on the Inequality of the Human Races* (1853–55). Gobineau, a French lawyer and diplomat, wrote a speculative history of civilizations and races – each divided into tribes and nations – that articulated an emerging consensus of the European reading public. Civilization, in Gobineau's view, was achieved through a mysterious combination of a "civilizing instinct" with a form of group-level culture and administration that brought stability, sociability, and the hatred of violence. He used the term "Aryan" to refer to Indo-European languages and the term "Caucasian" to refer to light-skinned Europeans. He identified white, yellow, and black races, arguing that each racial, linguistic, and civilizational group was fundamentally different and mutually repulsive, so that mixing them usually brought more harm than good. Overall, his interpretation of the nineteenth-century world was that race and civilization fit in a clear hierarchy that would not change. Gobineau was an amateur scholar but his publication restated in colorful language the assumptions that had already been adopted by leaders of public opinion in Western Europe and North America.[5]

The Great Divergence – as documented in Kenneth Pomeranz's study of the late-eighteenth-century decline in the wealth and productivity of Chinese economic centers as compared to those in Europe – expanded worldwide in the nineteenth century.[6] For the period after 1850, several additional nations joined Britain in global economic leadership in a pattern that Peter Gran calls "multilateralism" and that I have labeled an alliance of "pro-capitalist states" and "transnational proprietors' associations."[7] This amounts to an expansion to the global level of what I have called proprietors' associations. That is, collaboration among states that were themselves tightly allied to capitalistic proprietors' associations brought a mutually reinforcing application of free-trade policies by using gunboat diplomacy to respond to efforts by noncooperating states to limit trade and using police to repress domestic efforts at organization by industrial workers. European and North American leaders explained their economic growth with triumphal ideology of Western civilization. A series of mid-century conflicts appeared to confirm the rapid divergence separating the beneficiaries of capitalism, on one hand, from other communities that were vulnerable in comparison. The Mexican–American War (1846–48) ended with Mexican cession of its immense northwestern territories to the United States.[8] The 1848 rebellions in Europe began as republican uprisings by Sicilian commoners, then spread to become rebellions of wage workers against their conditions – though the rebellions failed.

Far more significant in terms of lives lost were three other great conflicts that reflected the emancipatory claims of the age. In the aftermath of the First Opium War, the Qing government raised taxes to pay its war debt, thereby contributing to the Taiping Rebellion of 1850. Hong Xiuquan, who envisioned himself as the brother of Jesus, raised massive support to challenge the Qing. In warfare with high casualties, the rebels seized Nanjing and ruled there from 1853 to 1864, while millions died in fighting in the Yangzi Valley until the empire reestablished control.[9] In India, halfway through the Taiping rebellion, Indian troops in the British army rebelled in an 1857 uprising that spread throughout north India. Britain sent in troops to repress the rebellion with ferocity: casualties of British soldiers were minimal, losses of Indian soldiers were larger, and losses of the Indian population reached into the millions in the course of two years of repression. London replaced the East India Company with a government headed by a viceroy and reporting to the India Office in London. Parliament thus established the British Raj, with direct or indirect imperial control throughout British territories of the Indian Ocean. Third, the Civil War of the United States raged from 1861 to 1865 as slave states of the South, great exporters of cotton to Britain, seceded to protect their system of slavery – while slaves in the South and free people in the North called for emancipation.[10] As with the other contemporaneous conflicts, casualties were immense. The three great conflicts were quite different, but all reflected major imbalances in society – all were tied to global and not just to local factors. The losses were of several million each in the Taiping and Indian uprisings, and most of a million in the US Civil War.[11]

Social Change: Nations Displace Empires, 1870–1970

The nineteenth and twentieth centuries brought successive institutional changes in government. Monarchies either adopted national constitutional governance or they collapsed. Great empires rose to a peak in the early twentieth century and then collapsed at midcentury. National governments, especially of republican form, grew until they became nearly universal. This displacement of empires by nations, in just over a century, is arguably an essential change in human social evolution.[12]

Especially after 1850, autocratic monarchs had to cede their power to elective parliaments that selected governing ministers. Such a transition had taken place in Britain and now took place in Spain, Denmark, Sweden, Japan, and Thailand. Monarchies were created or restored in the Netherlands, France, Brazil, Greece, Belgium, and more.[13] Existing monarchies expanded as constitutional monarchies in the unification of Italy (1860) and Germany (1870), and with the Meiji Restoration in Japan (1868). Nineteenth-century monarchies, however, were abolished even faster than they were created. In some cases, they became national republics. France created and overthrew four monarchical governments from 1793 to 1870; Brazil overthrew its empire in 1888 and the Qing dynasty in China ended in 1912. World War I brought the end to the monarchies of Russia, Austria, Germany, and (in 1924) the Ottoman Empire, yielding nearly a score of European republics. In other cases, imperial expansion by European monarchies brought the end to monarchies that had existed for centuries in Asia, Africa, and the Pacific. As European forces occupied Africa, mostly from 1880 to 1910, scores of African monarchies were made subservient, including the Sokoto Caliphate, the Asante empire, and the Zulu kingdom. In an extraordinary case, as the Dutch forces seized Bali in 1908, wave upon wave of members of the royal Bali family charged the repeating rifles of the Dutch forces, effectively committing suicide along with the end of their old order.[14]

The new wave of empire-building, also beginning about 1850, accompanied the global spread of capitalism, with results as shown in Map 9.2. One must remember that the institutions of capitalism had emerged in a challenge to empire during the sixteenth and seventeenth centuries, when merchants and proprietors struggled against monarchs and aristocrats. By the nineteenth century, however, capitalism wielded empire as its own tool. Nation and empire became fundamentally confused with each other as imperialism rose to a peak in the early twentieth century. Britain, with Ireland as an intimate colony, expanded further to seize territories in Africa, Asia, and the Pacific. In some of these territories, where there were large settler populations (for instance Canada and Australia), the British allowed local self-government; India and other British territories were governed from Britain. The Netherlands expanded its Asian empire from Java to include the whole of the Indonesian archipelago. Germany, almost as soon as it became a national monarchy, annexed colonial territories in Africa and the Pacific, and sought others in China and South America. Japan, similarly, launched military campaigns and annexed colonies in the decades after its redefinition as a national state.[15] Among republics, as for monarchies, powerful nation-states sought to conquer new lands. Republican France maintained control of colonies in the Caribbean, the Indian Ocean, and the Pacific, and gained control of large territories in Southeast Asia and much larger territories in Africa. For the United States and other large nations of the Americas, much of the territory under national rule, inhabited by indigenous peoples, was in practice in

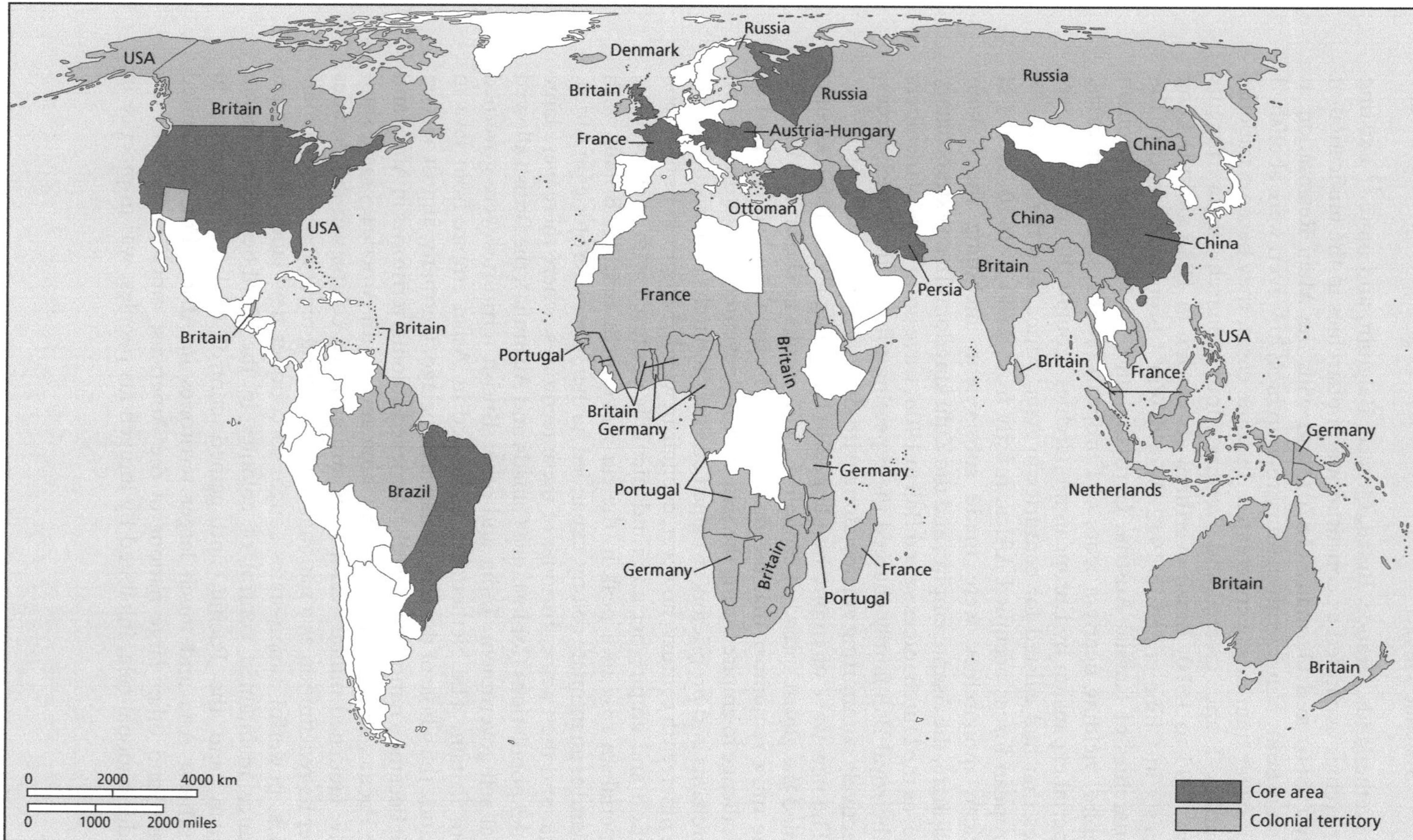

Map 9.2 Empires, ca. 1900. (Redrawn based on original image, courtesy of the University Library System, University of Pittsburgh; generated by Boris Michev, 2019.)

colonial status.[16] In these cases, a federal form of government permitted incorporating parts of the empire into the imperial homeland as states or provinces, with eventual recognitions of indigenous peoples as citizens, after a long delay. In addition, the United States gained an overseas empire with Alaska in 1867 and other territories from 1898.

World War I and World War II were at once wars among empires and wars among nations. The nations mostly survived and even expanded but, for empires, the future was more complex. Russia, China, and Iran, with centuries of imperial tradition, underwent social revolution and became great nations. The United States, Brazil, India, and Indonesia became great nations by breaking away from the empires of Britain, Portugal, and the Netherlands. Otherwise, most empires lost their colonial territories and the old imperial cores became national states: such was the case for Britain, France, Portugal, Spain, Turkey, Netherlands, Japan, and Italy. After a war of empires, the League of Nations formed in 1919; its successor, the United Nations, had to adjust to a world in which nations had at last become the norm.

Decolonization swept Asia in the 1940s, then Africa in the 1960s.[17] This dual wave of national political independence was also a powerful statement of antiracism, in that nonwhite people, who had been denied the vote and political office for the previous half-century, now had access to formal political participation. Overall, modern colonialism was at its peak from 1890 to 1940; by 1991, with the collapse of the Soviet Union, it was nearly gone. European powers had used arms and free-market ideology to create huge empires, only to see them dissolved in the aftermath of two world wars. Yet some imperial heritage persisted, as in the islands held as military bases by a few powers and in the wide usage of Indo-European languages, especially English, as second languages.[18] The emerging order of nations – "postcolonial globalization," in the words of A. G. Hopkins, brought new sorts of inequality, as nations shifted in their relative demographic, economic, and military influence.[19]

Growth and Its Limits, 1800–2020

The Anthropocene has brought two centuries of rapid growth in many areas of human experience. Here the effort is to identify a few patterns of growth that stand out in their significance for the Human System. I begin by noting the shift in sources and quantity of energy. The annual human consumption of energy increased by a factor of 25 from 1800 to 2015.[20] In 1800, firewood provided the great majority of global energy, accompanied by small but growing amounts of British coal. Petroleum, natural gas, and hydroelectric power entered the market in the late nineteenth century. As of 2020, coal, petroleum, and natural gas each provided just under one-third of energy consumed; energy from firewood, at 7% of the total, still exceeded the contribution of solar, nuclear, wind, and hydropower.[21] Animal power had also remained connected to other power sources – horse populations of cities peaked in the late nineteenth century.

World agricultural output grew at a rate exceeding the growth in population, suggesting that populations became better fed: wheat output rose by a factor of 14 from 1900 to 2000, up to 755 million metric tons per year. Several other global economic measures rose by a factor of about 30 from 1900 to 2000: the output of cocoa, a modest luxury, rose by a factor of 30, up to 3 million metric tons per year; worldwide steel production rose from 28 million tons per year in 1900 to 850 million tons per year in 2000, a factor of 30 in a century; and copper mine production rose at a similar rate, by a factor of 35 from 1900 to 2013. Tonnage of world shipping increased by a factor of 10 from 1800 to 1910. Then shipping tonnage increased by a factor of 26 from 1900 to 2000 – so that shipping tonnage increased by a factor of 260 in two centuries. The rate of expansion in shipping was greater than that in total production, because many goods were produced and consumed locally. Nevertheless, oceanic shipping capacity doubled from 1980 to 2018, reaching almost 2 billion dead-weight tons in capacity.[22]

The quality of economic life changed as well. In the early stages of industrial expansion, the small firms had little influence over their consumers: "competition" meant that profits rarely exceeded modest levels. By the late nineteenth century, firms became able to affect market prices. "Competition" was now directed at specific competitors and consumers rather than the market in general; firm profits rose, along with advertising expenditure and consumer prices. In the late twentieth century, as monopoly and oligopoly grew, prices and wages ceased to be set at equilibrium levels that cleared the market; they were set especially by the negotiating power of big buyers and sellers.

The issue of waste grew in importance with industrialization. The problem of urban sewage stood out, and sewerage systems began to be built in major European and North American cities in the mid-nineteenth century. By the early twentieth century, sewage treatment facilities had been implemented in all cities that had sewers, though sewerage systems were not able to keep up with the accelerating rate of urbanization. Solid waste, long neglected despite its volume, is now being analyzed: urban solid waste (household and industrial) for 2005 averaged 1.2 kg per person per day. The regional lows were 0.45 kg for India and 0.65 kg for Africa, while for the Organisation for Economic Co-operation and

Development (OECD) countries (20 wealthy countries) the average was 2.2 kg per person per day.

The human population grew by a factor of 8 from 1800 to 2020, from 1 billion to just under 8 billion.[23] Accounting both for the larger population and increased human lifespan, the number of years to be lived by persons alive in 2015 was 20 times greater than for those alive in 1800. Expectation of life at birth, worldwide, averaged roughly 25 years in 1800, a figure held that low especially because of high infant mortality. By 1900, the average global expectation of life at birth had risen to 35 years; by 2015 that figure had doubled to 72 years, so that average human lifespan nearly tripled in two centuries. The differences in lifespan between rich and poor countries declined with time, though the differences in income grew. Life expectancy in 2020 reached 84 years in Japan, 76 in Brazil, and 66 in Ethiopia. For child mortality, the changes are perhaps more remarkable: for 1800, it is estimated that no more than 56% of children worldwide lived beyond five years; by 2015, 96% of children were living beyond five years. Urban population grew spectacularly: cities held only 3% of total population in 1800, rising to 14% of total population in 1900, to 30% by 1950, and exceeded 50% of total population shortly after 2000. According to UN projections, as world population passes 8 billion in 2024, the global rural population will begin to decline.[24] Such a change is not unprecedented – rural populations declined in earlier moments of pandemic and climate shift – but we now appear to be headed for a permanent decline in rural population, even as the cities grow.

Population not only grew but also shifted from the Old World to the New World and from temperate zones to the tropics. Map 9.3 shows the era of large-scale migration from densely populated areas of Europe, Asia, and Africa, up to 1940. While enslaved Africans were the majority of overseas migrants from ca. 1600 to ca. 1850, the combination of steam shipping, expanding capitalism, and empire led to expanded migration of European and Asian populations up until World War II.[25] After 1940, population grew most rapidly in tropical regions: Africa, South Asia, Southeast Asia, and Latin America grew in relative population, at the expense of East Asia and Europe.[26] From 1820 to 2010, the European portion of world population declined from 21% to 11%, while the New World portion rose from 3% to 14%. As populations were redistributed, so also were the average earnings of workers. Recent estimates of the real income of wage workers, shown by region of the world in Table 9.1, increased over time, but with cases of both narrowing and widening of gaps among regions.

Changes during the era of postcolonial globalization included regional economic integration in place of the threads of empire, migration from ex-

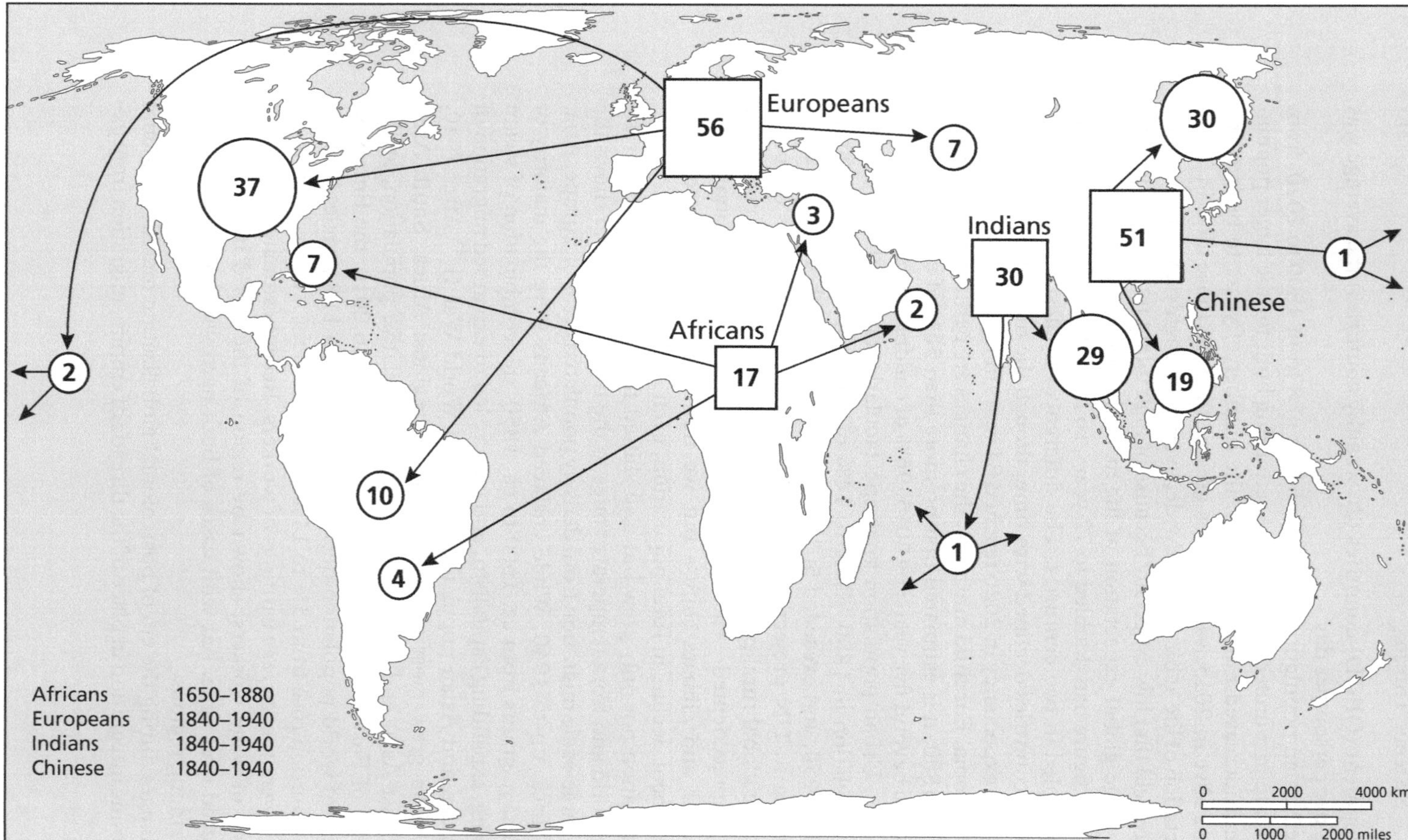

Map 9.3 Migration of Africans, Europeans, and Asians, 1650–1940. [Redrawn based on original image from Patrick Manning, *Migration in World History* (London: Routledge, 2012), 151; © 2015, 2013 Patrick Manning. Reproduced with permission of the licensee through PLSClear.]

Table 9.1 *Average real wages of building laborers, by region, in relative values based on comparable baskets of subsistence goods*

	W. Eur	N. Am	L Am	East Asia	S–SE Asia	Africa	World
1860s	11.5	43.9	6.6	2.8	4.7	3.1	6.3
1910s	16.9	77.2	10.6	4.3	3.7	4.4	12.5
1940s	26.2	75.8	14.4	5.2	4.3	6.2	13.8
1970s	49.4	145.5	21.3	19.4	4.7	6.1	26.4
1990s	105.2	168.1	29.9	26.9	8.5	11.0	35.3

Source: van Zanden, *How Was Life?*, 75, 80.

colonial lands to former imperial centers, a discourse of human rights in place of racial hierarchy, flows in capital from ex-colonies to former imperial centers, and trends toward equalization of health and education.[27] As examples of the fluctuations in national influence in the era of postcolonial globalization, after the USSR collapsed in 1991, Japan's rapid growth halted, while that of China accelerated. US population, while growing, went from 6% of the world total in 1960 to 4% in 2020. US economic output was 40% of global output in 1960 but, by 2020, the combined total output of the United States and China was 40% of the global total.

Pervasive growth, though relatively new as a general phenomenon, had become habit-forming. It became an ideological requirement. Yet growth varied by space and by time: it came late to the tropics and to the colonies. Reversals of growth never left the picture – storms, famines, economic depressions, and wars brought eras of decline and deprivation. Only in the 1970s did a vocal minority of analysts begin to argue that growth might not last as a general phenomenon.[28] Yet today, when the threats to growth and the dangers of growth appear everywhere, there remains a clear, global consensus: a desire and expectation for more growth.

Cui bono? Trade Unions and Institutional Fitness, 1880–1970

"Cui bono?" ("Who benefits?") – a question posed repeatedly by the Roman jurist Lucius Cassius – has shown its value in studies of law and society down to the present.[29] While such studies are usually conducted in the framework of individual-level analysis, I apply it as well to group-level analysis. I argue that the question of "Cui bono?" is central to the analysis of social institutions, in that it points toward questioning institutions' fitness in society. As I have argued, social evolution is somewhat analogous to biological evolution. That is, in biological evolution, the characteristics of an individual organism, if beneficial, yield the "fitness" of the organism, indicated by its thriving heritage of individual offspring within the species. For social evolution, the characteristics of an institution within a community, if beneficial, yield a "fitness" of the community through the heritage of benefits to a beneficiary population within the community.[30] In this case, I explore the success and failure of trade unions – their fitness – in the course of the century of their prominence. An analysis of this type could also be applied to such other social institutions as enterprises, governments, schools, ethnic groups, nations, religious organizations, and health organizations.

The additional complexity, for social evolution, is that a beneficiary population – for which the institution's fitness is to be assessed – is *not* uniquely defined, either in the population served or who makes the choice. First, the beneficiary population might be as small as directors of the institution itself or as large as the full community in which the institution exists; the beneficiary population, once identified, might be assessed through the number of its offspring (as is the case for assessing biological fitness), or through the "social welfare" of the population. Second, the individual or group that identifies the beneficiary population is open to variation. The "identifier" of the group's objectives and beneficiary population might consist simply of the directors of the institution itself; it might include all members of the institution; it could include all members of the community; or it could be a higher-level institution such as a state.[31]

Trade unions, new institutions of the nineteenth- and twentieth-century world, underwent growth and transformation through four waves of change. From 1850 to 1920, trade unions emerged along with industrial capitalism and global empire. Despite many defeats, they eventually gained large memberships and challenged employer control of labor relations. The scene changed in 1920: the International Labor Organization (ILO) arose as a mediator in labor relations; the Soviet Union emerged as a state formally allied to trade unions (followed later by other socialist states); and trade unions gained legal rights and access to social welfare systems in several countries. The end of World War II in 1945 brought a further sharp change, as the consequences of war strengthened the economic and ideological position of workers. Yet from the 1970s, in the fourth wave, trade unions declined sharply.

The most basic idea of trade unions was that workers should have collective representation in negotiations with their employer: that required a consensus among workers on the form of their representation – initially an informal group and later an institution – and a recognition, explicit or implicit, of the workers' voice by the employer. Trade unions, while commonly illegal in their early days, were increasingly supported by employees who felt the pressures from their conditions and chose to risk making common cause in submitting demands to their employer. Trade unionists sought co-workers with whom to unify and debate their objectives in changing wages, hours, job security and benefits. They further debated their beneficiary population – was it to be the wage workers of a given enterprise or such wider groups as their families, communities, or workers more generally?

Two problems complicate this analysis of the institutional fitness of trade unions: geographical dispersion and situational complexity.[32] Geographically, worker collectivities arose anyplace where there were groups of workers, free or slave, although those in the great industrial centers are best documented. Further, trade unions interacted with multiple levels of socioeconomic structure. For instance, as early as the great US railroad strike of 1877, trade unions engaged multiple employers, multiple communities, and provincial and national administration. The many situations encountered by trade unions required attention to such dynamics as business cycles, changing technology, solidarity of co-workers, and the specific workplace relations of farms, factories, transportation work, and office work. Trade unions steadily faced the issue of basic reproduction – sustaining and passing on their institutional structure. They relied on archives in paper and in the memories of members: oral tradition, meetings, songs, marches, newspapers, and conferences. If the workers had political rights, they could support electoral parties, such as labor or socialist parties; they could express anarchist views regardless of their rights.

With this complex context, trade unions faced the issue of fitness. Their beneficiary group was certain to include workers at the local level but might also include the communities in which the workers lived. Yet it might also be important for trade unions to seek alliances with consumers of the goods or services provided by their employing enterprise. The next question in fitness was that of who served as leaders of the institution, how they were selected and guided, and their role in determining the beneficiary population. For instance, employers commonly pressured trade union leaders to develop sympathy for employer outlooks at the expense of the workers and trade union members. One more issue in fitness was a trade union's relationship with state authorities. Were those legal, administrative, and police authorities hostile to trade unions? Were they neutral in labor conflicts or perhaps supporters of trade-union interests?

As of the opening of the twentieth century, trade unions in Western Europe and North America had done remarkably well in increasing membership of workers and support from their communities. Socialist (but also anarchist) ideology held trade unions together but also represented them in the political arena, to the degree that workers had civil rights and the franchise. Nevertheless, the scale of agricultural and industrial production, of transportation networks, mining and forestry, and white-collar work continued to grow, and with it the scale and ambition of trade unions. Corporations consolidated and resisted trade unions more fiercely in North America than in Europe, winning court battles yet also having to face national legislation that restricted the mergers and collusion of great firms. European trade unions and socialist parties, pursuing an electoral strategy, were often able to legalize trade unions. Corporations integrated their ownership and operations in horizontal and vertical terms, built their own police forces, hired private police, and called on the state for armed support in conflict with workers. In a 1914 Amsterdam meeting, the best-organized trade unions from North America and Western Europe met to form a new international organization, the International Federation of Trade Unions.

In the years after World War I, new chapters in trade-union history opened. In addition to the continuing transformation of capitalist enterprises, three institutional shifts took place: the creation of the International Labor Organization, the rise of the Soviet Union, a "workers' state," and legal rights to trade-union representation for workers in some countries. The questions of institutional fitness remained much the same, but the ability of trade unions to serve their constituency improved markedly, though inconsistently. In the period after 1920, while trade unions might be resisted bitterly in any given location of industry, they had gained formal recognition as part of an economic scene that featured alternating confrontation and conciliation in labor relations. Membership and recognition grew in times of prosperity and depression, yet with sudden changes. But ideological differences of workers and employers, often severe before World War I, became further hardened in the interwar years.

The ILO was intended to facilitate conciliation among employers, workers, and governments. Its governance provided representation for trade union associations, employer associations, and state representatives. Its activities included preparing model legislation to provide support for workers' rights, workplace insurance, and pensions worldwide; it remains a conciliatory force to this day.[33] While the ILO offered conciliation, the Russian Revolution and its supporters emphasized worker militancy. The world's socialist parties split into two groups beginning in 1919, when revolutionary Russia, under Bolshevik leadership since 1917, convened the Communist International in 1919. After the International

Federation of Trade Unions reestablished itself in 1919, the Red International of Trade Unions formed in 1920.[34]

In the interwar era, though it was a time of growing ideological contention, trade unions grew in strength, not only in the North Atlantic but also in China, India, Japan and its colonies, and Latin America. In the Soviet Union, workers maintained union membership and received benefits through their alliance with the government, but lost their ability to negotiate independently for their working conditions. Within two months after Adolph Hitler came to power in Germany, he abolished all trade unions, replacing them with a German Labor Front that included employers and workers. This was an early step in Hitler's reliance on militarism, anticommunism, anti-Semitism, and a desire for conquest that soon brought war. World War II broke out by stages: the Japanese invasion of China in 1937 initiated large-scale fighting. Brutality accelerated on every front as the war expanded, and the Allied opposition to the Axis powers responded with steadily greater emphasis on democratic rights and on antiracism.

At the end of 1945, as news spread of the Holocaust in Europe and of Japanese prisons in Asia, an unprecedented rejection of racial hierarchy united much of the world. This pro-worker, antiracist outlook provided an opportunity to build trade unions worldwide. From that perspective, the World Federation of Trade Unions (WFTU) formed in 1945, with the intention to include all trade unions regardless of ideology, in order to represent workers consistently within the International Labor Organization. But the divisions soon returned: the American Federation of Labor declined to join, especially because Soviet trade unions were included, and by 1949 the World Federation of Trade Unions had split: the competing International Confederation of Free Trade Unions, led by American and British affiliates, became strongest in the leading industrial countries; Catholic and other trade union federations remained separate. Nevertheless, although trade union organizations clashed bitterly, based on their ideological differences, the trade union movement reached the peak of its strength from the 1940s to the 1970s. This was a time of rapid economic growth, during which wage workers were recognized as important contributors to social welfare. Trade union membership reached high levels in Western Europe and in socialist countries; it grew rapidly in North America, Latin America, and Asia; and grew with a late start in Africa. Union recognition and collective bargaining were legally recognized in most countries, though collective bargaining agreements extended to a minority of wage workers. Substantial pension programs and health programs supported union members and in some cases most citizens, either through contracts with employers or in nationally supported systems.[35] While the changes were uneven, the postwar era of growth did bring about expansion in schools and public health programs in both rich countries and poor.

From the 1970s, trade unions declined steadily in membership and economic influence in most nations. It may be in some fashion that the success of trade unions in the postwar era created and exposed new internal weaknesses that combined with changing external conditions to weaken trade unions seriously. Trade unions lost community support; they lost worker support as well. Automation and shifting structures of industry, replacing full-time with part-time workers, reduced the numbers working under union contracts. Industrial production began to move from high-wage countries to low-wage countries, especially to Asia. Anti-union campaigns, led by corporate interests but gaining broader support, advanced multiple arguments against the efficiency and validity of trade unions.[36] Trade unions did not disappear, as they remained effective in certain industrial fields and in certain artisanal and public service fields, but trade unions following any of the numerous possible ideologies lost much of their strength in this transition, and the centrality of trade unionism as a social movement had apparently ended. Trade unions expanded in South Korea and elsewhere in Asia but did not achieve unity with those in the Atlantic. Additional problems in trade unions of many countries were their focus on male workers and on industrial and artisanal production, rather than on female workers, white-collar work, or service and retail work, so that trade unions became vulnerable to arguments that they represented only special interests rather than workers in general.

What has been the place of trade unionism in world history? What has been the place of socialism in trade unionism?[37] Each was an effort at ideological collectivism in contrast to the dominant individualism. What did this mean in practice? Whether unified or in disunity, trade unions and socialist parties brought important reforms in capitalist countries and built alternative systems in socialist countries. Trade unions, to this day, preserve influence in certain industries and national settings. But it became clear by 1980 that the trade-union movement had not succeeded in unifying the wage-labor class worldwide. Trade unions, on their own or in association with socialist parties, had lost strength and, while surviving, were marginalized: their effort to articulate the voice of humanity had fallen short. They may be compared with such other movements for social transformation as national liberation, national renewal, gender equality, and environmental protection.[38] Does the apparent weakness of all of these movements mean that attempts to organize group behavior are discredited as ineffective and that individualism is confirmed as the only framework for analysis and practice in human behavior?

Chaldean Detroit, Seen through Cultural Evolution

I turn now to a review of a contrasting analysis, conducted within the framework of cultural evolution. The analysis focuses on processes of voluntary cooperation within a community. My interest in exploring this study is to clarify what sort of group behavior it traces and to see whether or not the analysis takes account of the creation and functioning of social institutions. The authors, Natalie Henrich and Joseph Henrich, are a husband and wife team, an ethnographer and a cultural-evolution scholar, who have applied the logic of cultural evolution to an ethnographic study of the contemporary Chaldean community of Detroit.[39] This community, bilingual in English and Chaldean, is explored through the lens of cooperation and cultural evolution, based on a logic of reciprocity that has been developed in game theory. The main analysis of the Henrich and Henrich volume is to apply these behaviors to an ethnographic approach to the Chaldean community of Detroit.

The analysis focuses on "prosocial behaviors" of two types: cooperation and "altruistic punishment." The most basic level of cooperation occurs "when an individual incurs a cost to provide a benefit to another individual or people." A group of cooperating people, implicitly, is a population not otherwise bound together.[40] Examples include those involved in voting, food sharing, or recycling, where these activities are public goods, available to all. "In a public goods situation, individuals have an incentive to refrain from cooperating even though, in the long run, everyone will suffer from the loss of the joint benefit. In public goods situations, individuals can 'free-ride' by not cooperating while still reaping the benefits created by the contributions of others."[41] In altruistic punishment, "an individual pays a cost to inflict a cost on another individual in order to maintain a behavior, or norm, in a group" – basically, scolding those who misbehave. In an important assumption, the degree of cooperative and altruistic-punishment behavior in individuals is assumed to be influenced by genetic inheritance of those proclivities and by the marital and family patterns underlying genetic inheritance.

As I would state it, such a cooperative group is seen to exist not because of collective intentionality of a we-group, but because a cooperative norm arises within an I-group, enforced by individual-level altruistic punishment. Institutions do have a role in the analysis, though the term "institution" is mentioned but twice in the book. Henrich and Henrich argue that, "From the perspective of institutional design, cultural evolution should exploit our evolved reciprocal psychology by creating institutions that – among other things – partition cooperation into small, enduring social groups."[42] "To understand how culture solves the combinatorial explosion of reciprocity, it is important to recognize that ... cultural learning processes allow populations to rapidly adapt to novel situations without any genetic change.... [I]nstitutional structures can turn public goods into dyadic interactions."[43]

That is, Henrich and Henrich adopt, for cultural evolution, the approach to social institutions that is commonly applied in social sciences (and reaffirmed by Douglass North), which treats institutions simply as norms, not as groups, and treats them as exogenous rather than include them in the analysis. The analysis includes individuals within a world of institutions, without explaining the institutions. The book's later discussion of religion shows, however, that in addition to the prosocial world of I-groups, the Chaldeans also live in the world of we-groups, in that their religious and linguistic unity must be agreed upon and reaffirmed by each generation. Those who become Jehovah's Witnesses or who cease speaking Chaldean are gradually excluded, according to rules formulated by group members (identifying the beneficiary population) of a type that do not exist among I-groups. The churches require institutional consensus and structure that is not required for cooperation among voters or for recycling.[44]

My point is not that the analysis of Henrich and Henrich is erroneous or illegitimate. Indeed, I would hope to see more such studies, but I would hope to see them framed more explicitly in a discourse that explores the overlaps of cultural and social evolutionary theories. Henrich and Henrich have presented the Chaldeans as an example of a uniform ethnic group of the type theorized in cultural evolution for much earlier times. They make the case that the Chaldeans now live successfully, with their internal unity in a surrounding world of great diversity and huge numbers of institutions formed by collective intentionality. Yet by describing their analysis both in I-group and we-group terms, I have shown that, alongside the I-groups of their principal analysis, the same individuals also participate in we-group institutions of church and language, which the authors choose to take as descriptions and leave out of their analysis.

Transformations and Ideologies, 1970–2000

The world's postwar economic expansion – from 1946 to the mid-1970s – slowed, reached its end, and gave way to stagnation, reinforced by the early-1980s peak in interest rates.[45] A change in mood brought feelings of dissatisfaction from commoners who had benefited from decades of advances and who expected that the improvements in their lives would continue. For instance, the proliferation of independent nations built a hope, among those in the newly labeled Third World, that advances toward social and economic equality would continue at all levels. As go-slow policies took hold in domestic civil rights and in granting of independence, revolutionary nationalism and Black Power movements arose in protest, paralleled by a global wave of feminism. In China, the ferocity of the Cultural Revolution drew in part on dissatisfaction with the pace of revolutionary change and with persistence of social inequality. Debates over limited funding for schooling and health combined with ideological debates between proponents of capitalism and socialism. Clashes of national movements grew in antagonism, most persistently that between Palestinians and Israel in the regional wars of 1967 and 1973.

The Fruits of Decolonization. A global political jolt, especially in the years 1968–70, brought underlying tensions into the open but did not resolve them. Different struggles broke out in various regions, yet they somehow reinforced one another. The year 1968 is best known in Europe through the May rebellion of students and workers in France; it is best known in the United States for antiwar demonstrations from April to August. But events in Europe were tied as well to Germany and to Congo; events in the United States were tied as well to Civil Rights and feminism; and demonstrations and repression took place in Dakar and Mexico City. The Cultural Revolution continued to unfold in China, and debates about it fed disputes everywhere. Musical and visual artistic innovations played a role in every node of the global social contestation of 1968. While the voices and outlooks were varied and often contradictory, the events and debates of 1968 confirm a broadening popular discourse worldwide. Divisions were both expected and unexpected: workers who had most benefited in previous years were unlikely to rebel; workers at the margin found new techniques for protest. The proclamation of Earth Day in 1970 was in one sense an extension of protest to concern for the environment and, in another sense, a distraction from socioeconomic concerns.

The forces that had benefited from decolonization and social welfare policies, while they gained no great victories in the contestations of 1968–70, continued their campaigns. At the international level, the combination of ex-colonial and socialist governments maintained and expanded their majority of votes in the United Nations General Assembly. In the UN Security Council, the United States and its allies had the most votes but faced vetoes from the Soviet Union and, after 1972, from China. With this new balance, the anticolonial group began to prepare for creation of a New International Economic Order. It was their hope that United Nations agreements would limit the economic power of global corporations and the leading industrial nations.[46] Parallel trends arose among academics in Europe and North America, who launched the antigrowth Club of Rome, analysis of a steady-state economy, and efforts to model the global economy.[47] By the end of the 1980s, however, while campaigns for revolutionary nationalism had defeated Portuguese colonialism and the American war effort in Vietnam, the campaigns for feminism, trade union rights, and expanded education had not fared as well. Shifts in the financial world had enabled banks to extend loans to numerous nations that turned into devastating debts once interest rates reached peak levels; Structural Adjustment programs of the World Bank required nation after nation to cut back social services and investment in infrastructure. Within the ex-colonial camp, single-party governments, intended to limit neocolonial interference, protected political figures in their positions despite economic cutbacks. Trade unions, students, and community groups, calling for support from governments that were now in debt, found that governments feared them. Newspapers were shut down.

Environmental contestation opened another field of debate and analysis. Long-established conservation movements maintained an undercurrent of debate that could gain wide attention, as with Rachel Carson's 1962 analysis of the dangers of DDT for bird and other populations.[48] Individual cases arose, such as poisoning by chemical and radioactive waste, destruction of species, clear-cutting of forests, and excesses in mineral extraction, first in wealthy countries but very soon more broadly. In some cases, scientific and community consensus led to quick regulatory action. Chlorofluorocarbons (CFCs), used increasingly as refrigerants and propellants, were found in 1973 to be destructive of the atmospheric ozone layer. United States regulation from 1978, followed by international conventions of 1985 and 1994, led to an end in the commercial production of these chemicals soon after 2000.[49] James Lovelock, who first traced the role of CFCs in the atmosphere in 1973, went on to develop a broader ecological view. Lovelock, an inorganic chemist who compared Earth to Mars, developed a functionalist approach, arguing that Gaia was a planetary system in which plant life – especially the algae that occupied the oceans and tropical soils – used sunlight, water, and carbon dioxide to create energy and produce oxygen through photosynthesis, relying on feedback of various interactions linking the biosphere to the inorganic lithosphere and atmosphere, so as to create a stable temperature of between 10° and 20°C. Lovelock was not initially worried that the stability of Gaia might be overcome, but new evidence and more detailed analysis caused him to change his mind.

The Ideas that Became Neoliberalism. In contrast to the aforementioned movements of social critique and environmental concern, a corporation-oriented ideological outlook gained strength in the 1970s. In it, a rationalization of new developments in corporate governance combined with response to the postwar advances of trade unions, decolonization, and socialist governments. This outlook centered increasingly on attention to the economic vision of Friedrich A. Hayek.[50] Hayek's theoretical vision was that the system of open markets, free trade, and market determination of prices was wise not only for the economy but for all of society. Beginning in 1947, Hayek organized annual meetings in Switzerland, attracting well-established economists to discuss his vision of economic freedom – which meant that there could be no restrictions on market exchange. Hayek was a critic of Keynesianism, with its full-employment policy, and a critic of any form of centralized economy, especially socialist but also fascist.

He argued that prices and the market system, in human society, were the equivalent of natural selection in biological evolution. This was, therefore, a reductionist view in that it argued that prices and markets are adequate for explaining every arena of society. His arguments convinced economists, corporate leaders, and accompanying political figures that almost any government intervention in the economy led to decisions that were less efficient and effective than those that would be achieved through the price-making market. Hayek argued that a pristine price-and-market system in the economy was equivalent to a pristine natural selection system in genetic evolution. He thus adopted a timeless theory of biology at the very moment that biological theory was undergoing complex transformation, and he used it in an effort to impose a timeless price system in the economy.[51] His analysis of the market was limited to the general-equilibrium level, focused on single-level population thinking and neglecting either broader macroeconomics or more specific firm-level studies, at the moment when biological analysis was supplementing population thinking with analysis of multiple scales.[52]

Hayek's followers, inspired by his example, argued increasingly that regulation of corporations was inherently harmful to the economy. Hayek, extending this logic in his later work, argued that any intervention in the economy for the purpose of achieving justice was bound to do more economic harm than good. In this reasoning, he assumed that those who are disabled gained their disability at birth and from nature – he thus assumed that inequality brought by human discrimination was negligible.[53] Hayek had nothing to say on nature, aside from his analogy linking economic markets to natural selection; nothing to say on the limits to growth. His rejection of regulations led to the conclusion that economic leaders need not care for those who are lost in the inequalities of the economy.[54] Neoliberal ideology, as it grew from the basis of Hayek's analysis, treated private, profit-oriented, corporations as individuals; it was anticollectivist and antiregulation. It formally opposed groupthink while advancing corporate groupthink.[55] This ideology enabled powerful employers to believe with certainty that, in maximizing their profits and undermining competitors, they were providing unambiguous service to society as a whole, limited only by various groups of narrow special interests, such as self-interested workers or greedy regulators.

Implementation of Neoliberal Ideas. In the gradual shift in social relations during the 1970s and 1980s, Hayek's theories and policies became increasingly influential among corporate leaders. A 1980 Conservative electoral victory in the UK brought Margaret Thatcher to power, and brought policies of austerity that targeted social services, education, and trade unions. Thatcher rose to power with the phrase, "There Is No Alternative!" and made explicit her support for Hayek's economic vision.[56] Ronald Reagan became US president in 1981 with a similar program, successfully breaking the independent union of air traffic controllers in their 1981 strike. Opposition to Thatcher rose until 1982, when the Argentine military regime launched an attack on the Falkland Islands, claiming their return to Argentina as the Malvinas. A nationalistic response in Britain and a quick military victory reaffirmed the strength of Thatcher's government. The British mine workers were defeated in their 1984–86 strike, and legislation steadily weakened trade unions. Meanwhile, to protest the views of the postcolonial bloc, the United States and United Kingdom resigned from the United Nations cultural organization, UNESCO, weakening but not eliminating it.[57]

Another defense of corporate profitability followed a more devious path. In a campaign that began with the effort of tobacco firms to deny that smoking caused cancer, the development of public relations campaigns ended up fundamentally complicating scientific debates and public understanding of scientific results. Naomi Oreskes and Erik Conway have documented the four senior scientists who led in developing techniques for creating public doubt about established scientific results, to support their ideological preconceptions and to support corporations whose profitability would be reduced by regulations based on scientific knowledge.[58] Beginning in 1979, Frederick Seitz, a physicist and former president of the US National Academy of Science, worked with the tobacco industry to defend against the evidence of cancer caused by its use. A larger team joined with Seitz to throw doubt on the corporate causes of acid rain, depletion of the ozone layer, the negative effects of pesticides and, especially, the human sources of global warming. They carried out their campaigns more in the public arena than in scientific debate. But they set up "Potemkin villages," spurious research groups and think tanks, in which collaborators with some scientific qualification prepared attacks on established scientific results and on prominent individuals associated with those results. Their attacks on scientific consensus were carried out in major public media, notably *The Wall Street Journal*. These attacks were sufficient to bring elected officials into the discussion. In government hearings and in major media, the general public was faced with doubt on each major issue rather than with a set of scientific results that needed to be backed up with public policy. Corporate representatives emphasized the presumption that every issue had two sides, so that either side might cancel the arguments of the other. Over the decades, individual-level smoking declined substantially and, with it, the level of lung cancer, though many lives were still lost. Acid rain has declined, but with a heavy cost. But the broad campaign denying global warming has remained successful. Corporations and governments have given almost no support to reducing carbon emissions, despite increasingly disastrous storms.

A slightly more benign but perhaps more costly practice of the wealthy – both individuals and corporations – has been the escape from tax liability or simply from tax payment. In a related practice, Anthropocene-era technological transformations have taken place especially through substantial state investment or through reduction of corporate tax liabilities: railroads, electrification, water power, and telephone and internet networks have been developed and constructed in countries throughout the world, through the initiative of government investment or through public–private collaboration.[59] In these allocations, great quantities of resources have been granted to private firms, hidden from public view and free of taxation. The steady growth, during the past century, of tax havens has privileged not only individuals such as dictatorial rulers but especially the banks and corporations that facilitate their diversion of funds from standard accounting systems.[60] Privatization of public pension programs for workers, as we will see, became another technique for transferring wealth to elite individuals and institutions.

As Bas van Bavel argues, the current pattern of institutional sclerosis, economic inequality, and apparent decline in the United States and other economies is not unique. Instead, he shows how it evolved before in the cases of Britain, the Netherlands, Italy, Song China, and Abbasid Iraq.

Van Bavel has identified a long-term dynamic in the economic institutions of factor markets, in which expanding financial markets bring the most extreme inequality, followed by economic decline. Nevertheless, for application to the present, van Bavel's analysis of national units needs to be expanded to account for current global interaction, environmental degradation, and the new political system of nations, in which Peter Gran's "rise of the rich" characterizes global social history.[61]

Social movements arose periodically to challenge the practices of political and economic elites.[62] In the immense uprisings of 1989–92, social movements reached across borders to create continental and global contestation, at a scale exceeding the events of 1968–70. These movements led to displacement of governments, mostly by peaceful means, in roughly ten African countries, six East European countries, and the whole of the Soviet Union. At least as many other countries underwent major contestations, in which the existing government retained power only with difficulty. The governments most vulnerable to democratic contestation were single-party states, some ruled by communist parties, others by postcolonial single parties, and South Africa, ruled by a racialized elite with important capitalist holdings. Multiparty governments largely escaped challenges by demonstrators. The social movements, globally linked, brought a triumph in articulating global public support for democracy in government.

Yet the same wave of social movements also provided important opportunities for large-scale capital. The protestors, while they targeted mishandling of the public sector, seem not to have worried about the private sector. From the standpoint of large-scale capital, the democratization movements of 1989–92 provided an opportunity to expand private-sector financial holdings without, for the moment, becoming the object of massive protest of their accumulation. Most obviously, private capital gained through privatization of the state-held resources of the Soviet Union and Eastern Europe. The exception was South Africa, governed by both private and state capital, where the transfer of power to the African majority in 1994 did not include a significant effort to reallocate the wealth of the former governing elite.[63]

Viewing this transition in broader context, one may note that the project of privatizing pension programs had begun some two decades earlier in the leading capitalist countries. In the 1970s economist Milton Friedman joined those looking jealously at the pension funds that had been accumulated and held for workers in the postwar era. If pensions could be removed from ultra-safe government bonds and invested in speculative securities, financial firms would gain profits in handling the transactions. Friedman led a campaign for privatization of public and nonprofit pension funds that had considerable success not only in privatizing pension funds but also in depriving the contributors of their expected pensions; the campaign to weaken trade unions was thus linked to privatization of pensions.

Trade unions in capitalist countries fought against the privatization of pensions: they were largely victorious in western continental Europe but not in the UK. At the same time, Structural Adjustment programs of the 1980s, imposed by the World Bank on post-colonial nations, helped to restrict but also to discredit public spending. These developments may help to explain why the demands of the 1989–92 movements focused on political rather than economic demands. After the June 1990 crackdown on demonstrators in Tiananmen, Milton Friedman carried out one more venture for neoliberal activism. He visited China in about 1994 to make the case for privatization of the nation's pension funds, which had grown rapidly after the 1978 economic reforms. In his hope of transforming China into a fully capitalistic economic system, he won a partial victory, in that roughly one third of pension firms were privatized and subjected to the fluctuations of the market.[64]

A combination of factors had therefore greatly strengthened the financial sector of the global capitalist economy. Privatization of pension funds in the West and the East; privatization of state enterprises in the former Soviet bloc and thereafter in China; reduction of worker benefits in formerly communist countries; expansion of the internet and the option of accelerated stock trading; relatively successful campaigns to reduce state regulation of large firms; and tax breaks to encourage investment. Young men entering finance firms could find themselves to be rapidly enriched without much work; they began to believe in their own great value and in the essential importance of making the highest possible profit. Of course, there were many other factors at play in the world, but these elements help to emphasize the significance of ideology and inequality. The unregulated flows of capital led to a world-wide financial crisis in 2008; recovery from the crisis was achieved only by investing huge amounts of public funds in the failing banks and corporations.[65] Nevertheless, a neoliberal zeitgeist, privileging profit of major enterprises, maintained its dominance so that only minimal reforms of the financial system took place.

Environment and Society in the Twenty-first Century

Environmental historians J. R. McNeill and Peter Engelke emphasize the "Great Acceleration" since 1945, presenting it as two tales of drastic change.[66] First they review the Acceleration itself – the past three generations of growth in population, extraction, production, and waste, greatly exceeding all previous periods of growth. For topic after topic, they show the changes that have taken place in energy use, population size, biological diversity, and climate, in urban and economic life, and in the political contests of the Cold War. Their second story is the end of the Acceleration, to come within about fifty more years – that is, the end of population growth, a sharp decline in consumption of fossil fuels, and perhaps other contractions in the human order. This second story links the end of the Acceleration to continuation of the Anthropocene. That is, the expanded human footprint of the Anthropocene is here to stay. As they argue, the processes set in motion by human agency will continue, whether humans are still here or not.

Of the many dimensions of change in the natural world during the Acceleration, here are four that stand out, spreading their influence throughout the decolonizing world. Expanded extraction of the lithosphere, which began most obviously with coal mining in the eighteenth century, became fully manifest in the late twentieth century.[67] New earth-moving techniques expanded strip-mining to flatten mountains and destroy regional ecologies in order to reveal coal, copper, or aluminum ores. Dams interrupted most of the great rivers and many smaller rivers, usually with negative results for downstream regions. Concrete, assembled from a mixture of minerals, is poured or set in bricks into buildings and streets that are impenetrable to water, so that rainstorms, growing as temperatures rise, create urban floods. Second, stories of extinction, circulated since the loss of dodos and passenger pigeons, accelerate in all directions. The International Union for Conservation of Nature (IUCN) has conducted systematic analysis especially on mammalian and bird species, showing the extinction of 875 species between 1500 and 2012, with human agency appearing to be increasingly responsible for the extinction. Wild animals now account for a much smaller proportion of living animals than before. Recent attention focuses on the extinction of microorganisms, notably the bacteria that inhabit the gut of animals and participate in digestion. A third point is the expansion in knowledge of the natural world, combined nevertheless with corporate sequestration of that knowledge. James Lovelock's perspective on Gaia changed, as he wrote in 2006 on her "revenge."[68] Lovelock began to think of the life course of Gaia – she had existed for something over 2 billion years but had no more than another billion years of possible existence, because the steady heating of the sun would eventually overcome her processes. Fourth, in more immediate terms, the melting of polar ice may shift the direction of ocean currents, especially by cutting off the northward Atlantic current that keeps northwestern Europe relatively warm.

Yet when James Hansen, a physicist at the US National Aeronautics and Space Administration, expressed alarm in his 2003 report – "Can We Defuse the Global Warming Time Bomb?" – he was given specific instructions from US government officials that he was not to speak of climate change.[69] Further, the commodification of the environment proceeds: water prices rise while biological resources of medical value undergo patenting – and both of these factors reflect consumer desires as well as corporate profit.

The crisis in socioeconomic inequality tends to get less public attention than the environmental crisis, but when the two are put together, it becomes clear that humanity does indeed face serious difficulties. Thomas Piketty created much interest with his demonstration of the recent rapid growth in inequality, especially in wealthy countries.[70] Yet while ideological outlooks have shifted from the racial hierarchy of the nineteenth century to an argument that all humans are of inherent value, global society is far from implementing that vision. If all humans are of equal value, why should we leave so much human potential as wasted because of impoverishment? While economists have begun collecting data on inequality, studies of global inequality continue to be conducted at the national level. Aside from economic inequality, we lack data on inequalities in health status and social and political standing that are parallel to our data on environment. Scholars have collected many data on social conditions, and far greater quantities of data await in archives, but there has been little effort to integrate and analyze these data. The contrast is remarkable: while the analysis of global climate has advanced speedily since its 1958 beginning, the analysis of human society is only beginning now. Climatologists have theorized and collected data that show patterns of climate change as far back as millions of years ago. The Intergovernmental Panel on Climate Change (IPCC), created by the United Nations in 1988, gave its first report in 1990 and continues to update its reports. But for human society, despite the efforts of numerous groups of scholars, we have no broadly supported campaign to collect information on the changes in human society at global and local levels over time, especially patterns of inequality.[71]

How are we to face our global situation? Three overlapping arguments make the case that little can be done: human nature – humans are biologically unchangeable; inertia – human society changes only slowly; and denial – people choose to reject the unpleasant. James Hansen sadly accepted these views, writing of "scientific reticence," the analytical reluctance to speak forcefully in addressing environmental crisis.[72] Recent responses to disease, however, show the potential variance in human behavior. Consider the populations that have been facing HIV and Ebola. Denial, reinforcing illness and death, dominated South African HIV for a decade. But in West Africa, where combating Ebola meant giving up the treasured ritual of washing the body of a deceased person before burial, people learned in a hurry and survived.[73]

In a new standard for evaluating climate change, the IPCC publishes comparisons of worldwide temperature "anomalies" or variations each year, giving comparisons to the stable temperatures of the late nineteenth century. Temperatures rose during the twentieth century with a total rise of 1° Celsius before 2020. The IPCC and other groups now estimate that the temperature may rise by a total of 1.5° Celsius as soon as the mid 2030s; a subsequent rise of temperature to a total of 2 degrees increase is likely to raise sea level by 50 cm. Should we then seek to create similar calculations to monitor the impending social disasters? The next great social disasters are likely to be caused *both* by climatic disaster and by disastrous social policy, so that estimating their magnitude would be complex indeed.

In an approach that looks to the future in two ways, David Wallace-Wells has picked up where historians leave off.[74] First, he provides details on twelve types of frightening changes resulting from rising global temperature: the changes that have already taken place after a temperature increase of 1° Celsius and also the changes that will result once temperature has risen by 2° Celsius – the tipping point for climatic disaster as identified by the IPCC. Second, Wallace-Wells looks for ways to project how humans may respond to continuing environmental change. He begins this task by investigating popular culture: in a world of pervasive popular culture, how are the past and future represented in stories and in entertainment? He finds that stories – in novels, films, and games – center on individual heroes or sometimes on a tight group that can celebrate after victory, so that most humans are left out of the tale in end-of-the-world stories. As Wallace-Wells observes, "collective action, dramatically, is a snore." Do we leave it to a few experts to solve the problems? Nature has been portrayed as chaotic but, "as climate change expands, it ceases to be a story and becomes a setting." Who would heroes be? What would they do? In gaming, the game relies on a villain – but who bears moral responsibility for climate change? The shares are disputed: the wealthy, of course, are at fault, but consumers and producers at every level are complicit in environmental degradation. How should we balance their roles? Again, Wallace-Wells notes, "Complicity does not make for good drama." He appears to argue that popular culture, while valuable because it links so many people, provides us with little guidance in facing our crises. Should humans engage the long-postponed debate on the limits of growth? While that debate has barely started, in some ways it is already resolved. The McNeill-Engelke team and Wallace-Wells confirm that the Great Acceleration will not continue for long: there can be growth of some sorts for a few decades, but demographic and other sorts of human-focused growth are likely to come to an end.

The Three Crises

Environment. Human contribution to global warming, through greenhouse gases, is approaching its limits. Fossil fuels have not been exhausted as rapidly as was earlier predicted, but humans will surely exhaust them. That alone will slow the rise in temperature after a few decades, but not reverse it.[75] The big explosion in human population is now over (though migration will continue). Growth in populations of domestic animals, in land use, and in mineral extraction may also decline. In sum, the human impact on Gaia will stop accelerating, but is unlikely to decline. So the transformation of Earth will continue, but without such acceleration. Still, there is no going back to the world before. Climate, for instance, is rather certain to become far more erratic than in the past.

Socioeconomic Inequality. The Anthropocene has brought growth and dreams about more growth. In the same era, nineteenth-century movements for emancipation and twentieth-century movements for citizenship brought postwar expectations of social equality. The aggressive accumulation of wealth during the past century – especially since 1970 – conflicts deeply with the ever-advancing vision of social equality: the growth of that vision appears, for instance, in comparisons of North and South America, where the old gaps in literacy and life expectation are almost gone.[76] But even as schooling and health advanced, existing discrimination by race, religion, nationality and class continued to be reinforced, while certain factions in every portion of the world were able to gain disproportionate wealth and power. Was this "human nature" in action? The inequitable effects of environmental degradation reinforce the contradiction. Those now poor and living in the tropics will suffer especially from heat, flood, and storms. Refugees, driven from home, will seek to create new lives in unfamiliar lands, but will encounter resentment as much as welcome. Even with the rapid economic growth of China and India, there is no way to imagine implementing the vision of global equality without reallocating, back to the general population, the wealth that has been transferred from it during the lifetime of those now living.[77]

What to Do? The crises of the natural environment and of human socioeconomic inequality both require an effective response. The response must be broad-based and imaginative. But no adequate response is in sight. Supporters of neoliberalism, in particular, will simply be unable to provide help either for the environmental or socioeconomic crises. They would support either of the two projects only if they could be certain of gaining high profit levels and avoiding any regulation. They are interested in the general public but only as potential consumers. The long denial of the cancer-causing effects of tobacco can be seen as a rehearsal – a trial run – for campaigns of denial and refusal on a larger scale, undermining any effort to address social inequality or environmental degradation.

Among knowledge specialists, the situation is more nuanced. Experts in the natural sciences have accomplished much in diagnosis of environmental crisis. But their policy proposals, limited in scope, are mostly blocked by political figures and corporations. Experts in social sciences have done very little on diagnosis of the socioeconomic crisis; they have done especially little on policy, as they have not given up on the expectation of continued economic growth.[78] A further point is that social scientists have done little in taking up analysis of denial – how it is that leading elite figures and many among commoners choose to ignore or deny the current and expected future cost of social inequality and environmental degradation? How is it that corporations, officials, and many in the general public deny the clear evidence of these crises? Even more difficult will be the problem of developing adequate policies to overcome denial of environmental and social crises. How will people prepare themselves at all levels for global warming, social crisis, and the end of growth? Have humans reached the limits of individualism? Exploring group-based models of behavior is by no means guaranteed to solve our problems, but it does seem that a significant number of researchers should explore the possibilities. Can it be that the Human System has flaws within it that prevent it from diagnosing its own worst problems? More than attacks on individuals and political parties, we need an understanding of the functioning of the system, its strengths and weaknesses.

10 Hope for Adaptations

Ours is a time of great change – of that there is little doubt. One of my objectives in this book has been to invite readers to compare the current transition of the Human System to earlier systemic transitions, to see if any lessons emerge. As spoken language emerged long ago, humans created communities of perhaps 150 each to share their ideas and activities; these communities populated Earth. In the early Holocene, multiple communities combined to assemble societies; the societies developed the skills to create an incredible array of evolving institutions. The much-expanded Human System is now undergoing reorganization at a larger scale; we experience the details but cannot yet see the overall direction.

The previous chapter interprets the Anthropocene epoch by tracing institutional growth, accompanied by critique of individual-level analysis.[1] Empires expanded and collapsed into nations; corporations and other institutions of capitalism grew relentlessly at national and global levels; international organizations of many other types paralleled corporations; big and small nations wrestled in a shifting hierarchy. Accelerating growth reinforced inequality and deepened environmental crisis. In all of these transformations, new networks and new hierarchies expanded and interconnected.

This chapter continues my analysis of the period from 1800 to the present: it introduces three more networks – worldwide social networks – and shows that they require rethinking of the Human System and its problems. The first two networks are popular culture worldwide and knowledge worldwide – the latter including both general and specialized knowledge. These networks have grown throughout the Anthropocene and are now of formidable strength. The third, democratic discourse worldwide, arose in the late twentieth century along with decolonization. I explore these networks through the lens of group behavior. As I argue, the combination of popular cultural connection, expanding knowledge, and global debate presents a significant counterweight to the apparently relentless expansion of hierarchy and inequality in human affairs. If we consider the full array of structural elements, including these three, the results change the picture of the Human System and its trajectory. The debates shared across the internet scrutinize organization by hierarchy or by network; they challenge the fitness and functioning of institutions from the family to the corporation. The debates among nations get the most attention in news media organized in national terms; but similar debates are carried on within nations, within communities, and across borders of every sort. Altogether, we find ourselves with an emerging tribune for exchange of views at every level.

Groups, of course, are composed of individuals and cannot function without individuals. The issue in group theory is whether the decision to form a group adds something new to the character of behavior.[2] We-groups, created by agreement, exist in a complex world. There are still individuals who are not associated with groups, or persons within an institution who do not follow the institutional norms and objectives. Groups and institutions must survive periodic assessment of their effectiveness, perhaps by the beneficiary population initially designated to benefit from the institution's work. Discussions of the institutional priorities may become ideological and heated. Out of these discussions arise calls for regulation and even elimination of the institution; results may include redefinition of the beneficiary population. Yet the extra effort of participating in group efforts may pay off in amplified achievements.

This chapter proposes an analysis of the new and expanded institutions of the Anthropocene epoch. Chapter 9 has already explored corporations, nations, international organizations, trade unions, and other new institutions. In this chapter, the first section adds the rise of popular culture and its institutions, while the second section explores knowledge – the rapid growth in institutions of general knowledge and specialized knowledge. The third section, focusing on networks of the Anthropocene, considers the interaction of the full range of institutions discussed in both chapters. The fourth section presents one more institutional change: the recent formulation of global democratic discourse. This inclusive framework for exchange of ideas, having survived its initial tests, shows potential as a forum for debate on the full range of social choices. At the conclusion of this chapter, I return to three of humanity's big choices – decisions on modification of the Human System itself and its ties to the natural world. Certainly, the new institutions and networks add important dynamics to the Human System. Will their combination facilitate the much-needed reform of the Human System?

What ideology would people turn to in order to set social priorities enabling a creative and energetic response to the dangers that humans face? We may pose the question from the standpoint of individuals, each seeking a way to contribute; or to groups large and small, seeking similarly to do their part. We may also pose the question from the standpoint of the Human System, asking what inherent strengths and resources it has that may be valuable in facing such a test. For the moment, however, two major outlooks dominate the ideological scene: the approaches of employers – corporations, financiers, and the powerful capitalistic institutions built up around them – and the policies of great powers, which remain suspicious of social movements. These outlooks focus narrowly on short-term interests of restricted populations and are poorly prepared to address issues of the general interest. Meanwhile, popular culture rose to the peak of its influence at the beginning of the twenty-first century yet had gained no attachment to any forceful ideology that appealed especially to the common people of the world. In particular, most groups continue to ask for growth in their resources, though growth is sure to cause more difficulties than it will resolve, both for humanity and for Earth.

Popular Culture

Today's group-level culture expands well beyond the limits of the ethnicities of old, beyond the great families and elites, because of the new connections and new technology of life today.[3] In an early Anthropocene example that linked popular culture and social movements, antislavery movements arose around the Atlantic as a result of ideas arising from inherent ideas of freedom, the experience of participating in the world market, migration and encounters with new communities, rebellion of the enslaved, and the development of oral and visual media of communication. That is, antislavery movements were networks of humanitarian institutions and informal groups that campaigned for the end of slavery. Two memorable images of that age became iconic in the rise of popular culture: the image of the slave ship "Brookes," showing hundreds of captives chained into place; and a medallion with the image of a kneeling captive, asking, "Am I not a man and a brother?"[4] These images brought new supporters to the alliance of those in slavery, free blacks, and supportive whites. For a whole century, songs, novels, dances, and oratory summarized the lessons of the slave market, exploitation of slave labor and of slave persons, made the case for a principled abolition of enslavement, and strengthened the communities that became the African diaspora.[5] During this century, enslavement continued and expanded in Africa, Asia, and even the Americas, yet was ultimately beaten back to the level of an uncommon abuse. Meanwhile, the campaign for emancipation ended up strengthening certain Christian and Islamic communities. For instance, in the United States, the black church became a powerful and continuing sociopolitical institution.

Literacy became a central impetus to expanding popular culture. Especially in English language but also in many other languages, teachers and expanded schooling dramatically raised the number who could read newspapers. Short stories and serials in the newspapers provided steps toward the expanding market for full-scale novels. The printed word brought entertainment but also brought discussion of local and broader social issues. The trade in books and newspapers expanded steadily; those who were literate read aloud to families and friends. Through colonization and later through the internet, English spread widely as a second language and as a medium of global popular culture. One advance in the availability of printed texts was the invention in England of the Stanhope hand press, with a series of modified versions beginning in 1800. Nile Green documents the expansion of printing in Qajar Iran from 1820.[6] He traces Iranian men who began active careers of publishing works in Arabic and Persian. Mirza Salih Shirazi traveled to London and Mirza Zayn al-'Abidin traveled to St. Petersburg; each studied printing for years and returned with a Stanhope printer to Tabriz, printing religious and secular texts. More broadly, Green shows that such printing in Arabic characters expanded elsewhere at the same time, as presses were set up in Cairo and Lucknow, India.

Migration, facilitated by steamships, opened new cultural interconnections. European migration shot up with the Irish exodus from the 1847 famine, after which migrants from every part of the continent settled in Europe, the Americas, and elsewhere; millions of Russians, many of them ex-serfs, moved eastward. Almost equal numbers of migrants went from India to Southeast Asia and the coasts of every ocean. Chinese migrants settled in Southeast Asia alongside those from India but also moved west and north in East Asia, and overseas. Migration declined from the 1930s to 1950s (except for immense migrations of soldiers and refugees in wartime), then expanded again but in new directions. For migrants in whatever time and place, music, dress, games, sport, and other dimensions of popular culture accompanied them. Music and dress traveled with particular ease, crossing the boundaries of language. In the early 1870s, the Fisk Jubilee Singers traveled throughout the United States and to England to convey their version of black culture. They joined the ranks of such international celebrities as the Swedish operatic singer Jenny Lind and the African American Shakespearian actor Ira Aldridge. Sport arose from schools and other men's organizations – soccer football, wrestling, and boxing. In an elite and nationalized recognition of the growing importance of sport, the modern Olympic Games were convened in Athens in 1896.

In the twentieth century, cosmopolitan popular culture seized upon a succession of new media. The African diaspora – breaking away from the heritage of slavery and putting down urban roots for claims on equality – continued to play a leading role in developing new forms and in linking those cultural traditions to a determined and ultimately successful campaign for citizenship in nations of Africa and abroad.[7] The musical traditions of jazz and blues arose at the end of the nineteenth century in the United States – along with rhumba in Cuba, samba in Brazil, highlife in West Africa, and Swahili music in East Africa – and the beginnings of recorded music spread the new forms widely from 1900. Telephone, radio, television, portable devices, tapes, CDs, personal computers, the internet, and mobile phones brought endless revolution to communication. With time and the wider audiences resulting from film and electronic media, celebrities gained still greater fame: Pablo Picasso and Charlie Chaplin were among the greatest celebrities of the early twentieth century; the football striker Pelé and the boxer Muhammad Ali gained truly global recognition at the end of the century. In film, strong national traditions arose in Europe, Asia, and the Americas in the 1920s, though Hollywood gained an early lead. With time, however, Hollywood shared its space with Bollywood (films from Bombay) and Nollywood (videos from Nigeria); soap operas, if invented in the United States, prospered in Mexico, Brazil, and Egypt.

In medium after medium, young people took advantage of their wider contacts to draw on cultural styles from their peers rather than their parents. In dress, dance, music, and speech, young people stretched across the boundaries of family and ethnicity; as a result the rate of innovation rose, as did the rate of obsolescence of previous fads. Visual art was perhaps as important as music, film, and literature as a medium of expression by common people around the world, but it could not be copied and distributed as easily. The works themselves were of various sizes and media, including large installations. By the twenty-first century, however, electronic images were easily created and distributed, so that images of works could be seen widely, and ideas and appreciations could develop in interaction. Terry Smith has emphasized the twenty-first-century explosion in biennial exhibitions of visual art, by artists who have yet to make sales to major museums but whose work conveys a sense of the planetary moment.[8] From these artists may come the visual symbols to inspire future generations as they take up new challenges.

In every medium, performers had to share their place and especially their purse with entrepreneurs – producers, publicists, and other sorts of owners. As audiences grew, each medium in popular culture developed a substantial capitalistic dimension. Under some circumstances, the artists maintained the upper hand; in other situations, the marketing team, the producers, or the owners of the firm had the greatest control of the entertainment product and of the earnings. The audience too gained in influence. Consider the changes within music: in the times of live performance, the audience heard what was selected by the performers. In the days of records, the audience could purchase a range of performances by artists of interest. With today's equipment, the audience can shift at will among a wide range of performances.

What is the meaning of this popular culture? While popular culture has its fashions, cohorts, and cliques, it is surely far more open to new stimuli from unfamiliar sources than the culture of the past. One important consequence is that, with greater interchange and familiarity among communities, the fears and hatreds that earlier separated communities into hostile and antagonistic camps declined significantly: for instance, the merciless Pacific warfare of Americans and Japanese gave way to close alliance, not only political but cultural.[9] In this regard, we can distinguish the I-groups of audiences that experience multiple fashions from the we-groups that develop among those devoted to specific cultural expression. Either way, young people make their own cultural choices, rather than accepting molds provided by their parents. Both individualism and group identity took on new forms, nurtured by expanding popular culture.

Knowledge: Sources and Destinations

During the Anthropocene, human knowledge has grown more rapidly than any other category of our existence.[10] We tend to be pleased rather than fearful that knowledge exists and is expanding. Our fears, instead, are about the allocation and control of knowledge: is there secret knowledge that enables some to control or benefit preferentially? Are we creating the right sort of knowledge? Can we continue to accelerate the creation of knowledge? One is tempted to argue, *yes* – in contrast to the limits on our ecological, material, and social resources, there are few indications that we are running out of resources with which to produce knowledge.

For this discussion, I have chosen to divide knowledge into general and specialized knowledge, where general knowledge is of potential interest to everyone, while specialized knowledge is for specific purposes and requires specific disciplinary training and practice. But the two overlap and have changed with time: literacy, for instance, once an expert specialization, is now expected of people everywhere as part of general knowledge.

General Knowledge. The practical knowledge of common people, buttressed by their ideology, comprises a huge and expanding accumulation of information. Yet it is not coordinated or catalogued in the same way as specialized knowledge. One way to begin a discussion of general knowledge is in terms of its sources. That is, one may consider what a person learns from his or her parents, cohort or age group, community, or religion. Literacy is most often associated with school, and there one may ask whether the instruction is in the student's first language or a second language. Schooling for children and young adults may depend on choices by parents; very commonly it draws on a curriculum set by a national government, including a patriotic emphasis. At an adult level, general knowledge may be gained through the paths of popular culture, the informal networks of work, commerce, migration, but also the continuing messages of empire. General knowledge may begin with factual information and the knowledge necessary to maintain a household or pursue an occupation, but also includes cultural values and ideological precepts that articulate social priorities. To what degree is one equipped to challenge old beliefs or to accommodate to surprising rituals? Is it the case that the inculcation of general knowledge in populations of commoners includes philosophy that is helpful in making sense of the very rapid pace at which the world is changing? Such notions might include an understanding of positivism – the idea (prominent from the nineteenth century) of breaking each problem down into its smallest logical pieces and solving them separately, allowing only one piece to change at a time. But they might also include an understanding of interactive or dialectical thinking – the more current effort to keep track of multiple changes as they influence each other. Further, such thinking might include attention to the inevitability of multiple perspectives on any issue, as each observer comes to a problem with different experience. The recognition of multiple perspectives, in turn, presents difficulties for an attempt to establish an irrefutable truth on a given problem. In addition, because the social situations of commoners, like those of others, include multiple scales of aggregation, they should develop skills in understanding patterns characteristic of the various scales.

At any level, the quantity of available knowledge expanded beyond the ability of any person to absorb it. Libraries and the internet became overloaded; each new medium also resulted in destruction of old knowledge. On one hand, impressive indices and search engines have made it possible to find specifics, narrowing immense numbers of files to get to particular details. On the other hand, current search systems are not very good at presenting users with authoritative summaries of general background. Wikipedia, while an admirable compromise between housing specialized knowledge and making simplifications for broad audiences, also makes clear to users its limitations. The very quantity of available information has opened the gates for those who steal protected information or who circulate false information to steal from or to mislead others. Still, the presence of a well-informed general public might become a powerful force for making sense of specialized knowledge to formulate beneficial steps for humanity. Yet the general public and knowledge specialists appear to be quite distant from one another as groups, though they are not uncommonly the same people when seen as individuals. But to pose one more question on this topic: do popular culture and general knowledge give people experience with interpreting group behavior?

Specialized Disciplines and Their Technologies. Specialized knowledge has grown disproportionately during the Anthropocene. In the days of Newton, the study of physics was limited to mechanics and astronomy, but electricity, magnetism, and thermodynamics joined in the nineteenth century and quantum mechanics in the twentieth. Chemistry expanded as new chemical elements were discovered; the rise of atomic theory led to organic chemistry, then to biochemistry and vitamins. In this nineteenth-century context, the geographical and social distribution of expanded scientific knowledge became and remained very tightly limited to elite strata in Europe and North America, at least in the experimental areas identified here. On the other hand, field studies in geology and various biological fields, as well as observations in astronomy, continued to be recorded in many parts of the world. Engineering fields changed their character in response to changes in natural science: new findings in physics, chemistry, and mathematics enabled engineering to move from practical tinkering to scientifically based analysis. Formal study began in fields of civil, mechanical, chemical, and electrical engineering; biological and aeronautical engineering came later. From today's point of view, it may be that specialists in medicine, because they work with humans around the planet, have done the best so far in developing and sharing knowledge worldwide.

Studies of human society and the natural world, previously overlapping in significant degree, became separated from each other increasingly during the course of the eighteenth and nineteenth centuries. For instance, specialized disciplines formed more gradually in the social sciences than in the natural sciences. Economics (previously political economy) grew in the early nineteenth century, and later developed a theory of the firm, a theory of output in the national economy, and input–output analysis. Sociology took form with the academic studies of Emile Durkheim and Max Weber. Physical anthropology, led by Paul Broca in France, focused on the issue of race, seeking to discern essential differences among human groups; in this work the field became close to the eugenics theories of Francis Galton in psychology, which sought to demonstrate lineal inheritance of intelligence. Cultural and social anthropology began through the work of two lawyers, Edward Tylor in England and Lewis Henry Morgan in the United States. They conducted valuable empirical studies but fit them into a simplified stage theory of human development.

The arts and humanities stand out, in that their many fields of study and practice have existed for millennia. In contrast to the natural sciences, the literature and visual art of ancient times remain of high value and interest to today's audiences. As a result, specialized knowledge in the arts and humanities did not fit easily into the grid of academic disciplines modeled on the natural sciences.[11] Further, advanced study of arts and humanities took place in workshops and other inherited institutions, certainly not in laboratories. Ultimately, however, arts and humanities specialists gained connection to the university system. Thus, practitioners of oral history and tradition in Africa met up with academic historians of Africa. The oral texts were not only recorded and critiqued, but both sides of this collaboration also joined in creating an innovative scholarly field. With time, the new media and techniques of the electronic age had their effect: activities in "digital humanities" began to arise in the 1980s, with results that were influential in numerous fields, especially computer visual art, so that new technology and conceptualization could connect to ancient arts. Critique of literature and art, in the world of today, accounts more systematically for social and cultural differences than in past times.

Institutions and Ideology in Specialized Knowledge. The big question about specialized knowledge, in my view, is whether it is to be directed toward special interests or toward the general interest; whether it is to be held in secret by corporate owners or made available to all who wish to share it. One can see contending influences in play. The sheer size of powerful corporations, even when they publish promises to act in support of public interests, leads one to understand that there will be many restrictions on the flow of knowledge. The governmental institution of patents, granting legal monopolies to the recognized creators of inventions, has been very influential in limiting flows of knowledge and in the path of economic change. This concept has been generalized more recently to the notion of intellectual property, which defines owners of intellectual creation (either individual or institutional) and attempts to protect their interests – possibly slowing the circulation of information as a result. In another application of patents, pharmaceutical companies have been able to identify natural constituents used in popular remedies, and to patent and price them in such a fashion that the originating communities lose all access to their remedies.[12]

Meanwhile, throughout the process of university expansion, the geographical and social distribution of new knowledge has undergone steady transformation – to university and public libraries and, especially, to access through the internet. Specialists in knowledge have devoted energies to reaching the general public, seeking to convey the methods and the results of their analysis. In the field of biological evolution, Stephen Jay Gould, a key figure in the expansion of biological theory in the 1970s, was outstanding not only as a researcher but as one able to convey new debates and results to a wide audience. Academic journals in the natural sciences are requiring, increasingly, that contributors of articles provide access to their original data along with clear statements of their methodology. In addition, science writers – journalists focusing on scientific topics – produce a steady stream of books and articles with news on recent research. This same process of outreach to general audiences is evident in the social sciences, where authors of key contributions write up their results in a form accessible to general readers.[13]

Can specialized knowledge fit with the interest of common people? Despite the ongoing efforts to disseminate new research knowledge, in schools and colleges and to the general public, there remains a substantial gap between the specialization of knowledge experts and the daily lives of citizens. The growth of global democratic discourse through popular culture and social movements raises the question of what relations do exist and could exist among knowledge specialists and the general public. The quantity of new knowledge and its availability are impressive. But linking this knowledge to the social and environmental problems we now face is not automatic. Will it be possible to provide information to the common people of Earth that will improve their ability to formulate their views on the big questions that affect us all?

Changing Institutions and Networks, Small and Large

In thinking of the social institutions of today, I emphasize large-scale institutions with global reach, as it is easier to survey them: nations, corporations, international organizations, and big cultural institutions. Yet I note that the large-scale institutions exist only because of the smaller institutions that support them. It is equally important, therefore, to keep in mind the changes in the functioning of localized teams, ethnic groups, religious communities, and workplaces. Their reliance on changing technology (electronic communication and datasets) and changing social organization (expanded migration and commercialization) has been a part of institutional change throughout the Human System.

Decolonization – the world's shift from governance by empires to nations – unfolded over more than two centuries but ended in a rush, following immediately on the final expansion of empires. Map 10.1 distinguishes those regions that gained independence from empires early (as in the Americas), or late (as for Africa), also noting regions where empires did not undergo decolonization (as in parts of Europe and Asia.) Nations have become the principal container for social and political debate and reform. The issues of popular culture, systems of knowledge, capitalist transformation, and ecological reform are debated much more in national than in imperial context. While some nations have existed for centuries, the twentieth century brought a cascade of national movements, as people everywhere demanded citizenship in a recognized national government. The long campaigns of people in Africa and the African diaspora to gain citizenship and nationhood relied especially on popular culture in music, literature, dress, photography. These creations and their audiences built the reputation of each national group and of African peoples as a whole.[14]

Once the world came to be subdivided everywhere by national units – by about 1970 – the dynamics of international relations changed. The process unfolded in the face of objections that the newly independent peoples were "not ready for self-government," but these claims from shrinking empires were forgotten well before 2000. The Soviets and Americans held on as great powers but not precisely as empires. Decolonization had brought the addition of another 150 nations by the end of the twentieth century. With time, additional benefits of national units began to appear – benefits that grew in value as industrial and electronic technology improved. A dynamic of global tolerance and apparently friendly competition arose with international sports events, popular-culture competitions, and beauty contests. Within nations, the national state became a forum for political struggle on such issues as regional and ethnic divisions, the problems of class differentiation, and the question of how to provide adequate resources for health and education. Latin America, for instance, met some of its worst days with the military dictatorships of the 1960s–1980s, but thereafter made remarkable progress in setting up political systems that allowed for alternation of power and recognition of the value of most groups within the nation.

Nevertheless, the differences between big and small nations remain a curious artifact of historical experience. Huge nations that were

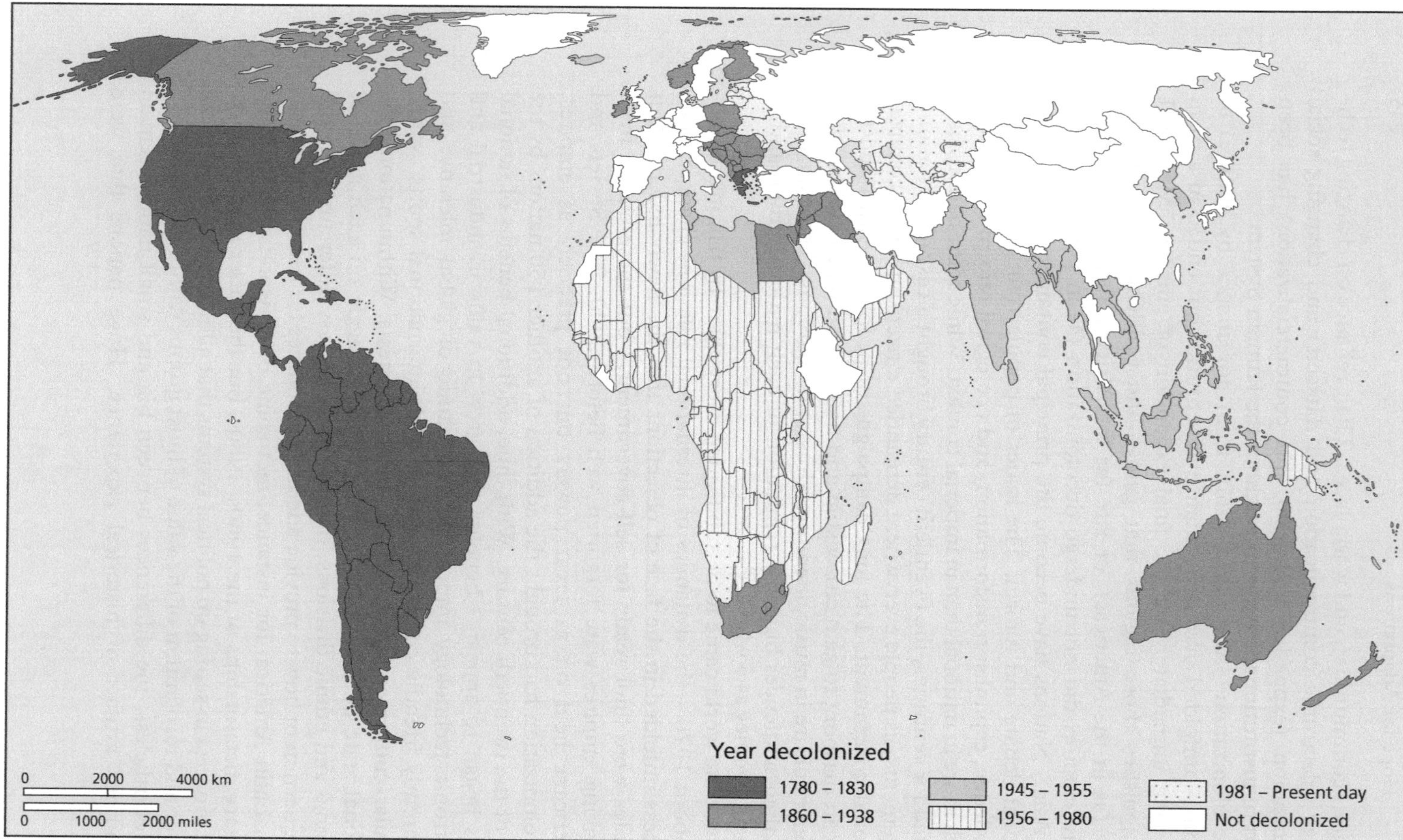

Map 10.1 Decolonization. (Redrawn based on original image, courtesy of the University Library System, University of Pittsburgh; generated by Daniel Andrus, 2019.)

formerly imperial centers have disproportionate global influence: China, Russia, India, Brazil, and Iran. Other huge nations were previously colonies: the United States, Mexico, Indonesia, Pakistan, and Nigeria. The most populous seven nations in 2020 include 4 billion persons, half of the world's population; the next most populous seven include another billion people.[15] The former imperial homelands – Britain, France, Germany, Japan, Italy, Spain, the Netherlands, and Turkey – while smaller still in population, preserve significant influence. Given this imbalance in populations, one can only wonder what will be the future of human political relations: these inherited inequalities will not soon disappear. Nevertheless, the existence of nationhood makes it much more difficult to simply oppress or neglect the smaller nations and their peoples.[16]

The great corporations of today, clearly descended from the founding firms of the late nineteenth century, are transformed in size, structure, goods and services produced, and in their pervasive influence. The leading industries continue to shift from decade to decade – since the 1970s, economic leadership has shifted from automobiles to petroleum to finance to electronics, along with other leading firms in retail sales and in health care. While corporations formally register their home in a single nation, in fact the large corporations are distributed widely, gathering resources in some areas, focusing production and marketing in others, hiding financial resources in certain nations, paying minimal taxes in others, and participating in stock markets in New York and on every continent. Corporations invest in large-scale public relations campaigns to argue that they provide service to all; nevertheless, their principal focus remains that of maximizing profit and market share. Despite the size and power of great corporations, they operate successfully only through their collaboration with immense numbers of much smaller firms, down to the level of single proprietorships.

International organizations, while they have existed for centuries (most obviously through the two millennia of the Catholic Church), expanded dramatically in the second half of the twentieth century. The United Nations created the Security Council as an arena for the big powers and the General Assembly as an arena for all member states, in which smaller and weaker nations predominated. This led to codification, by the UN and its numerous bodies, of a great number of international standards for behavior. These were violated as often as respected, but they did change the nature of global affairs. Nationalism was still a threat in many ways, in that it led to oppression and exclusion rather commonly and to genocide in several tragic cases. Yet nationalism also brought multinational social movements, as in pan-Arab and pan-African nationalism, although these serious attempts to create larger federations were unsuccessful especially because of big-power fear and interference. Religious organizations include the Catholic Church, the World Council of Churches, and the Organization of the Islamic Conference. Philanthropic foundations have expanded along with the profitability of the corporations that created them.[17] International service organizations, relying heavily on voluntary contributions include Médecins sans frontières in health, Amnesty International in human rights, and the World Wildlife Fund. All of these institutions, most of them recently created, have become a central factor in international relations.

Culture, at the level of group interaction, involves actions of representation, modeling, and innovation to portray the surrounding world. The initial representations of the world by the Founders of speaking communities, expressed in words or in other media, have developed with time into a remarkable range of institutions in culture and knowledge. The results of this representational activity yield entertainment and knowledge, often in a mix of the two. Cultural institutions are those that support the creation and dissemination of entertainment and knowledge. They include public and private schools, universities, museums, research institutes, entertainment institutions, and media corporations, but also armies. In these institutions, we commonly divide knowledge into natural sciences (including studies of health and engineering), social sciences (emphasizing analysis of human behavior), and the arts and humanities (representing the world in a wide range of media).

While this book focuses on issues of individual and group behavior in biology and social science, it would be of interest to extend its discussion to the arts and humanities. In particular, it is striking that the arts and humanities address both individuals and groups among humans, moving across the two perspectives with apparent ease. Artistic composition and performance are somewhat different when completed by individuals and groups – solo music versus choirs and orchestras. One can distinguish between informal groups (jams) and those that are rehearsed through formal agreement (bands). The work in arts and humanities is not analytical to the same degree as that in social or natural sciences. But the production of a play or a film involves complex group interaction as well as the starring efforts of individuals. The arts and humanities show that we have much to learn on the combination of individual and group behavior. More broadly, popular culture, expanding knowledge, and (as I will argue) global democratic discourse have emerged as informal, global-scale I-networks that exchange experience and ideas among people in all stations. The emerging discourse has the potential to facilitate a consensus at an unprecedented scale – or to lead to costly, horrendous debates. There is some possibility that these three global networks may crystallize into global institutions with clear membership and purpose. As yet, the purposes, the membership, and the norms of any institutional form have yet to be set. Yet the scale of these networks suggests that big changes are afoot.

Global Democratic Discourse

How can members of local institutions and communities gain the attention of those in a position to make global decisions? One way is to contribute to the hierarchical discussions of which their institutions form a part, for instance by petitioning local structures of government and religion. Another means of expression is through the network of social movements, in which people sharing common concerns reach across boundaries to make their case in regional and global debate. In the latter instance, social movements circumvent hierarchies. While they are often structured informally as I-groups rather than as the we-groups of social institutions, social movements play a central role in the global interactions of social institutions.[18]

I begin my exploration of global democratic discourse with the example of colonized and indigenous peoples. In Map 10.2, I show the dates at which colonization began in various parts of the world: in it I seek to privilege the perspective of the colonized peoples rather than the imperial conqueror. The peoples of most world regions underwent colonization, beginning at times from the fifteenth to twentieth centuries. All such peoples lost their sovereignty, though at varying moments. Some lost their lands as well – especially those in the Americas, Australia, and Pacific Islands, and also those enslaved or otherwise recruited for their labor. All were flooded with the impact of the expanding capitalist system, and many were outnumbered by settlers arriving from Europe, Asia, and Africa. In almost no cases, however, did the colonized peoples actually disappear. The renewed Caribbean interest in Taino culture, beginning in the 1980s, was a recognition of continuing presence of the ancestral Taino people.[19] In a related but not exactly identical issue, kingdoms and communities, large and small, were overrun and incorporated into larger colonial units, especially in Africa during the twentieth century, but earlier as well in Siberia, India, Southeast Asia, and the Americas. When nations were created, in the nineteenth and twentieth centuries, indigenous peoples were included within them, but they usually did not gain full sets of rights.

Despite these centuries of losses, indigenous peoples found ways, with time, to draw on nationhood and international organizations. They created social movements that criticized the oppressive, long-neglected

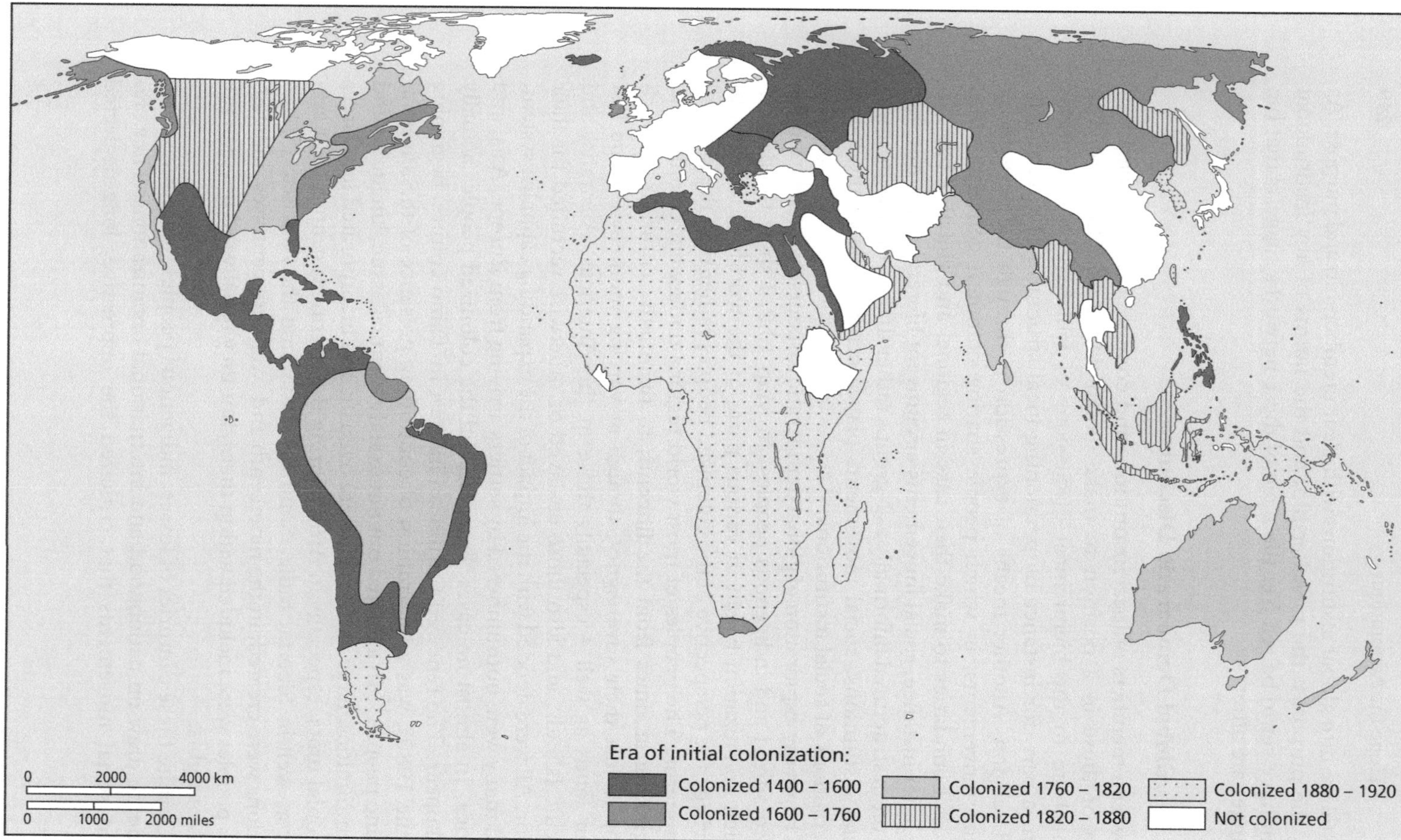

Map 10.2 Colonized peoples over time, by era of regional colonization. (Redrawn based on original image, courtesy of the University Library System, University of Pittsburgh; generated by Boris Michev, 2019.)

conditions that had restricted them. Indigenous peoples, despite their small numbers and the losses of their rights, their land, and many of their children to the great powers, forcefully claimed full recognition of their community, their language, their lands, and their cultural production. Remarkably, they gained significant support from other groups, and began to win concessions. The indigenous people's movement was signaled by the seizure of Alcatraz Island in the San Francisco Bay by the Native American Movement in 1968 and reached a global peak with the Chiapas uprising of 1994. Those who had been the weakest and had lost the most were able to achieve various sorts of formal recognition, as in the creation of a formal negotiating framework between the peoples of the polar region and the national states that governed them.[20] In this and other fashions, popular voices made themselves heard in a way that brought the concerns of communities and other local-level institutions to the attention of others around the world. These campaigns put pressure on the most powerful institutions to find adjustments that could be made in response.

Such social movements are the organized and active parts of a larger phenomenon that I call "global democratic discourse." This global discourse has arisen from three principal sources: the practice of popular culture for purposes of entertainment, the expansion of both general and specialized knowledge, and the campaigns of movements for social change.[21] I trace the evolution of such a global discourse out of localized discussions, linked in various ways: through connections in commerce that have moved commodities to new lands, migration that has introduced travelers to new communities, communication media that have reached across social lines in text and audio, and the formal schooling of the young. Most prominently, the emerging forms of popular culture have stimulated new feelings through music, dress, dance, theatre, film, and the internet. Global democratic discourse relies on the ancestral culture of each community, but it relies even more on the novelty of reaching beyond the boundaries of home. It encompasses nations, societies, communities, and organizations to sustain a discourse on human priorities. It draws on the long history of local debate and governance, in some cases invoking the well-known Athenian model of democracy.

An implicit premise of new global discourse is the notion of human equality. This vision has gained wide support – perhaps majority support – among the people of the planet. It contests the hierarchy that is so evident in the leading institutions of society. On the other hand, the notion of equality is vague and is not embodied in any effective institutions or any explicit ideology. In contrast, well-organized corporate and financial interests, backed by other institutions and with great influence in governments, argue that their own interests must be attended to first. They assert that the welfare of humanity is best addressed by ensuring the welfare of the most powerful interests. Special interests of various sorts – corporate, financial, government officials, landed aristocrats, and other big institutions – greedily take care of themselves. For general populations of all sorts, they recognize the greed but see no strong basis for developing alternative approaches. The various efforts at popular reform – expanded schooling, consumer protection, environmental protection, assistance for the poor – meet with instinctive and forceful attacks from the established special interests. In between are coalitions of groups seeking to provide valuable social services through alliance with the corporate interests.[22]

Still, one can see elements of an emerging ideology in the assertions, in popular culture, of the equality of individual humans. The reasons for this growing vision of human equality need to be explored. I argue that we are in the midst of a great shift in human social organization: the emergence of global democracy as a forum for discussion on social priorities. While this is very much a modern and contemporary transformation, it is also important that we compare our current transformation with the biggest social transformations of earlier times. For instance, empires arose three thousand years ago as extensions of societies, yet they were commonly oppressive and were unable to expand beyond a regional scale. Nations have arisen recently to displace empires, also as extensions of societies, and we may ask whether they indicate a return to the societies of old or a step toward global democracy – or both. Nations, for instance, appear to be more open than empires to recognizing both individual and group rights within their borders. Prasenjit Duara, in considering principles for construction of a new world order, suggests that priority be given to the values inherited from religions based in eastern Asia rather than those of Abrahamic religions, as the latter, dominating for the past several centuries, have brought more conflict than consensus.[23]

If a global forum is coming into existence, the discourse has many topics and concerns to debate. The Anthropocene epoch has provoked many crises. Hierarchy and network, earlier out of balance, have become even more inconsistent. The dilemmas of higher productivity, tighter global connections, capability for war, further invasion of Gaia, and great expansion in knowledge face us all. Grasping elites are of obvious concern – they have generated world conquest, untold riches, and impoverishment of whole continents. Elite determination to govern economy and society from the top down have disrupted the reproduction of the Human System by distorting processes for regulating institutions.[24] Nevertheless, to stigmatize the elites alone as scapegoat for the crises of today is to hide from other imbalances: it may be that almost all humans are complicit in undermining the smooth and equitable functioning of the Human System.

How greatly have social relations changed since the opening of the Anthropocene? One way to consider this question is to compare two great sets of social movements – in the North Atlantic (1789–92) and the planet (1989–92).[25] In one sense they reveal clear parallels, as social movements arose in each age to contest elite dominance, seeking to establish democratic structures. In the eighteenth century, the French uprising brought revolutionary change for a period of time, though much of it was reversed in 20 years of warfare. Within the same decade, powerful if often unsuccessful social movements for emancipation and governmental reform arose in the Caribbean, North America, Ireland, and Poland; the 1792 antislavery petition campaign, in which thousands of British subjects signified their opposition to slave trade, launched a campaign that lasted half a century. Two centuries later, in 1989, demonstrators in Tiananmen, Berlin, Cape Town, and many other areas of the world, in contact with one another, claimed recognition of democratic rights. By the end of 1990, unpopular governments had stepped down in Europe, Africa, and others were under pressure. The 1793 execution of the king of France accompanied the invasion of France by European monarchies; the 1990 seizure of Kuwait by Iraq brought a US-led invasion of Iraq. In each case the wars lasted 20 years. Among the differences were that the twentieth-century social movements were in closer mutual contact and were more likely to bring change in government and society. The United States was a hegemonic power in the latter case, but not quite a Napoleonic empire.

The results of great social movements were complex, in the twentieth century as in the eighteenth. In the eighteenth century, Britain ultimately vanquished France; in the twentieth, the United States vanquished the Soviet Union. The communist regimes of Russia and Eastern Europe fell while that in China reaffirmed its dominance; in South Africa, the Communist Party, a participant in the anti-apartheid movement, became part of the new constitutional government in 1994. The eighteenth-century executions by the Jacobin government of Revolutionary France, followed by the extinction of Poland and the crushing of Irish revolt, were more than matched by the later episodes of genocide in Cambodia, Rwanda, and collapse in Yugoslavia. Antagonism arose in the Islamic world, followed by terroristic attacks – especially the attacks of September 11, 2001 in the United States and then more broadly. The United States responded to these tensions, again, by launching war on Iraq in 2003. The United States, ignoring worldwide antiwar demonstrations, invaded, declared victory, then fell into a long and bloody stalemate.

At a larger scale, one may argue that global democratic discourse successfully averted atomic war. Even 80 years after the two bombs exploded over Hiroshima and Nagasaki, no government or army has dared to launch a nuclear weapon. The mechanism for preventing atomic war is not clear, but it does seem plausible that global popular opinion has been central to restraining the firing of these weapons.

Global democratic discourse, arguably an innovation of recent times, requires more thorough analysis. Perhaps one will learn more about global democratic discourse through investigation at a more intimate level, exploring debates on the divisions among common people. The many divisive issues that had long existed – racial classification, ethnic identity, gender roles, religious toleration, and justification of violence – are now debated more publicly through film, music, and internet comment. Painful issues of violence and sexual exploitation, especially directed toward children and women, are being brought into the open in communities throughout the world. At worst, these reports and debates reveal that cruelty and mistreatment remain unexpectedly widespread within humanity. At best, these same reports and debates suggest that people of many groups are gradually developing common understandings and a global discourse that, while uncomfortable, can nevertheless be direct and informative.

One of the remarkable institutions to develop out of such debates was the series of "truth and reconciliation" commissions that collected and publicized information on brutal national conflicts in Latin America and then in Africa. These public discussions could not resolve the anger, hatred, and fear that remained in each case, yet they set the goal of some form of reconciliation and perhaps contributed to its achievement. In its various institutional forms, global discourse highlights the cost of social inequality – as measured in loss of individual development, social productivity, and in the potential for unleashing new ideas if more people suffered less from inequality. Yet the discourse alone has yet to reduce decisively the level of inequality. South Africa provides a challenging example: on one hand, a long struggle to overcome racial hierarchy resulted in a peaceful transition and promises to build equality. On the other hand, the passage of a quarter century shows as much of the creation of new sorts of inequality as it does of constructing national unity.

I have highlighted three global-level structures in formation – the framework for popular culture, the exchange of general and specialized knowledge, and the global democratic discourse – each resulting from a level of communication and exchange far exceeding any from earlier times. They provide a basis for coordinating efforts to face the three great crises of social inequality, environmental degradation, and policy for facing the first two. The most obvious solutions to global choices, however, might do more harm than good. In particular, a focus on forming a world government, under today's circumstances, would lead mainly to expanded hierarchy and further inequality, as existing hierarchies in governance extend their influence. If there is no preexisting popular consensus, world government will be open to seizure by special interests. In contrast, formulating a global discourse provides a context for the emergence of a consensual, global social structure. The essential next step, as I see it, is not a technological quick-fix but the discussion and distillation of knowledge-based ideologies: a set of ideologies appropriate to workable social priorities for humanity. Humanity, in recent times, has arguably made more steps toward creating a global discourse than toward forming a world government, and that is as it should be.

Choices

The Human System, to preserve itself, must continue to reproduce itself, transforming matter and energy and processing information. The system conducts these activities in combinations of networks and hierarchies at levels of community, society, and now at the level of global democratic discourse. The system ingests and extrudes matter, energy, and information, and grows as a result. While a thorough investigation of the choices to be made for the Human System would involve exploring each of the elements just listed, I restrict myself in this conclusion to posing three questions on choices about the operation of the system.

Is growth necessary for the Human System? While the Human System has evidently grown in extent and density for most of its existence, it would not necessarily be unable to survive without growth. The question of growth, however, is usually posed in terms of national domestic product rather than in system-wide terms. Indeed, does "growth" refer only to economic growth, or does it also refer to growth in knowledge or levels of social welfare? In any case, it is clear that political figures and economists tend systematically to assume that economic growth is necessary in order to achieve social welfare and political satisfaction for national units. Nevertheless, given that growth in the Human System is clearly threatening Gaia and the stability of the Earth's systems, it seems worthwhile to explore the question of whether growth is necessary. Further, since analysts are certain that population will level off and fossil fuel consumption will decline, there are strong arguments that GDP too will level off or decline on the warmer planet of our grandchildren. Do we wait for the end of growth to happen or seek to prepare ourselves for it?

Is social inequality a necessary aspect of the Human System? Social scientists generally accept that inequality is an inherent characteristic of human society. That is, there is a common impression among social scientists that there exists a sort of happy medium in the level of inequality within societies – the society is too oppressive if inequality is too great, but the society provides insufficient incentive to innovate if inequality is too low. This view of inequality appears to be related to the thesis that growth is necessary for economic welfare. But what if it were agreed that the level of inequality were too high: what would be appropriate measures to reduce it? Commonly, the social science literature presents war and economic depression as the main processes for reducing levels of inequality (though there are studies of Scandinavian tax systems as alternatives). Do we have any other ways, within existing social science, of setting up the problem of how to address inequality? One possibility is to conduct analyses of human capital. That is, how much is currently invested in each person, rich or poor? What is the productivity of the capital invested in each person – could it be that overall productivity would rise if the existing investments in human capital were applied in a more equitable fashion?

Third, in what way should the operation of social institutions be regulated? This is a question about conscious governance of big institutions, such as the possibility of government regulation of financial institutions. Yet it is also a question about small and local institutions. Who are the intended beneficiaries of each institution? What role do beneficiaries play in articulating the effects of institutions on their welfare? If, as has been theorized in this book, social evolution includes an archive of institutions and a system of social selection that retains only the valuable institutions, then it is worth asking how social selection has been working in recent times. Is it possible, for instance, that mechanisms of social selection have been weakened so that they no longer eliminate institutions that are harmful to the system?[26] Are there other processes for regulating institutions – aside from self-regulation by institutional directors and government regulation – by which institutions could be adjusted to fit better to the needs of their beneficiary populations? More study of social institutions, the groups that constitute them, and the wider groups influenced by them – from local to global levels – would help to clarify their functioning, success, failure, and oppression.

Appendix: Frameworks for Analysis

Six frameworks of knowledge and analysis, summarized here, provide assumptions, principles, and resources on processes of human evolution. The first and broadest framework is that of the system – beginning with its general conceptualization and accompanied by examples including the Earth System, Gaia, living systems, and the Human System. The second framework, on evolution, summarizes the parallel though very different processes of biological, cultural, and social evolution: these are the basis of much of the analysis in this book. The remaining four frameworks are steadily more specific but address issues of importance in the understanding of overall human evolution. Human migration, studied with the best available methodology, must be documented in order to reveal the various ways in which humans have populated and repopulated Earth. Individual behavior and group behavior of humans, as shown in the fourth framework, can now be compared in detail because of recent advances in the conceptualization of human groups through "collective intentionality" – this framework assists readers in recognizing when a text assumes humans are acting as individuals, when they are acting in groups, and when groups and individuals are mixed. Fifth, social institutions, a central emphasis in this book, are listed here under the assumption that they are created only by the "collective intentionality" of groups of people that agree to create and preserve them. The same section presents a related list of social networks and hierarchies, including the overall Human System. Sixth, emotions are assumed to arise at the individual level, giving support to behaviors of altruism, violence, or other sentiments and actions; further emotions may interact with social institutions, giving rise to "human nature" as we see it acted out in communities.

The concise summaries of six frameworks presented here are intended to guide the reader to major topics, concepts, terms, and processes of analysis. Most of the processes identified in these frameworks go back millions of years – even to the days of *Homo erectus*. Yet I have also assumed that, about 70,000 years ago, there was a relatively rapid and innovative rise of spoken language. As a result, a deeper level of specificity in communication enabled humans to create new social structures and to make far more efficient use of their many accumulated skills and capabilities. As a result, certain processes described arose only with the emergence of speech: these include social evolution, cross-community migration, conscious group behavior, human nature at the group level, and more. Human history after 70,000 years ago centered on the development of these new practices and on integrating them into the preceding. human practices.

1. Systems and Coevolution

General Systems. A system is a complex of interacting elements; systems are open to and interact with their environments (Bertalanffy). Closed systems are self-contained; open systems ingest and extrude matter and information. Systems are generally self-regulating in that they self-correct through feedback. System thinking includes both part-to-whole and whole-to-part thinking about making connections among the various elements so that they fit together as a whole. **Subsystems,** within an overall system, perform distinctive functions.

Earth System. The term "Earth System" refers to Earth's interacting physical, chemical, and biological processes. The system consists of the land, oceans, atmosphere, and poles. It includes the planet's natural cycles – the carbon, water, nitrogen, phosphorus, sulfur, and other cycles – and deep Earth processes (Ruddiman). Life too is an integral part of the Earth System. **Gaia** theory (Lovelock) proposes that all organisms and their inorganic surroundings on Earth are closely integrated to form a single and self-regulating complex system, maintaining the conditions for life on the planet.

Living Systems. Living systems function at any of eight levels: they are open systems composed of 20 critical subsystems that process inputs, throughputs, and outputs of various forms of matter–energy and information. Two of these subsystems – reproducer and boundary – process both matter–energy and information. Eight of the subsystems process only matter–energy. The other 10 subsystems process information only (Miller). Communities, confederations, societies, and networks of combinations of these are social subsystems, each containing functional sub-subsystems that perform relevant tasks. (In 1990, Miller added the 20th subsystem to his 1978 list of 19: the "timer," the tenth subsystem for processing information.) **Coevolution** consists of interactions among systems or within systems. Coevolution was initially defined as the interplay in the phenotype of species that were in close contact (Ehrlich). With time the notion of coevolution came to be generalized to refer to interaction between processes of biological and cultural evolution (Boyd and Richerson). In the history of evolution, occasionally, a new level of system emerges – such as from cell to organism and from organism to group, in biological terms (Maynard Smith). Here I argue that the Human System is a new level among living systems.

Human System. The *Human System* is a historically formed totality of the processes carried out by humans. It took form through initial links among nearby communities; thereafter, the system expanded progressively to encompass links among confederations and societies. As hierarchies emerged, the Human System incorporated links among hierarchical levels. Still, the Human System was more than the encompassing envelope: it included the vertical and horizontal links interacting at every scale down to the local. In the framework of human evolution, coevolution is further generalized to include significant interactions within an evolving system.

References. Bertalanffy, *General Systems*; Boyd and Richerson, *Evolutionary Process*; Ehrlich, "Co-evolution"; Lovelock, *Gaia*; Manning, *Methods*; Maynard Smith, *Transitions in Evolution*; Miller, *Living Systems*; Ruddiman, *Earth Transformed.*

2. Evolution: Biological, Cultural, and Social

Three processes – biological, cultural, and social evolution – are united by the overarching analytical framework of Darwinian thought. At the core of Darwin's analysis, biological evolution relies on variation in characteristics, selection of characteristics through a struggle for existence, and inheritance of characteristics – as these factors are judged by the standard of fitness that is the degree of reproduction of individuals and the species. Darwin's initial phenotypical analysis of individual organisms has now been extended to other scales of life, giving us genetic and molecular details that provide deep new insights yet do more to reaffirm than to revise Darwin's original insights. In recent years, analysts have been turning to exploring questions that Darwin raised in his later books on the descent of humanity and on emotions in animals and humans. It remains plausible that Darwin's founding principles are valuable as well at these additional levels (Darwin). Beyond biology,

Table A1.1 *Assumptions in various theories of evolution*

Evolution type	Biological	Cultural	Social
Unit of evolution	Organism	Organism in dual inheritance	Community or society
Variation: Locus	DNA in organism; Epigenetics	• DNA in organism • Brain in organism	Brains and institutions in social group
Source	Random	• Random • Individual choice	Group choice
Selection: mode	Natural selection	• Personal selection • Inclusive selection • Multilevel selection	Social selection and regulation
Inheritance: Retention locus	Inheritance (DNA); Epigenetics	• Inheritance (DNA) • Brain in dual inheritance	"Archive" (distributed social retention)
Reproduction locus	DNA and proteins	• DNA and proteins • Brain in reciprocity and social learning	Rituals
Fitness measure	Reproduction of offspring (of gene or organism)	• Personal fitness of individuals • Inclusive fitness of individuals • Multilevel fitness of groups	Social fitness: strengthening of social group (at selected scale)

Prepared by the author. NB: For simplicity, the table assumes neo-Darwinian selection processes, abstracting from the more complex processes (for instance, epigenetics) known to be part of evolutionary change.

I extend Darwinian logic to cultural and social evolution. That is, I concur with the basic arguments of the growing field of cultural evolution, in which analysts use analogies to Darwinian principles to explore processes of social learning that are archived in human brains paired with the evolution of population characteristics in the genome. In addition, I advance a theory of human social evolution, arguing that new formulations in the study of group behavior in humans may also benefit from analogies to Darwinian reasoning. To confirm the parallels of these first three processes of change, Table A1.1 summarizes elements of biological, cultural, and social evolution in a framework that highlights the Darwinian framework. The initial column lists the various Darwinian elements of evolution; in this description of the table, I discuss these elements for each of the last three columns, from top to bottom.

Biological Evolution. The term applies to all species. It centers on the genetic change in species through natural selection but has been extended to include ontogenic change or life-course development of species (Gould). In *biological evolution*, the *unit of evolution* is the individual organism. *Variation* takes place in the DNA of the organism, but also takes place through epigenetic change; the nature of genetic variation is primarily random change. *Selection* among biological variations takes place through natural selection, a term that encompasses a wide range of potential changes. *Inheritance* takes place through DNA, as the genome (in association with epigenetics) both retains any successful variations and reproduces them through mitosis. *Fitness* of any variation is assessed by the level of biological reproduction of organisms carrying the variation. **Infectious disease** arises from biological evolution of microorganisms, bacteria or viruses, that occupy plants and animals as hosts (Diamond).

Cultural Evolution. The term has been developed for study of humans, though it may be applied to primates and other animal species. Within humans, the theory of cultural evolution focuses on individual learning, where it is known as "dual heritage" in that genetic and cultural evolutionary processes coevolve. In cultural evolution, the *unit of evolution* is again the individual organism, now including both biological evolution in the genome and cultural evolution in the brain (Boyd and Richerson). The *variation* takes place in the genome and the brain of the organism; the source of variation is random change in the genome and individual choice in learning or reciprocal behavior, archived in the brain. *Selection* can take place according to three processes, in which genetic selection is paired with (1) personal selection through individual learning; (2) inclusive selection on learning or reciprocal behavior; and (3) multilevel selection on learning and reciprocal behavior. *Inheritance* of innovations is archived and retained in the genome and the brain of the individual; the dual inheritance is conveyed genetically through DNA and culturally by exchange from brain to brain across the generations through learning practices. *Fitness* may be determined at three levels: (1) the personal rate of reproduction of organisms carrying the cultural variation and associated biological variation; (2) the rate of reproduction of organisms included in the behavior of an individual organism through inclusive fitness; and (3) the rate of reproduction of a group of organisms through the aforementioned processes. These three scales of cultural analysis are argued to be equivalent to each other (McElreath and Boyd). This analysis of cultural evolution has been given empirical and historical application in two groups of studies: analysis of social learning in mid-Pleistocene hominin populations and studies of *Homo sapiens* at more recent times and larger scale, for which the term "ultrasociety" is employed. **Culture in the framework of cultural evolution:** This includes accretions to learning by individuals, which constitute cultural elements (Boyd and Richerson). These acts of learning, whether by trial-and-error or by imitation, are related to human imagination and ideas. **Uniformity in groups relying on multilevel selection**: Note that, in the framework of cultural evolution, there is a premium on uniformity within communities and variety among communities, to facilitate selection of the fittest communities; hence migration brings dangers (Turchin).

Social Evolution. The term applies to the study of conscious, speaking humans. In *social evolution*, the *unit of evolution* is a social group, especially a community or society, for which we trace the Darwinian changes through the evolution of social institutions based on groups of people (Tuomela). *Variation* appears in the minds of innovators (as individuals and in groups) and in the institutions they propose; the source of variation is the choice by members of a we-group to create or modify a social institution. *Selection* takes place through "social selection," an analog to natural selection, in which new institutions survive or fail through various processes according to their social fitness. In addition, social selection is more complex than the dichotomous choice of success or failure: institutions may also undergo *regulation* and reform, in which institutional members or social groups modify existing institutions (Campbell). *Inheritance* of an institution takes place through two associated institutions, which handle preservation and reproduction of information on the institution. For preservation, an *archive* retains information on the institution, where the archive is distributed across brains and ultimately material recording of information. For reproduction, *rituals* pass information from the archive to the next generation through ritual, rehearsal, and initiation. *Fitness* of any variation in institutions is assessed as strengthening of the relevant social group: such social fitness and strengthening however, can be assessed according to varying standards. That is, standards of social fitness are open to debate among members of the social group. Social fitness can be measured through the number of individuals sharing the institution; it can also be measured through changing levels of social welfare per individual. Further, social fitness might be assessed as the contribution of an institution to the welfare of a whole community or to the welfare of a subgroup, such as an elite. **Culture in the framework of social selection:** In this framework, in contrast to that of cultural evolution, "culture" consists of the exchange of representations among individuals and the creation of group representations. Group-level culture is created in a world of social institutions; it offers commentary on social institutions and the natural world. **Diversity in groups relying on social selection:** In the framework of social evolution, there is a premium on variety within communities, to facilitate innovation and institutions adding to the fitness of each community by expanding its social welfare; hence migration brings benefits.

References. Biological evolution: Darwin, *Origin of Species*; Diamond, *Guns, Germs, and Steel*; Gould, *Ontogeny and Phylogeny*. Cultural evolution: Boyd and Richerson, *Origin and Evolution of Cultures*; McElreath and Boyd, *Mathematical Models*; Turchin, *Ultra Society*. Social evolution: Campbell, "Biological and Social Evolution"; Manning, *Methods*; Tuomela, *Social Ontology*.

3. Human Migration

Human migration relies on certain basic impulses, yet it has changed in character along with social transformations. Human flexibility in adaptation to habitats makes the species prominent worldwide. Humans have repopulated the Earth in characteristic fashions as social, ecological, and economic conditions change. Migration fills in new gaps as they develop. Human migration takes place through two processes: **general migration**, diffusion of population to extend its habitat or occupy a new habitat (Dingle); and **cross-community migration**, to link communities and occupy a new habitat (Manning, *Migration in World History*). Documentation and analysis of migration strengthen the spatial understanding of the Human System. **How to document migration:** there are four principal disciplines for documenting migration: historical linguistics, genetics, archaeology, and written records.

Historical Linguistics. Example: The distribution of Nilo-Saharan languages confirms that they began in the mid-Nile Valley and spread northwest to the Sahara and the Niger Valley but also southward into Central and East Africa (Ehret). **Method:** Through lexical and grammatical comparisons, languages are classified into phyla – families of unilineal descent of languages from their ancestor, that go back more than 15,000 years. Geographic analysis of families and subfamilies permits estimation of the geographical homeland for each language group. This tradition has advanced a list of 14 language phyla which include virtually all human languages in hierarchical classifications intended to reflect the divergence and descent of languages within each phylum (Greenberg; Ruhlen). (For details, see the online resources associated with this book at www.cambridge.org/Humanity.) Combining this set of language phyla with the results of recent whole-genome ancient DNA analysis, one gains strong support for the conclusion that the 14 phyla have been persistent over very long periods of time, in some cases reaching back to the earliest days of language. I will rely on this presumption in my migratory analysis. While linguists commonly assume that the rate of language change is such that it becomes impossible to trace language groups back further than 10,000 years, I think that the correlation of language-group homelands with genomic data is high enough that it would be a mistake not to explore in detail the hypothesis of deep persistence of language phyla, as I do in Chapters 4, 5, and 6. **Dispute and inconsistency in language classification:** The categorization of 14 language phyla remains in dispute by linguists who specialize in certain regions, notably the Americas, Southeast Asia, Oceania, and much of Eurasia. As a result of these disputes, linguistic specialists on Africa see four distinctive phyla accounting for all languages on the continent (except for recent immigrants), while linguistic specialists on the Americas identify more than twenty separate language groups plus numerous unrelated isolates for the hemisphere (Greenberg, Turner, Zegura et al.). These two approaches are definitely at variance: the methodological disputes in historical linguistics are in serious need of resolution. I documented the methodological discrepancies in language classification and made a plea that historical linguists should hammer out a methodological consensus (Manning, "*Homo sapiens*"). Here I reaffirm that plea, arguing that linguists should confer and make agreements in four areas: the basic principles of language classification; a review of classifications with attention to creating worldwide standards for language classification; and criteria for estimating the ordinal and chronological time of language divergences. Fourth, they should debate their model of language differentiation – will a tree model of language differentiation be adequate, or is it necessary to substitute another model? As author of these words, I am not a linguist and do not have the skills to analyze syntax or the relationships among lexical elements. But I do have world-historical experience in comparing datasets, and I am clear on the differences separating archaeology and genetics from linguistics. In the former two, individual differences and debates among scholars do not undermine a worldwide consensus on methodology; in the latter case, the subfield of historical linguistics lies low in priority within linguistics overall and isolated camps of specialists follow local standards and criteria that are quite inconsistent with each other. The results inhibit use of language data in analyses where consistent language analysis could add greatly to understanding of the human past. As an indication that the rejection of large-scale language groups among linguists continues to advance, I note that the Ethnologue website, in its comprehensive listing, has deleted several large groups and replaced them with smaller groups since my 2006 review of the site. It seems to me that the denial of relationships among languages by scholars working on Amerind and Indo-Pacific is a latter-day reenactment of the rejection of any debate on the origin of language by the Linguistic Society of Paris in 1866.

Genetics. Example: Mitochondrial DNA (Mt-DNA) study of individuals of worldwide ancestry showed the greatest genetic variation among people of African descent, confirming that humans have lived longest in Africa, as shown in 1987 by Rebecca Cann, Mark Stoneking, and Allan C. Wilson (Cann et al.). **Method:** Analysis of Mt-DNA (for female ancestry), Y-chromosome (for male ancestry), or somatic chromosomes (not gender-specific) enables tracing of alleles over time. **Problems in genetic analysis:** (1) when DNA samples are taken from humans today, there is the problem of classifying the present-day sample source and projecting that genome to classifications of past populations. Both classification and projection require consultation with social scientists (Olson). Ancient DNA clarifies but does not resolve this problem, though pairing modern and ancient DNA studies should provide helpful variance (Reich). (2) **Problems in the molecular clock**: Genomic analysis gives definitive ordinal results, giving the order in which genetic divergence has occurred, but does not give the specific chronologies required for comparison to other data. Assumptions of constant rates of genetic change have shown to be undependable: continuing research must explore ways to advance and verify molecular clocks.

Archaeology. Examples. Analysis of content of Strontium isotopes in bones and teeth of individuals from archaeological sites can give an indication of where they were born or lived earlier, by linking the individual isotope ratios to those known from worldwide mapping (Keita). Archaeology is a labor-intensive, multidisciplinary, and costly field of study (Mithen). Yet the growing knowledge of Denisovans is an example of advances in archaeology, as are new dates for human settlement in Australia and South America (Douka). The historic focus on Europe and the Levant has expanded impressively to other regions, yet deeper investment in archaeological study would be most valuable. For Africa, in particular, work on paleontology has developed impressively, but there is continuing need for archaeological study of migrations and settlements of late Pleistocene and Holocene Africa (Tattersall).

Written Records. Examples. On the assumption that human migration has been a consistent set of processes throughout our history, written records can be linked to other sorts of evidence on human displacement to yield a more comprehensive picture (Hoerder). The ample written documentation on migration began with writing itself and includes transcriptions of oral accounts. Such records range from ancient chronicles of wandering groups to early modern oceanic migrations to urbanization in our own time.

References. Cann, Stoneking, and Wilson, "Mitochondrial DNA"; Dingle, *Migration*; Douka, "Age Estimates"; Ehret, *History and Testimony*; Greenberg, *Languages of Africa*; Greenberg, Turner, Zegura et al., "Settlement"; Hoerder, *Cultures in Contact*; Keita, "Geochemical Method"; Manning, "*Homo sapiens*"; Manning, *Migration in World History*; Mithen, *After the Ice*; Olson, *Mapping*; Reich, *Who We Are*; Ruhlen, *Origin of Language*; Tattersall, *Masters*.

4. Individual and Group Behavior

Behavior in humans is analyzed in two basic frameworks: behavior as individuals or behavior in groups. **Individual behavior:** Biological analysis is study of individual organisms as representatives of a species. Individual motivation, action, and reaction has been the framework for most detailed study of human behavior in economics, sociology, and psychology. It has been assumed that the behavior of human groups is best explained by the logic of individuals within those groups: actions of groups are explained through actions of individuals. **Group behavior:** Recent research has developed a parallel logic of group behavior, identifying distinctive behavior for human groups. Assuming that this work goes ahead, future social analysis should give attention to the balance and interaction of individual and group behavior.

Individual Behavior

- **Individual Behavior Defined**. An individual makes decisions based on self-interest without any agreement with others or influence by an authority.
- **Population Analysis of Individuals.** An individual may be a member of a population. Populations (including subpopulations) are accumulations of individuals. They are analyzed according to the distribution of individual characteristics in a population. (Thus, population genetics addresses the frequency of alleles in a population.) Population behavior arises from the dynamics and accumulations of individual behavior.

Table A4.1 *Game: Prisoner's Dilemma*

Game: Prisoner's Dilemma		Agent 2	
		Cooperate	Defect
Agent 1	Cooperate	2.2	0.3
	Defect	3.0	1.1

Prepared by the author, based on Tuomela, *Social Ontology*, 186.

- **Individual-Level Game Theory**: Prisoner's Dilemma. Individual agents pursue their own self-interest without the direction of an authority. Payoffs are set and agents cannot communicate with or influence each other. Each agent selects a move, *cooperate* or *defect*; aware of the payoff they will receive. Irrespective of the opponent's choice, *defect* gives a higher payoff than *cooperate*. Yet both would be better off if each chooses *cooperate* (Axelrod; McElreath and Boyd).
- **Individual-Level Cooperation and Reciprocity**. It was long argued that the most advantageous strategy in the Prisoner's Dilemma is the "tit-for-tat" strategy, in which a player first defects and thereafter matches the move of the other player, so that cooperation based on reciprocity could grow from this base (Axelrod). Later analysis showed that a slightly different strategy is best: in any case, the results show that human cooperative behavior could gradually arise from individual motivation.
- **Methodological Individualism.** This is social science analysis interpreting large-sale social structures and behavior in terms of the interaction of individuals (Udehn).

Group Behavior

- **Group Behavior Defined**. Group behavior arises from collections of individuals and patterns of their behavior, which may differ according to whether the individuals recognize each other, respond to each other, or share common intentions. Human group behavior relies on "collective intentionality," in which individuals agree and determine a joint intention (collectively), of which a we-intention is an individual dimension. Group behavior presumes the preexistence and continuation of individual behavior (Tuomela).
- **Types of Groups.** Participation in group-level cooperation does not eliminate individual-level cooperation: the two coevolve with each other

and with other factors. Groups may take the form of an **I-group** or a **we-group**. An **I-group** is a population, as for individual-level populations, in which each participant acts on individual motivations and initiative. A **we-group** is a group of members unified and bounded by their collective intentionality, by which is meant their shared objective, recognition of their common interest, and agreement to act for the interest of a group.

Table A4.2 *Game: Hi–Lo*

		Agent 2	
		A	B
Agent 1	A	2,2	0,0
	B	0,0	1,1

Prepared by the author, based on Tuomela, *Social Ontology*, 187.

- **Group-Level Game Theory.** For group games, agents are members of a team. They communicate before but not during the game. Each agent selects a label, A or B, aware of the full payoff structure of the game. To maximize the payoff, each player wants it to be the case that they both choose the same label. As each player makes the choice that is best for the group, a unique outcome results, avoiding the impasse resulting from individual reasoning (Bacharach).

Also note Primal Contest – a zero-level, minimal-assumption game, in which groups struggle for advantage (Elias, 122–28).

- **Types of Group-level Cooperation and Reciprocity.** Cases include individuals in I-mode in group context; individuals in I-mode or we-mode intending to do or believe something together; individuals with collective commitment acting as a group in we-mode. Each of these three modes of analyzing group behavior has made some progress in showing that it can lead to human cooperation and reciprocity.
- **Methodology of Collective Intentionality.** Philosophical and social science analysts, using this approach, are now interpreting large-scale social structures and behavior in terms of the interaction of groups. Group theory, while analytically complex, shows that groups can be more efficient than individuals for reaching and implementing decisions (Tuomela).

References: Individual behavior: Axelrod, *Evolution of Cooperation*; Hamilton, "Genetical Evolution"; Maynard Smith, *Game Theory*; McElreath and Boyd, *Mathematical Models*; Udehn, *Methodological*

Individualism. Group behavior: Bacharach, *Beyond Individual Choice*; Elias, *What Is Sociology?*; Manning, *Methods*; Tuomela, *Social Ontology*.

5. Institutions and Social Networks

As defined here, institutions require membership of persons who agree to the objectives of a group and who join in its institutional practice. Institutions generally require members to follow certain norms, but the norms themselves are not sufficient to create or sustain an institution. Coherent definition of an institution must account for the creation of institutions, their constituents, their dynamics, their reproduction, their modification, and their role in society broadly. Institutions rely in different ways on *I-group* and *we-group* behavior.

Institutions as Constituted by Groups. Social institutions are central to the view of social evolution presented here. A social institution is an organizational form, a *we-group* with explicit or implicit objectives, involving human activities, behavior, and norms, constructed and supported by members of one or more groups, requiring we-mode and we-attitude in identity and activities. The institution defines ground rules on how to act for a collective item with a signified symbolic or social status in a collectivity. Specific institutions have dynamics arising from the character of their activities (Tuomela). To repeat, I define institutions in terms of their membership and their common accord rather than in terms of the results of institutional activity.

We-groups are characterized by "collective intentionality"; members share a group objective, are able to specify that they share common interests, and make individual commitments to the group. We-groups permit shared commitments to action, common beliefs, and collective emotions. The "irreducibility" of we-groups means that, in game-theory context, a we-group gains efficiency through its ability to make decisions that are not accessible to I-groups. That is, in games where an I-group cannot select between alternative equilibria as to which is best for individuals in the group, a we-group can identify, among alternative equilibria, a choice that is best for the group. We-groups allow for existence of hierarchy within the group; we-groups may function with direction from outside. We-groups function for specified purposes – outside of those purposes, members may exercise their individual wills.

Individuals in Institutions. *I-groups* are groups of individuals in which participants act on their individual interests. This type of group behavior evolves by processes of cultural evolution; it can develop cooperation and reciprocity according to mechanisms described by

Robert Axelrod (Axelrod). Individuals are assumed to respond to institutions but not to create them or influence them (Udehn). Groups can rarely function as a basis for biological evolution among animals (Williams). But groups can function as a basis for cultural evolution among humans (Boyd and Richerson) and as a basis for social evolution among humans (Tuomela).

Representation in Groups. Acts of representation are the main source of innovations in institutions. Representation is the modeling, formalization, and interpretation that is created in response to an aspect of the world as it is or might be. Representation is a complex mix of individual creativity with exchange and reflection among groups. At one level, representation takes place within the mind of an individual – internal and conceptual formulation in the individual mind to represent a thing, a process, or an idea. At another level, the exchange of representations generates discourse, debate, and innovation – especially through speech within a we-group context. Observation of group-level cultural practice inspires responses from individuals and groups. Representation is commonly the modeling and interpreting of one aspect of the world for expression in another arena: representations thus cross boundaries of material culture and expressive representation in visual art, dress, dance, or song, with new media for representation developed steadily through the course of human experience.

Social Networks. *Social Network* is defined as a set of entities – actors, organizations, or locations, for example – and the ties that exist among them. **Social networks and individuals**: Social networks most commonly assume individual-level actors. Individual-level social networks are constituted of individual human nodes and links among them. Individuals enter the network space at birth and leave it at death or upon migration; they have links to family members and other individuals. Some individual nodes are more central than others; information can be passed through the network from node to node (McLean). **Social networks as institutions:** In cases in which members of a network express collective intentionality and a common goal, the network is an institution. Links can extend, directly and indirectly, to individuals and groups of nodes at great distance, with attenuated intensity. Networks of networks may be formed, as each network forms a node in the larger network of networks.

Hierarchies. A hierarchy can be seen as a specialized network in which the nodes are placed at varying scales or ranks. Individuals or groups may be ranked according to the scale of the hierarchy. A group formed of a collection of nodes may have a hierarchy within it, with certain nodes being in ruling or highly influential roles; these hierarchies are operative within the boundaries of the hierarchical group. If the hierarchy is in I-mode, the individuals at the various levels retain individual-level motivation. If the hierarchy is in we-mode, the individuals agree to accept and act out the role assigned to them in the hierarchy.

Social Institutions Discussed in the Text. Each of these institutions relies on we-groups and changes through social evolution. These are examples of institutions in each period: more institutions could be explored for each time period. See the Index to locate references, in the text, to the social institutions listed here.

- **Pleistocene institutions:** ceramics, confederation, community, fishing, language (speech), marriage, migration, religion, ritual, workshop
- **Holocene institutions:** agriculture, animal husbandry, capitalistic institutions [enterprise, factory, proprietors' associations, transnational proprietors' association or multilateralism], ceramics, cities, commercial institutions [banking, money, ports], cultural institutions [literacy, republic of letters, schooling, workshops], fishing, governance institution [empire, monarchy, republic, state], judiciary, metallurgy, military force, nation, religion and large-scale religion, ritual initiation, slavery, society, town, water supply
- **Anthropocene institutions:** agriculture, capitalistic institutions [banking, corporation, factory, proprietors' association, transnational proprietors' association or multilateralism, trade unions], cities, commercial institutions [money, ports], cultural institutions [literacy, schooling, knowledge institutions, universities], ethnic groups, governance institutions [empire, monarchy, nation, republic, state, international organizations], religion and large-scale religion, slavery, water supply

Networks and Hierarchies Discussed in the Text. Each of the structures listed in this section can be understood as a network (linking people across space to perform a function), a hierarchy (ranking people by scale to perform a function), or both. In some of them, people in the network or hierarchy function as independent individuals (I or I-group); networks and hierarchies based on I-groups are implicit or informal groups. In others, people in the network or hierarchy function in an agreed-upon group (we or we-group); networks and hierarchies relying on we-groups are also classified as institutions. These are examples of networks and hierarchies in each period: more could be explored for each time period. See the Index to locate references, in the text, to the networks and hierarchies listed here.

- **Pleistocene:** community networks (I), confederation of communities (I), Human System (I), marriage network (we), migration (I, we), religious networks (I), ritual networks (I)

- **Holocene:** capitalistic networks (I, we), cities (I, we), commercial networks (I), cultural networks (I), empire (I, we), Human System (I), military force (I, we), monarchy (I, we), nation (we), religious networks (I), large-scale religion (we), republic of letters (I), ritual initiation (we), schooling (I, we), slavery (I, we), society (we), state (I, we), town (I, we), water supply (we)
- **Anthropocene:** capitalistic networks [banks, corporations, trade unions], cities (I, we), ethnic groups (we), cultural networks [literacy (I, we), networks of knowledge (I, we), schooling (I, we), universities (I, we)], Human System (I, we), governance networks [empire (I, we), international organization (I, we), nation (I, we)], large-scale religion (we), slavery (I, we), water supply (I, we)

***References*:** Axelrod, *Evolution of Cooperation*; Boyd and Richerson, *Origin and Evolution of Cultures*; McLean, *Culture in Networks*; Manning, *Methods*; Tuomela, *Social Ontology*; Udehn, *Methodological Individualism*; Williams, *Adaptation and Natural Selection.*

6. Emotion and Human Nature

Emotions in Individuals. Emotions are functional mental states in individuals that are typically caused by sensory inputs, that typically cause behavioral outputs, and that also cause changes, and can be caused by, other mental states like perceptions, memories, attention, and so forth. Major topics within emotions: common lists of six basic emotions, such as happiness, sadness, fear, anger, surprise and disgust. Yet it is commonly unclear whether they are defined at the phenotypical or genotypical levels (Adolphs and Anderson). What are the processes linking genetic emotions to their expression, and to the subjective recognition of the emotions, and to reports on them? At the phenotypical level, there are numerous studies of individual emotions or behavior, especially violence and cooperation. These center on efforts to determine whether the behavior is inherent and unchanging, or whether it changes with social circumstances. Theorists of genetic emotions assume an emotional state in the context of stimuli, volition.

Emotions in Groups. Emotional expression is conditioned by individual characteristics, as these are expressed through institutional dynamics, ideology, and configurations of social fitness. An institution may therefore develop and express group emotions (Barrett).

Human Nature: Biological Aspects. The term "human nature" is widely and variously used. While it is included in the title of many books and articles, only rarely does it seem to be a specific object of research (Degler). In the work of Lopreato, "human nature" refers to patterns of phenotypical or directly observable human characteristics and human behavior (Lopreato). Human nature, while arguably observable in its expression, is widely understood to be based on underlying biological makeup. Researchers into human nature assume it is a multiscale phenomenon, ranging from the genetic level to the level of behavior and expression, but with no consensus on the details of the scales and the mechanisms.

Human Nature: Social Aspects. My hypothesis: Biological emotions exist (with their typology to be set by the Adolphs–Anderson process); modification of emotional capabilities with increased brain capacity; cultural evolution develops attitudes as well as learning; behavioral predispositions arise at the phenotypical level; expression of emotions as described by Barrett. Social conditions – especially the functioning of institutions – encourage certain behavior (Pinker; Wilson). As a result, a zeitgeist may result as a common expression of outlook and ideology linked to certain institutions and to behavior and emotions that are central to those institutions, which have been reinforced within the institutions by interplay of individual outlook and the patterns of the institution. Multiple *zeitgeisten* might exist, sustained by different sets of institutions. Human nature becomes more group based as institutions expand.

References: Adolphs and Anderson, *Neuroscience*; Barrett, *How Emotions Are Made*; Degler, *In Search of Human Nature*; Lopreato, *Human Nature*; Pinker, *Blank Slate*; Wilson, *Human Nature*.

Notes

THE HUMAN SYSTEM

1. James Lovelock, *Gaia, A New Look at Life on Earth* (Oxford: Oxford University Press, 1979). Lynn Margulis contributed substantially to development of the theory of Gaia.
2. For a survey of historical methods at the level of the Human System, see Patrick Manning, *Methods for Human History: Studying Social, Cultural, and Biological Evolution* (New York: Palgrave Macmillan, 2020).
3. For a thoughtful argument that the biggest problem in human affairs comes from the earliest days, see James Burke and Robert Ornstein, *The Axemaker's Gift: A Double-Edged History of Human Culture* (New York: G. P. Putnam's Sons).
4. Theodosius Dobzhansky, *Genetics and the Origin of Species* (New York: Columbia University Press 1937), 126–27.
5. Fields making active contributions include population genetics, paleontology, sociology, cultural evolution, cultural anthropology, evolutionary linguistics, psychology, history, and philosophy.
6. Charles Darwin, *On the Origin of Species by Means of Natural Selection* (London: John Murray, 1859), 145.
7. Stephen Jay Gould, *Bully for Brontosaurus* (New York: W. W. Norton, 1991), 63–65, as cited in Alex Mesoudi, *Cultural Evolution: How Darwinian Theory Can Explain Human Culture and Synthesize the Social Sciences* (Chicago: University of Chicago Press, 2011), 99–100. Mesoudi notes that genetic material is transmitted across species boundaries in bacteria and plants by viruses.
8. Herbert Spencer, "Progress: Its Law and Cause," *Humboldt Library of Popular Science Literature*, No. 17 (New York: J. Fitzgerald, 1881), 236.
9. Edward B. Tylor, *Primitive Culture*, 2 vols. (London: J. Murray, 1871), I:1. It appeared the same year as Darwin's *Descent of Man*, but Tylor gave no attention to *Descent of Man* either then or later.
10. In later chapters, I show that the meaning of "culture" can change at individual and group levels.
11. Tylor, *Primitive Culture,* 2–10. Spencer, in his essay on progress, treated "culture" in a fashion anticipating Tylor, referring to groups within modern societies. Spencer, "Progress."
12. Spencer's approach was at once universal and vague (though some argue that Spencer was really focused on human society rather than on other scales). Tylor and other social scientists treated society as quite separate from biology

but agreed with Spencer in assuming unspecified progress rather than a specific process of change.

13. Karl Marx's theory of social change through class conflict was closer to specifying a mechanism of change, although theorists in the Marxian tradition, in practice, often treated changes among stages as inherent. Economic theory, arising at much the same time, focused its attention on individual consumers and firms.
14. Donald T. Campbell, "On the Conflicts between Biological and Social Evolution and between Psychology and Moral Tradition" (presidential address, American Psychological Association), *American Psychologist* (December 1975): 1103–26.
15. The Appendix: Frameworks for Analysis provides a review of my analytical assumptions, methods within six categories, and references to methodological texts.
16. For a recent survey of strengths and concerns in studies of world history, see Patrick Manning, "Locating Africans on the World Stage: A Problem in World History," *Journal of World History* 26 (2015): 605–37. For an earlier overview of method and interpretation, see Patrick Manning, *Navigating World History: Historians Create a Global Past* (New York: Palgrave Macmillan, 2003).
17. Jared Diamond, *Guns, Germs, and Steel: The Fates of Human Societies* (New York: Norton, 1997); John R. McNeill and William H. McNeill, *The Human Web: A Bird's-Eye View of Human History* (New York: W. W. Norton, 2003); David Christian, *Maps of Time: An Introduction to Big History* (Berkeley: University of California Press, 2004); Fred Spier, *Big History and the Future of Humanity*, 2nd ed. (Malden, MA: John Wiley & Sons, 2015); Andrew Shryock and Daniel Lord Smail, *Deep History: The Architecture of Past and Present* (Berkeley: University of California Press, 2011); Christopher Chase-Dunn and Bruce Lerro, *Social Change: Globalization from the Stone Age to the Present* (Boulder, CO: Paradigm Publishers, 2014); and Yuval Noah Harari, *Sapiens: A Brief History of Humankind* (New York: Harper, 2015).
18. Christian defined time periods in terms of human technology (Paleolithic and Neolithic) while I have defined time periods in terms of geological epochs (Pleistocene and Holocene). McNeill and McNeill, *Human Web*; Christian, *Maps of Time*, 292.
19. For the key work in this approach, see George C. Williams, *Adaptation and Natural Selection: A Critique of Some Current Evolutionary Thought* (Princeton, NJ: Princeton University Press, 1966).
20. This approach to cultural evolution of "dual inheritance" and "multilevel selection," while it has yet to advance specific mechanisms for the creation of large-scale institutions, shows no interest in exploring a distinctive mechanism of social evolution. For the most forceful statements of this position, see Alex Mesoudi, *Cultural Evolution*, 25–54, and Robert Boyd and Peter J. Richerson, "Culture and the Evolution of Human Cooperation," *Philosophical Transactions of the Royal Society* 364 (2009): 3281–88.
21. For an individual-level definition of culture in the context of cultural evolution, see Robert Boyd and Peter J. Richerson, *Culture and the Evolutionary Process* (Chicago: University of Chicago Press, 1985), 1–2; for a group-level approach to culture in an influential theory of social evolution, see Campbell,

"On the Conflicts between Biological and Social Evolution"; for a group-level approach to culture from the standpoint of humanities disciplines, see Felipe Fernández-Armesto, *A Foot in the River: Why Our Lives Change – and the Limits of Evolution* (Oxford: Oxford University Press, 2015),148.

22. The analysis addresses individual human behavior (Chapters 2, 6, and 9); group behavior of humans (Chapters 3 through 10); participation in networks (Chapters 3, 6, 7, 8, and 10); and participation in hierarchies (Chapters 6, 7, 8, and 9).

BIOLOGICAL AND CULTURAL EVOLUTION

1. Douglas Johnson and David Anderson (eds.), *The Ecology of Survival: Case Studies from Northeast African History* (London: Lester Crook, 1988).
2. On climate, John L. Brooke, *Climate Change and the Course of Global History: A Rough Journey* (New York: Cambridge University Press, 2014); on paleontology and epigenetics, Ian Tattersall, *Masters of the Planet: The Search for Our Human Origins* (New York: Palgrave Macmillan, 2012); on epigenetics, Gary Felsenfeld, "A Brief History of Epigenetics," *Cold Spring Harbor Perspectives in Biology* (2014) DOI: 10.1101/cshperspect.a018200; and on genomics, David Reich, *Who We Are and How We Got Here: Ancient DNA and the New Science of the Human Past* (New York: Pantheon, 2018).
3. On cultural evolution and multilevel selection, Robert Boyd and Peter J. Richerson, *The Origin and Evolution of Cultures* (New York: Oxford University Press, 2005); and Okasha, *Evolution and the Levels of Selection.* On the size of hominin groups, Leslie C. Aiello and R. I. M. Dunbar, "Neocortex Size, Group Size, and the Evolution of Language," *Current Anthropology* 34 (1993): 184–93. On visual communication, Michael Tomasello, *A Natural History of Human Thinking* (Cambridge, MA: Harvard University Press, 2014). On reasoning underlying language capability, Derek Bickerton, *Language and Species* (Chicago: University of Chicago Press, 1981); W. Tecumseh Fitch, *The Evolution of Language* (Cambridge: Cambridge University Press, 2010), 297–385; and Robert Berwick and Noam Chomsky, *Why Only Us: Language and Evolution* (Cambridge, MA: MIT Press, 2016). On emotions and human nature, Joseph Lopreato, *Human Nature and Biocultural Evolution* (Boston: Allen & Unwin, 1984) and Ralph Adolphs and David J. Anderson, *The Neuroscience of Emotion: A New Synthesis* (Princeton, NJ: Princeton University Press, 2018).
4. Recent work argues that bipedalism began not in open savanna but in a Late Miocene wooded environment in which individuals alternated between walking and climbing trees. Brigitte Senut, M. Pickford, D. Gommery, and L. Ségan, "Palaeoenvironments and the Origin of Hominid Bipedalism," *Historical Biology* 30 (2018): 284–96.
5. Chris Stringer and Peter Andrews, *The Complete World of Human Evolution* (London: Thames & Hudson, 2005).
6. *Australopithecus* was followed by the genera *Paranthropus* and perhaps *Kenyanthropus*; the former persisted until 1 million years ago.

7. On the debates over breadth of classification, see Tattersall, *Masters of the Planet*, 82–89; and Jeffrey H. Schwartz, "What's Real about Human Evolution: Received Wisdom, Assumptions, and Scenarios," in Schwartz (ed.), *Rethinking Human Evolution* (Cambridge, MA: MIT Press, 2018), 61–91.
8. Cann, Stoneking, and Wilson, "Mitochondrial DNA and Human Evolution," *Nature* 325 (1987): 31–36; Reich, *Who We Are and How We Got Here*. Also on biological evolution, see Manning, *Methods*.
9. It is worth emphasizing that biological evolution continues for all species, right up to the present. See Edmund Russell, *Evolutionary History: Uniting History and Biology to Understand Life on Earth* (New York: Cambridge University Press, 2011).
10. Patrick Manning and Aubrey Hillman. "Climate as a Factor in Migration and Social Change, 200,000 to 5000 years ago," in American Historical Association annual meeting, New Orleans, January 5, 2013.
11. The individuals known as African *Homo erectus* are also known as *Homo ergaster*. (Tattersall, *Masters of the Planet*, 131).
12. Head louse survives from complete body louse; pubic louse arrived after loss of body hair. The two lice parted ways 3 to 4 million years ago. See David L. Reed, Jessica E. Ligh, Julie M. Allen, and Jeremy J. Kirchman, "Pair of Lice Lost or Paradise Regained: The Evolutionary History of Anthropoid Primate Lice," *BMC Biology* 5 (2007), DOI: 10.1186/1741-7007-5-7; D. M. Sandgathe, H. L. Dibble, P. Goldberg et al., "Timing of the Appearance of Habitual Fire Use," *PNAS* (2011), DOI: 10.173/pnas.1106759108.
13. Tattersall, classifying this individual as *Homo ergaster* rather than African *Homo erectus*, argues that "it seems clear that the radically new *Homo ergaster* must have originated in some behaviorally sophisticated australopith isolate. And the chances are that, far from being propelled by circumstances, this new body build just happened to open up hugely advantageous new possibilities for its possessors in a new and expanding environmental setting" (Tattersall, *Masters of the Planet*, 117). On epigenetics, see Manning, *Methods*.
14. Tattersall, *Masters of the Planet*, 98.
15. Ibid., 115.
16. Compare the two sections of Figure 2.1.
17. The vast majority of *Homo heidelbergensis* sites have been found since 1996. A related site is the fossils from Gran Dolina in Spain, which date to 800,000 years old, and may be *Homo heidelbergensis* or a different species, *Homo antecessor*. In addition, a species labeled *Homo naledi* has recently been discovered in South Africa.
18. On their possible use of fire, see Joop Goudsblom, *Fire and Civilization* (London: Allen Lane, 1992) and James C. Scott, *Against the Grain: A Deep History of the Earliest States* (New Haven, CT: Yale University Press, 2017), 3–5, 37–42.
19. Tattersall, *Masters of the Planet*, 157.
20. Chris Stringer, *In Search of the Neanderthals: Solving the Puzzle of Human Origins* (London: Thames & Hudson, 1993).
21. Dates for Denisovan occupants of the cave are estimated to range from 195,000 years ago to 52,000 years ago. Katerina Douka et al., "Age

Estimates for Hominin Fossils and the Onset of the Upper Paleolithic at Denisova Cave," *Nature* 565 (2019): 640–44.

22. J. Jeffrey Werner and Pamela R. Willoughby, "Middle Stone Age Technology and Cultural Evolution at Magubike Rockshelter, Southern Tanzania," *African Archaeological Review* 34 (2017): 249–73.
23. Chris Stringer, "The Origin and Evolution of *Homo sapiens*," *Philosophical Transactions of the Royal Society B* (2016), https://doi.org/10.1098/rstb.2015.0237.
24. Tattersall, *Masters of the Planet*, 185.
25. Ibid., 207–08.
26. The *Homo floriensis* remains, discovered in 2003, are now argued to be most closely related to *Homo habilis*, based on similarity in stature and brain capacity. Further research in West, Central, and northwestern Africa may reveal additional groups that descended from *Homo heidelbergensis*.
27. Boyd and Richerson, *Culture and the Evolutionary Process*, 1–2. For a clear summary of many analytical advances in cultural evolution, see Richard McElreath and Robert Boyd, *Mathematical Models of Social Evolution: A Guide for the Perplexed* (Chicago: University of Chicago, 2007).
28. Boyd and Richerson, *Culture and the Evolutionary Process*, 1–2. To provide broader context, I use the term "individual-level culture" in the context of cultural evolution, where it means individual-level learning and transmission of knowledge. The general term "culture" applies most commonly to creative processes that go far beyond transmission of behavior. To emphasize the distinction, I use the term "group-level culture" to refer to group-level cultural representation, which involves creation, reception, exchange, and interpretation of knowledge and materials.
29. Albert Bandura and Richard H. Walters, *Social Learning and Personality Development* (New York: Holt, Rinehart and Winston, 1963); Albert Bandura, *Social Learning Theory* (Morristown, NJ: General Learning Press, 1971); Bandura, "Self-efficacy: Toward a Unifying Theory of Behavioral Change," *Psychological Review* 84 (1977): 191–215.
30. Such a study is discussed in Chapter 9. See Natalie Henrich and Joseph Henrich. *Why Humans Cooperate: A Cultural and Evolutionary Explanation* (Oxford: Oxford University Press, 2007).
31. Boyd and Richerson, *Culture and the Evolutionary Process*, 172–202. On cultural evolution, see Manning, *Methods*.
32. Open to continuing discussion is whether the effectiveness of changes in cultural evolution is determined by a type of "cultural fitness" or whether it is determined by the genetic fitness of the biological traits that coevolve in association with cultural evolution.
33. On inclusive fitness, see W. D. Hamilton, "The Genetical Evolution of Social Behaviour," *Journal of Theoretical Biology* 7 (1964): 1–52; McElreath and Boyd, *Mathematical Models*, 78–82.
34. On the Price Equation, see John Maynard Smith and George R. Price, "The Logic of Animal Conflict," *Nature* 246, 5427 (1973): 15–18; McElreath and Boyd, *Mathematical Models*, 228–32. For a focus on the work of Price, see Oren Harman, "On the Importance of the Parvenu: The Amazing Case of

George Price in Evolutionary Biology," in Oren Harman and Michael R. Dietrich, eds., *Outsider Scientists: Routes to Innovation in Biology* (Chicago: University of Chicago Press, 2013), 312–30.

35. On multilevel selection, McElreath and Boyd emphasize that, "the effect of selection can always be decomposed into effects at different levels and can always be conceptualized as acting only at the individual level – the so-called *personal fitness* approach. Moreover, these approaches are completely equivalent." McElreath and Boyd, *Mathematical Models*, 224–28. See also the appendix on Frameworks for Analysis, part 2. See also Manning, *Methods*.
36. Joseph Henrich has displayed a table showing a succession of parent-to-child transmissions of various traits, assuming a fairly high rate of successful transmission, and argues that this process leads at a certain point to a self-reinforcing expansion of new behavior. Joseph Henrich, *The Secret of our Success: How Culture Is Driving Human Evolution, Domesticating Our Species, and Making Us Smarter* (Princeton, NJ: Princeton University Press, 2016), 58–61; Michael Tomasello, Ann Cale. Kruger, and Hilary Horn Ratner, "Cultural Learning," *Behavioral and Brain Sciences*, 16, 3 (1993): 495–511. DOI:10.1017/S0140525X0003123X.
37. The process in human innovation, however, involves nothing as focused on human agency as a decision to cross the river. Henrich, *Secret of Our Success*, 57, 285.
38. Henrich, *Secret of Our Success*, 292–94.
39. Francisco Jose Ayala and Theodosius Dobzhansky (eds.), *Studies in the Philosophy of Biology: Reduction and Related Problems* (London: Macmillan, 1974); Novartis Foundation Symposium, *The Limits of Reductionism in Biology* (New York: John Wiley & Sons, 1998); Michel Morange, *A History of Molecular Biology*, trans. Matthew Cobb (Cambridge, MA: Harvard University Press, 1998), 5, 243–46. Also on reductionism, see Manning, *Methods*.
40. As Tattersall argues, "But it seems overwhelmingly likely that – like all of our other unique attributes of structure – our new cognitive ability was acquired as a byproduct of the hugely ramifying genetic accident that resulted in the appearance of *Homo sapiens* as a distinctive entity … The addition of the neural ingredient that predisposed our species for symbolic thought was simply one passive consequence of the developmental reorganization that gave rise to anatomically recognizable *Homo sapiens* some 200 thousand years ago." The change begins not as an adaptation to a lifestyle "but as ***ex***aptions: features that must necessarily be co-opted *post hoc* into a new use." Tattersall, *Masters of the Planet*, 209–10.
41. In fact, I have been a bit more explicit about this early stage than were Berwick and Chomsky; I find the model to be promising both for language and, more broadly, for conceptualizing early stages in cultural evolution. Berwick and Chomsky, *Why Only Us*, 111–12. See also Bickerton, *Language and Species*, 130–68; and Fitch, *Evolution of Language*, 87–118. Also on evolutionary linguistics, see Manning, *Methods*.
42. Berwick and Chomsky, *Why Only Us*, 10. For more on Chomsky's analysis, see W. Tecumseh Fitch, "Noam Chomsky and the Biology of Language," in Oren Harman and Michael R. Dietrich, *Outsider Scientists: Routes to Innovation*

in Biology (Chicago: University of Chicago Press, 2013), 201–22; and Noam Chomsky, *The Minimalist Program* (Cambridge, MA: MIT Press, 1995).
43. Berwick and Chomsky, *Why Only Us*, 149.
44. Tattersall gives general support to this strategy: "So perhaps we shouldn't be looking for one single major 'keystone' acquisition. Instead, the extraordinary properties of the human brain are likely emergent, resulting from a relatively tiny – and altogether accidental – addition or modification to a complex structure that was already, and exaptively, almost prepared for symbolic thought." Tattersall, *Masters of the Planet*, 223–24.
45. The term "social learning" has gradually expanded in scope, as it has come to be understood that animals as well as humans learn through both imitation and instruction. Tomasello and his colleagues sought to introduce the term "cultural learning" to indicate a form of learning at a higher level, perhaps restricted to humans, in which learning focuses on mutual understanding. Tomasello et al. "Cultural learning."
46. Michael Tomasello, "Human Culture in Evolutionary Perspective," in M. Gelfand (ed.), *Advances in Culture and Psychology* (Oxford: Oxford University Press, 2011), 5–51; Tomasello, *Becoming Human: A Theory of Ontogeny* (Cambridge, MA: The Belknap Press of Harvard University Press, 2019).
47. Aiello and Dunbar, "Neocortex Size."
48. Dunbar and Sosis, examining contemporary and recent human groups, found that groups centered on sizes of 15, 50, 150, and some larger groups. They reaffirm the importance of groups of 150 but note that humans also have characteristic groups of different sizes. Robin I. M. Dunbar and Richard Sosis, "Optimising Human Community Sizes," *Evolution and Human Behavior* 39 (2018): 106–11.
49. The 1973 Nobel Prize in Physiology or Medicine was awarded to three founding figures in ethology, the study of animal behavior in biological content. Richard W. Burkhardt, Jr., *Patterns of Behavior: Konrad Lorenz, Niko Tinbergen, and the Founding of Ethology* (Chicago: University of Chicago Press, 2005); Irenäus Eibl-Eibesfeldt, *Human Ethology* (New York: Aldine de Gruyter, 1989).
50. Adolphs and Anderson, *Neuroscience*, 40–41. For a broader review of ethology, see Eibl-Eibesfeldt, *Human Ethology*; see also Jonathan H. Turner, *Human Emotions: a sociological theory* (London: Routledge, 2007). Also on emotions, see Manning, *Methods*.
51. The list includes sociability, valence, persistence, generalization, global coordination, automaticity, and social communication. Adolphs and Anderson, *Neuroscience*, 65.
52. For instance, Barrett argues that words create categorizations in the minds of children, generating interactions that lead to definition of emotions and types of behavior in specific social context. Barrett too explores the emotions of animals but leaves her analysis outside the formal study of ethology. Lisa Feldman Barrett, *How Emotions are Made: The Secret Life of the Brain* (Boston: Houghton Mifflin Harcourt, 2018), 169–70.

53. In an insightful speculation, Felipe Fernández-Armesto has suggested that imagination, meaning the ability to anticipate the next step of prey in flight from hunters, benefited from selection in humans and served as a basis for what later became group culture. Fernández-Armesto, *A Foot in the River*, 148.
54. Tattersall, *Masters of the Planet*, 188–93.
55. Sally McBrearty and Alison Brooks, "The Revolution that Wasn't: A New Interpretation of the Origin of Modern Human Behavior," *Journal of Human Evolution* 39 (2000): 453–563.
56. Stephen Shennan argues that stone tool technology can be learned by observation, not requiring language. See Stephen Shennan, *Genes, Memes and Human History: Darwinian Archaeology and Cultural Evolution* (New York: Thames and Hudson, 2003), 37–42; see also Robert L. Kelly, *The Fifth Beginning: What Six Million Years of Human History Can Tell Us about Our Future* (Berkeley: University of California Press, 2016).
57. Aiello and Dunbar. "Neocortex Size."
58. Henrich, *Secret of Our Success*, 295.

SPEECH AND SOCIAL EVOLUTION

1. McBrearty and Brooks, "The Revolution that Wasn't." For a follow-up, providing even earlier examples of tool development, see the commentary on work of McBrearty and Brooks by Patrick Randolph-Quinney and Anthony Sinclair, "African Tools Push Back the Origins of Human Technological Innovation," *The Conversation* (March 16, 2018). See also Sally McBrearty, "Down with the Revolution," in Paul Mellars, Katie Boyle, Ofer Bar-Yosef, and Chris Stringer (eds.), *Rethinking the Human Revolution* (Oxford: Oxbow Books, 2007), 133–51.
2. Richard Potts et al. "Environmental Dynamics during the Onset of the Middle Stone Age in Eastern Africa." *Science* 360 (2018): 86–90; Alison S. Brooks et al. "Long-Distance Stone Transport and Pigment Use in the Earliest Middle Stone Age." *Science* (March 15, 2018).
3. Cara Roure Johnson and Sally McBrearty, "500,000 Year Old Blades from the Kapthurin Formation, Kenya," *Journal of Human Evolution* 58 (2010): 193.
4. Founders: this term is to distinguish the initial community of speaking humans from the larger population of *Homo sapiens* of which they formed a part. In addition, the term fits with the use of the term "founders" by Ernst Mayr in his explanation of how small populations of distinctive groups might grow rapidly. William B. Provine, "Ernst Mayr: Genetics and Speciation," *Genetics* 167, 3 (2004): 1041–46.
5. Understandably, terms on language overlap and conflict. Within the confines of this book, I use terms as follows: (1) "language" refers to any and all forms of language – spoken or internal, syntactical or not; (2) "i-language" refers to a logical but unspoken language; (3) "visual communication" refers to communication by gestures; (4) "pre-language" refers to spoken but non-syntactic language; (5) "proto-human" refers to the presumed original, syntactic language; (6) "speech," "syntactic speech," and "e-language" refer to syntactic language. In other usage, Bickerton's "protolanguage" refers to nonsyntactic speech; Berwick

and Chomsky refer to "the mental organ language" or "i-language" to identify its internal use and its reliance on universal grammar, as contrasted with "e-language." In historical linguistics, "proto-human" refers to the presumed ancestral, syntactic language; similarly, proto-Khoesan refers to an ancestral and perhaps reconstructed language. For a broad overview, see Fitch, *Evolution of Language*, 399–432; see also Berwick and Chomsky, *Why Only Us*, 57–65.

6. Philip Lieberman, "The Evolution of Human Speech: Its Anatomical and Neural Bases," *Current Anthropology* 48, 1 (2007), 39–66.
7. Potts et al., "Environmental dynamics."
8. For instance, as Richerson and Boyd said, "paleoanthropologists have no idea when human language evolved." They consider various possibilities, but conclude, "People might have been mute until comparatively recent times." Richerson and Boyd, *Not By Genes Alone*, 144.
9. One could imagine that there was more than one community of Founders, but it appears that only one such language tradition survived. Ehret, personal communication. For his list of world deictics, including widespread use of "*na" for "this," see www.sscnet.ucla.edu/history/ehret/World_deictics.pdf. In earlier work, Merritt Ruhlen sought to identify elements of the proto-human language, proposing instances beginning with *aq'wa* for "water" and *tik* for "finger" or "one." Merritt Ruhlen, *The Origin of Language: Tracing the Evolution of the Mother Tongue* (New York: John Wiley & Sons, 1994), 107–18.
10. Atkinson's analysis of variety in language phonemes confirms that languages began in Africa and suggests that languages began in southwestern Africa. I argue that he may rely on a misunderstanding of the geographical homeland of the Khoesan languages which, according to Ehret, began in modern Tanzania. Tattersall notes the possibility that a Southern African drought may have cut short the promising developments found at Blombos Cave. Quentin D. Atkinson, "Phonemic Diversity Supports a Serial Founder Effect Model of Language Expansion from Africa," *Science* 332, 6027 (2011): 347; Christopher Ehret. "Early Humans: Tools, Language, and Culture," in David Christian (ed.), *The Cambridge World History*. Vol. 1. *Introducing World History, to 10,000 BCE* (Cambridge: Cambridge University Press, 2015), 346–47; Tattersall, *Masters of the Planet*, 201.
11. Noam Chomsky, *Syntactic Structures* ('s-Gravenhage: Mouton, 1957).
12. Zaccarella and Friedrici claim identification of the Merge function; Reich argues that single mutations are very unlikely sources of phenotypical change. Emiliano Zaccarella and Angela D. Friederici, "Merge in the Human Brain: A Sub-Region Based on Functional Investigation in the Left Pars Opercularis," *Frontiers in Psychology* (November 27, 2015), https://doi.org/10.3389/fpsyg.2015.01818; Reich, *Who We Are*.
13. Lieberman, "The Evolution of Human Speech."
14. Tomasello, *Origins of Human Communication*, 169–241. Similarly, Bickerton has emphasized the possibility that a verbal pre-language or protolanguage had developed previously, relying on a lexicon without syntax. Bickerton, *Language and Species*, 145–69; Fitch, *Evolution of Language*, 401–14.
15. Berwick and Chomsky, *Why Only Us*, 64, 78–82.

16. Tattersall also makes brief reference to the possibility that children may have been the initiators of speech. "Indeed, I am greatly entertained by the notion that the first language was the invention of children, who are typically much more receptive to new ideas than adults are." Tattersall, *Masters of the Planet*, 220. Fitch describes the case of dispersed Nicaraguan deaf children who, when gathered together in a group reaching 400, were rapidly able to develop their own sign language. The example emphasizes the importance of an institution to hold speakers together and shows that a syntactic language can develop rapidly under that condition. Fitch, *Evolution of Language*, 379–80, 419.
17. Robin Dunbar, *Grooming, Gossip, and the Evolution of Language* (London: Faber and Faber, 1996), 148–51.
18. A fuller definition of "institution" appears in a subsequent section of this chapter. Searle and then Tuomela anticipated my characterization of language as the first institution. More precisely, Tuomela labeled language as "the most fundamental, underlying social institution." John R. Searle, *The Construction of Social Reality* (New York: The Free Press, 1995); Tuomela, *Philosophy of Social Practices: A Collective Acceptance View* (Cambridge: Cambridge University Press, 2002), 159. Earlier contributions by John Rawls, John R. Searle, Joseph Ransdell, and Anthony Giddens launched the study of constitutive rules, including for speech: see Frank Hindriks, "Constitutive Rule, Language, and Ontology," *Erkenntnis* 71 (2009): 253–57, 260–61, 267–69.
19. Nevertheless, UNESCO documents decline to offer an approximate number of speakers necessary to sustain a language. Following the reasoning of Aiello and Dunbar, I assume that a human speaking community required an average of 150 members. "Language Vitality and Endangerment," UNESCO Programme Safeguarding of Endangered Languages," Paris, March 10–12, 2003. www.unesco.org/new/fileadmin/MULTIMEDIA/HQ/CLT/pdf/ Language_vitality_and_endangerment_EN.pdf (accessed July 19, 2018).
20. These basic terms have tended to be retained in speech.
21. In the words of Saussure: "Language is not complete in any speaker; it exists only within a collectivity ... only by virtue of a sort of contract signed by members of a community." Ferdinand de Saussure, trans. Wade Baskin, *Course in General Linguistics* (New York: McGraw Hill, 1966), 14.
22. On common human community sizes, Dunbar and Sosis have analyzed small communities among Hutterite and other American utopian communities of the nineteenth century and Israeli kibbutzim of the twentieth century. They conclude that the results confirm previous common group sizes of 15, 50, 150, 500, and 1000. Dunbar and Solis argue that these results for recent agricultural societies should be valid for earlier hunter-gatherer communities. Dunbar and Sosis, "Optimising Human Community Sizes." Also on group size, see Manning, *Methods*.
23. Tuomela makes this distinction and uses a team approach to game theory to demonstrate that the behavior of a we-group cannot be reduced or simplified into that of an I-group. Raimo Tuomela, *Social Ontology: Collective Intentionality and Group Agents* (Oxford: Oxford University Press, 2013), 183–84, 194–95.
24. Bickerton has treated pidgins and creoles as stages along the way to the initial creation of syntactic speech. Nevertheless, he should have been clearer about

the difference between a pidgin as an initial and untutored assemblage of words and the subsequent and common case in which a pidgin is simplified cross-community communication among people who are fully fluent in their native, syntactic language. Bickerton, *Language and Species*, 118–26; Fitch, *Evolution of Language* I, 406.

25. Talcott Parsons noted that the term "institution" is commonly used in two ways. He contrasted "organizations and other collectivities" with "property, contract, and authority" as "complexes of normative rules and principles ... to regulate social action and relationships" – Parsons' analysis was strictly within the latter definition. If the latter, normative definition is relevant for contemporary sociology, the former definition and its focus on membership is more appropriate for study of social evolution, especially its early stages. The "new institutionalism" of Douglass C. North also treats institutions as norms: "Institutions are the humanly devised constraints that structure human interaction." According to sociologist Jonathan Turner, "Institutional analysis is, therefore, inherently evolutionary because it explores how humans create population-wide structures and cultural systems that enable them to survive in the environment, often an environment of their own making." Yet his analysis of institutional evolution, while highly detailed, is principally snapshots of institutional norms in successive time periods. In contrast, according to Raimo Tuomela, "Social institutions typically have as their general goal or at least function to create order in society by solving coordination problems and collective action problems between individual and collective rationality." Talcott Parsons, "Social Structure and the Symbolic Media of Interchange," in Peter M. Blau, *Approaches to the Study of Social Structure* (New York: The Free Press, 1975), 97; Douglass C. North, "Economic Performance through Time," Nobel Prize Lecture, 1993; Jonathan H. Turner, *Human institutions: A Theory of Societal Evolution* (Lanham, MD: Rowman & Littlefield, 2003), 5; Tuomela, *Social Ontology*, 215–15; see also Joseph Henrich, "Cooperation, Punishment, and the Evolution of Human Institutions," *Science* 312 (2006): 60–61; and Manning, *Methods*.
26. The analysis of institutions as groups does also address institutional norms but does so through endogenous analysis of institutional dynamics rather than treat institutions as exogenous factors for which norms are assumed. For clarification of this point in the work of North, see Lars Udehn, *Methodological Individualism: Background, History, and Meaning* (London: Routledge, 2001).
27. For a restatement of these terms and their use, see the appendix on Frameworks for Analysis, part 4.
28. Nevertheless, in a we-group, the implementation of decisions is carried out by the individual action of members. Tuomela, *Social Ontology*, 46–51.
29. In a different way, the discipline of cultural evolution also focuses on changes in groups and variation among groups. In that framework, the individual process of cultural evolution gradually builds cooperative mentality among individuals so that, over many generations, they form ethnic groups that could be called I-groups. Collaboration and commonality are emphasized, so that each group may hope to displace competing groups. (Migration, by introducing differences to a group, is seen in cultural evolutionary perspective as weakening group solidarity.) Social evolution, in contrast, focuses on

agreements, creation and reform of institutions within communities. It can take place more rapidly and yields variation within linguistic groups as a basis for generating innovations that can be adopted by the group – and can be exchanged with others through further migration.

30. Ernst Mayr, *Evolution and the Diversity of Life* (Cambridge, MA: Harvard University Press, 1975), 26–29; Ronald A. Fisher, *The Genetical Theory of Natural Selection* (Oxford: Clarendon Press, 1930). The full set of these aspects became the neo-Darwinian synthesis; the rise of epigenetics in the 1970s added another dimension to biological theory.
31. "Population thinking" may be defined as analysis performed at a single level on a unit including many analytically identical elements, tracing their patterns of interaction. Ludwig Boltzmann developed this thinking especially in the thermodynamics of gases; Léon Walras applied it in his general-equilibrium economic analysis; and Alfred Lotka's work founded population theory in demography. Most importantly here, Ronald Fisher established the field of population genetics. Population thinking became central for biological evolution and, later, for cultural evolution. But for biological evolution, an additional layer was added to analysis with the theorization of epigenetics and its study of life-course development by members of a population. For social evolution, population thinking is very important but it is supplemented by analysis of the logic of collective intentionality, allowing for differences among individuals so that members of populations form we-groups to carry out selected functions.
32. In the biological equivalent, DNA is the archive; it reproduces itself chemically, and it reproduces the organism in a process beginning with RNA and creation of proteins.
33. In a well-known example, the sickle cell trait in blood cells, while it may bring anemia, also protects against malaria. Natural selection works at micro, macro, and environmental levels to achieve this result.
34. The specific nature of feedback to the cultural archive, both conscious and unconscious, requires further theorization and analysis. The issue surely involves a balance of nature and nurture.
35. Institutions to be discussed in this chapter include language, community, family, rituals, migration, and initiation of youths. In addition, the archive for sustaining knowledge about social institutions may be seen as an institution, as may the record of practices for archiving and replicating other institutions.
36. Research and conceptualization continue to specify examples of archives within this framework. For a collection of studies of institutions in evolutionary perspective, see Kaori Kawai (ed.), *Institutions: The Evolution of Human Sociality*, trans. Minako Sato (Kyoto: Kyoto University Press, 2017 [2013]).
37. In contrast, theorists of individual-level cultural evolution argue that the human brain is at once the source of cultural behavior and the repository of the information that is passed on to the next generation.
38. For Holocene times (see Chapter 6), where we will consider hierarchical institutions, some institutions will have directors with interests distinct from those of other members.

39. R. I. M. Dunbar, "Mind the Bonding Gap: Constraints on the Evolution of Hominin Societies," in Stephen Shennan (ed.), *Pattern and Process in Cultural Evolution* (Berkeley: University of California Press, 2009), 223–34.
40. This hypothesis of the creation of communities out of families, with a 5- to 10-fold increase in unit size, should ultimately be testable. But archaeology is not well placed to conduct tests, as residential groups and material culture presumably did not change right away. Periodic gatherings and ceremonies would have been the main activities of the communities.
41. Patrick Manning, with Tiffany Trimmer, *Migration in World History*, 2nd ed. (London: Routledge, 2012), 6–13.
42. Patrick Manning, "Cross-Community Migration: A Distinctive Human Pattern." *Social Evolution and History* 5, 2 (2006): 24–54.
43. James G. Miller, *Living Systems* (New York: McGraw-Hill, 1978). Miller's work anticipated, in important ways, the synthesis by Maynard Smith and Szathmáry, *Major Transitions in Evolution*.
44. Boyd and Richerson, *Culture and the Evolutionary Process*, 172–203.
45. Miller, *Living Systems*, xxiv–xxvi, 1–4, 30–34.
46. Miller, *Living Systems*, 18, 32, 67, 333.
47. To give an example that arose much later in history, the *market* emerged as an institution that dispersed decisions on exchange of commodities, rather than attempt to centralize them at level of the society.
48. To study networks, systems theorists developed the term "complex systems" to analyze systems that "can be analyzed into many components having relatively many relations among them, so that the behavior of each component depends on the behavior of others." Herbert Simon, "Near Decomposability and Complexity: How a Mind Resides in a Brain," in H. J. Morowitz and J. L. Singer (eds.), *The Mind, the Brain, and Complex Adaptive Systems* (Santa Fe Institute Studies in the Sciences of Complexity, Proceedings Vol. XXII, Reading, MA: Addison Wesley, 1995), 26.
49. On networks, see Paul McLean, *Culture in Networks* (Cambridge: Polity Press, 2017); and this volume's appendix on Frameworks for Analysis, part 5. The related issue of hierarchy, while significant in conceptualization at this early stage, was to gain significance in social relations at later stages.
50. Miller uses the term "nets of communication" rather than "network," but devotes attention to these links in discussing the *internal transducer* and *channel and net* subsystems. Miller, *Living Systems*, 333–36.
51. I have labeled these institutions according to anthropological descriptions of their purposes. Tuomela lists institutions according to the levels of individual and collective intentionality. Tuomela, *Social Ontology*, 6.
52. Aiello and Dunbar, "Neocortex Size."
53. The concept of coevolution was developed by Ehrlich and Raven to refer to interplay in the phenotype of species that were in close contact but the term became generalized thereafter; Stiner and Feeley-Harnik recently reviewed coevolutionary processes involving humans. Paul R. Ehrlich and Peter H. Raven, "Butterflies and Plants: A Study in Coevolution," *Evolution* 18 (1964): 586–608; Mary C. Stiner and Gillian Feeley-Harnik, "Energy and Ecosystems," in Shryock and Smail, *Deep History*, 78–102; see also William

H. Durham, *Coevolution: Genes, Culture, and Human Diversity* (Stanford: Stanford University Press, 1991).

54. There is reason for equivalent concern for drift in social evolution.
55. A firmer focus on coevolution should help to limit the tendency of biologists to explain the maximum range of changes in terms of biological evolution, as well as the tendency of cultural evolution scholars to explain all changes in human behavior through the mechanisms they analyze. Scholars focusing on social evolution, as their theories gain in strength, will need to be alert to parallel tendencies.
56. Adolphs and Anderson, *Neuroscience of Emotion*; Barrett, *How Emotions Are Made*. See also Manning, *Methods*.
57. "Our emotions also play a role in the subordination of our self-interest for the good of the group." Kent Flannery and Joyce Marcus, *The Creation of Inequality: How Our Prehistoric Ancestors Set the Stage for Monarchy, Slavery, and Empire* (Cambridge, MA: Harvard University Press, 2012), 561. Tuomela argues that groups may express "collective emotions." See Tuomela, *Social Ontology*, 6, 260–62; see also the appendix on Frameworks for Analysis, part 6.
58. As I term it, these are I-groups, held together by family loyalties rather than by institutional agreements. Peter Turchin, *Ultra Society: How 10,000 Years of War Made Humans the Greatest Cooperators on Earth* (Chaplin, CT: Beresta Books, 2015), 61; Peter J. Richerson and Robert Boyd, "The Evolution of Subjective Commitment to Groups: A Tribal Instincts Hypothesis," in Randolph M. Neese (ed.), *Evolution and the Capacity for Commitment* (New York: Russell Sage Foundation, 2001), 186–220.
59. As I term it, these are we-groups, held together by agreement more than by inherent similarity.
60. On the continuing biological evolution of humans and the influence of humans on biological evolution of other species, see Russell, *Evolutionary History*, 6–30, 85–102.

SYSTEMIC EXPANSION

1. McBrearty and Brooks, "The Revolution that Wasn't."
2. For a confirmation of this point from a genomic standpoint, see Reich, *Who We Are*, 207–08.
3. It would be valuable if a comprehensive review could be assembled – perhaps for times before 1000 CE – surveying studies of mitochondrial DNA, Y-chromosome, somatic DNA, and ancient whole genomes.
4. The date at which articulated language began is not known. But genetic and archaeological data confirm substantial out-migration from Northeast Africa to elsewhere in Africa and Asia by 65,000 years ago.
5. S. H. Ambrose, "Late Pleistocene Human Population Bottlenecks, Volcanic Winter, and Differentiation of Modern Humans," *Journal of Human Evolution* 34, 6 (1998): 623–51. On the methodology used to identify population size and bottlenecks from genomic data, see Henry Harpending and Alan Rogers, "Genetic Perspectives on Human origins and Differentiation," *Annual Review of Genomics and Human Genetics* 1 (2000): 361–85; and

Heng Li and Richard Durbin, "Inference of Human Population History from Individual Whole-Genome Sequences," *Nature* 475 (2011): 493–96.

6. Ehret, "Early Humans," 346–47. For a review of the debates on the interpretation of bottlenecks and N_e (effective population size) in genomic data, see M. Husemann, F. E. Zachós, R. J. Paxton, and J. C. Habel, "Effective Population Size in Ecology and Evolution," *Heredity* 117 (2016): 191–92, and the associated special issue.
7. On the variety of ecologies in Northeast Africa in recent times, see Johnson and Anderson, *Ecology of Survival.*
8. They contrast primate and human social orders in that, while primates are led by alpha males, for humans, "the alphas are invisible supernatural beings ... betas are invisible ancestors who do the bidding of the alphas and protect their living descendants from harm .... No living human can be considered more than a gamma within this system." Flannery and Marcus, *Creation of Inequality*, 59.
9. For my initial survey of human migration based on language data, see P. Manning, "*Homo sapiens* Populates the Earth: A Provisional Synthesis, Privileging Linguistic Data," *Journal of World History* 17 (2006): 115–58.
10. The site includes an introduction and guide to the site, maps showing language distributions and homelands for 14 phyla, classification of phyla and subfamilies by tree diagrams in spreadsheets, and methods and citations.
11. Appendix: Frameworks for Analysis, part 3, "Human Migration," this volume.
12. Joseph H. Greenberg, *Studies in African Linguistic Classification* (New Haven, CT: Compass Publishing Company, 1955); Greenberg, *The Languages of Africa* (Bloomington: Indiana University Press, 1963); Greenberg (ed.), *Universals of Language*, 2nd ed. (Cambridge, MA: MIT Press, 1965); William Croft, *Joseph Harold Greenberg, 1915–2001: A Biographical Memoir* (Washington, DC: National Academy of Sciences, 2007), 23.
13. Critics of Greenberg's analysis of American languages rejected his large-scale categorization. On the other hand, proponents of a proposed Nostratic superfamily of languages, by including Afroasiatic in a family based in Eurasia, were likely confusing evidence of the initial proto-human language with that for its northern subgroup, which Greenberg classified as Eurasiatic. For debates on American languages, see Joseph H. Greenberg, Christy G. Turner II, Stephen L. Zegura et al., "The Settlement of the Americas: A Comparison of the Linguistic, Dental, and Genetic Evidence [and Comments and Reply]," *Current Anthropology* 27 (1986): 477–97; and Greenberg, *The Languages of the Americas* (Stanford: Stanford University Press, 1987). For Nostratic versus Eurasiatic, see Aharon Dolgopolsky, with an introduction by Colin Renfrew, *The Nostratic Macrofamily and Linguistic Palaeontology* (Cambridge: McDonald Institute for Archaeological Research, 1998); and Greenberg, *Indo-European and Its Nearest Neighbors: The Eurasiatic Language Family*, 2 vols. (Stanford: Stanford University Press, 2000). On the confusion of Nostratic and proto-human classifications, see Christopher Ehret, "Nostratic - or proto-Human?" in Renfrew, *Nostratic Macrofamily.*

14. A pattern known today in Northeast Africa is that of age grades, in which both males and females have undergone community-wide initiation, every several years, forming all the children into a cohort, and similarly forming a cohort for those graduating from the initial cohort. Social responsibilities are allocated throughout the social order, by age and by sex, in a decentralized but effective system of social order. It is not now known when this sort of social organization was invented.
15. J. E. G. Sutton, "The Aquatic Civilization of Middle Africa," *Journal of African History* 15 (1974): 527–46.
16. Study of the genome of body lice, associated with clothing, showed that the lice species originated at least 83,000 years ago, in Africa. The African origin suggests that clothing was initially worn for display rather than for warmth. Melissa A. Toups, Andrew Kitchen, Jessica E. Light, and David L. Reed, "Origin of Clothing Lice Indicates Early Clothing Use by Anatomically Modern Humans in Africa," *Molecular Biology and Evolution* 28 (2011): 29–32; for a study on earlier lice, see D. L. Reed et al., "Pair of Lice Lost."
17. Ehret, "Early Humans," 348–57. Genetic data reaffirm both the outlines and the specifics of his analysis. See also Patrick Manning, "Migration," in David Christian (ed.), *Cambridge World History,"* 1: 277–310; Jibril Hirbo, A. Ranciaro, and S. A. Tishkoff, "Population Structure and Migration in Africa: Correlations between Archaeological, Linguistic and Genetic Data," *Causes and Consequences of Human Migration: An Evolutionary Perspective,* C.B.C. (2009): 135–71. 10.1017/CBO9781139003308.011; Michael C. Campbell and Sarah A. Tishkoff, "The Evolution of Human Genetic and Phenotypic Variation in Africa," *Current Biology* 20 (2010): R166–73.
18. Reich notes the great importance of obtaining additional genomic data on Africa. Reich, *Who We Are,* 207–09.
19. See also Ehret, "Khoesan Languages,"; and J. Lachance et al., "Evolutionary History and Adaptation from High-Coverage Whole-Genome Sequences of Diverse African Hunter-Gatherers," *Cell* 150 (2012), 457–69. As for the language group of those who settled in the forest region, this question has not been resolved yet, as the languages of Bantu-speaking settlers of the Niger–Kordofanian family displaced earlier languages during their Holocene-era immigration.
20. Ehret, "Early Humans"; Ehret, *Proto-Afroasiatic.*
21. See also Greenberg, *Languages of Africa,* 6–49, 149–60; Kay Williamson and Roger Blench, "Niger-Congo," in Heine and Nurse, *African Languages,* 11–42.
22. Christopher Ehret, "Africa from 48,000 to 9500 BCE," in David Christian (ed.), *Cambridge World History,* 1: 362–93.
23. For an overview of human and primate kinship that gives insights into the early process of forming explicitly bounded families, see Thomas R. Trautmann, Gillian Feeley-Harnik, and John C. Mitani, "Deep Kinship," in Shryock and Smail, *Deep History,* 160–88.
24. Lopreato labeled these contrasting motivations as Homologous affiliation and Heterologous affiliation. Lopreato, *Human Nature,* 343.
25. Marlize Lombard and Laurel Phillipson, "Indications of Bow and Arrow Use 64,000 years ago in KwaZulu-Natal, South Africa," *Antiquity* 84 (2010): 635–48.

26. On reed craft, see Paul Johnstone, *The Sea-Craft of Prehistory* (Cambridge, MA: Harvard University Press, 1980), 9–16. The oldest archaeologically known watercraft for Africa is a dugout canoe from today's Niger, but it is only 6000–8000 years old. Abubakar Garba, "The Architecture and Chemistry of a Dug-Out: the Dufuna Canoe in Ethno-archaeological Perspective," *Berichte des Sonderforschungsbereichs* 268 (1996): 193–200.
27. On the logic of networks, see Chapter 3; Appendix: Frameworks for Analysis, part 5, "Institutions and Methods"; see also Flannery and Marcus, *Creation of Inequality*, 24–25, 33–35, 54–55.
28. One exception is the analysis of Australian genomes, which includes what may be a 2% contribution of a group that had moved earlier from Africa to Asia. My hypothesis is that such a group, if it existed, would have moved no further than South Arabia, and would have been incorporated into the later migration. For a summary of preceding arguments in support of multiple dispersals from Africa, see Saioa López et al., "Human Dispersal Out of Africa: A Lasting Debate," *Evolutionary Bioinformation Online* 11, Suppl 2 (2015): 57–68.
29. Greenberg's view is conveyed by Croft's comment on Greenberg's last years: "He planned to turn to a southern group consisting most likely of Niger-Kordofanian, Nilo-Saharan, Elamo-Dravidian, Indo-Pacific, and Australian." William Croft, *Joseph Harold Greenberg, 1915–2001: A Biographical Memoir* (Washington, DC: National Academy of Sciences, 2007), 23.
30. This pattern would explain the relatively high proportion of Denisovan DNA in today's Papua New Guinea. According to my narrative, interbreeding of *Homo sapiens* and Denisovans in Central Asia would not have been able to take place more than 45,000 years ago. Current research has yet to document Denisovan specimens in Southeast or South Asia; it does not yet document Denisovan ancestry in South Asian populations. Mayukh Mondal, Jaume Bertranpetit, and Oscar Lao, "Approximate Bayesian Computation with Deep Learning Supports a Third Archaic Introgression in Asia and Oceania," *Nature Communications* 10, 246 (16 January, 2019).
31. Manning and Hillman, "Climate as a Factor in Migration and Social Change."
32. The complexity of the Indo-Pacific homeland is discussed in the text that follows.
33. Later migration along the Danube and the Mediterranean coast at roughly 40,000 years ago, confirmed archaeologically, tends to confirm this logic for the Indian Ocean coast and its rivers.
34. The Yangzi River, also flowing from the Himalayas, runs a more northerly course through temperate China; the initial wave of tropical migrants did not reach this far.
35. These are also known as Sino-Tibetan languages. See also George van Driem, "Trans-Himalayan," in Thomas Owen-Smith and Nathan W. Hill (eds.), *Trans-Himalayan Linguistics: Historical and Descriptive Linguistics of the Himalayan area* (Berlin: de Gruyter, 2014), 11–40.
36. Austric languages are commonly known by their principal constituent groups: Austronesian, Austroasiatic, Hmong-Mien, and Dai-Kadai.

37. Chris Clarkson's group (University of Queensland) now asserts that human remains from North Australia can be dated to 65,000 years ago. *New York Times*, July 19, 2017.
38. See also Joseph H. Greenberg, "The Indo-Pacific Hypothesis (1971)," in Greenberg (edited and introduced by William Croft), *Genetic Linguistics: Essays on Theory and Method* (Oxford: Oxford University Press, 2018), 193–276.
39. Jim Allen and James F. O'Connell, "Getting from Sunda to Sahul," in Geoffrey Clark, Foss Leach, and Sue O'Connor (eds.), *Islands of Inquiry: Colonisation, Seafaring, and the Archaeology of Maritime Landscapes* (Canberra: Australian National University Press, 2009), 31–46.
40. Johnstone, *Sea-Craft of Prehistory*.
41. Dating of images in these caves in was carried out through uranium-series analysis of coralloid speleothems associated with the image of a banteng or wild ox (Kalimantan, age 40,000 years) and with the image of a babirusa or pig-deer (Sulawesi, age 35,000 years). African and Australian paintings have so far been dated as early as 29,000 years ago. Maxime Aubert et al., "Paleolithic Cave Art in Borneo," *Nature* 564 (2018): 254–57; Aubert et al., "Pleistocene Cave Art from Sulawesi, Indonesia," *Nature* 514 (2014): 223–27; and Ehret, "Early Humans," 358. For an argument that such visual representations were the work of shamans, see Jean Clottes, trans. Olivier Y. Martin and Robert T. Martin, *What Is Paleolithic Art? Cave Paintings and the Dawn of Human Creativity* (Chicago: University of Chicago Press, 2016).
42. The exception here is the brief settlement of *Homo sapiens* (presumably not yet speaking) in Qafzeh Cave in Israel, about 115,000 years ago.
43. On the influence of humans on their habitats, see Russell, *Evolutionary History*, 42–53.
44. Manning and Hillman, "Climate as a Factor in Migration and Social Change."
45. John J. Shea and Matthew L. Sisk, "Complex Projectile Technology and *Homo sapiens* Dispersal into Western Eurasia," *PaleoAnthropology* 10 (2010): 100–22.
46. E. J. Michael Witzel, *The Origin of the World's Mythologies* (Oxford: Oxford University Press, 2012), 5–6, 357–73.
47. Witzel proposes that the Laurasian mythology emerged before 20,000 years ago and after 60,000 years ago; I propose 45,000 years ago as the moment and some point at the southern end of the Kyber Pass route as the place. It might be that the challenge of the new environment was part of the reason for developing a more elaborate myth. Adoption of a revised myth must have involved social conflict, in that the religious institutions preserving the preceding Gondwana mythology now had to be regulated and replaced. Perhaps the new mythology arose as a challenge to the old; perhaps the old mythology failed to perform its function in changing social circumstances. Among the questions was whether the criteria for social fitness of a mythology changed along with the emergence of Laurasian mythology. Witzel, *World's*

Mythologies, 105–85. On myth as a social charter, see Flannery and Marcus, *Creation of Inequality*, 56.

48. This process of interbreeding was similar to the interbreeding that must be presumed for the African populations in the same or slightly earlier times.
49. Since Kartvelian languages have not shown any convincing relationships to North Caucasian or to Eurasiatic, it may be that they resulted from a separate migration from the south – for instance, by Elamo-Dravidian-speakers rather than Trans-Himalayan speakers.
50. North Caucasian and Basque languages seem to be related to Trans-Himalayan languages; there are remnant Trans-Himalayan languages in modern Pakistan and Afghanistan. Ruhlen, *Origin of Language*, 74, 164–66, 191–93.
51. Qiaomei Fu et al., "DNA Analysis of an Early Modern Human from Tianyuan Cave, China," *PNAS* 110 (2013): 2223–27.
52. Ruhlen, *Origin of Language*, 49, 164–66.
53. Kambiz Kamrani, "Tianyuan Man Genome Reveals the Nuances of Asian Prehistory," *Anthropology.net – Blog, Physical Anthropology* (October 14, 2017).
54. Erlandson cites human remains on Okinawa and Miyako islands in the Ryukyu archipelago, dated from 29,000 to 17,000 years ago, with voyages presumably from Japan or Taiwan. They might have been Indo-Pacific mariners. Jon Erlandson, "Ancient Immigrants: Archaeology and Maritime Migrants," in Jan Lucassen, Leo Lucassen, and Patrick Manning (eds.), *Migration History in World History: Multidisciplinary Approaches* (Leiden: Brill, 2010), 201–02.
55. Johnstone, *Sea-craft of Prehistory*, 26–43; Harri Luukkanen and William F. Fitzhugh, *The Bark Canoes and Skin Boats of Northern Eurasia* (Washington, DC: Smithsonian Institution Scholarly Press, 2018).
56. Location of the human–canine encounter has yet to be confirmed, but I argue that Northeast Asia is a likely place. Laura R. Botigué et al., "Ancient European Dog Genomes Reveal Continuity since the Early Neolithic," *Nature Communications* 8, 16082 (2017).
57. Karen E. Carr, "Who Invented Sewing? History of Clothing." Quatr.us Study Guides, June 8, 2017. January 14, 2018. https://quatr.us/central-asia/invented-sewing-history-clothing.htm

PRODUCTION AND CONFEDERATION

1. V. Gordon Childe, *What Happened in History* (Harmondsworth, England: Pelican Books, 1942). Jared Diamond also emphasized the impact of agricultural production, as early as the Younger Dryas but usually later. Diamond, *Guns, Germs, and Steel.*
2. Archaeological records might ultimately provide information on the movement of humans to zones of refuge during the Last Glacial Maximum. To my knowledge, neither genetic data nor language distributions have yet provided evidence of these refugee movements. On variation of temperatures at low and high latitudes, see Shaun Marcott et al., "A Reconstruction of Regional

and Global Temperature for the Past 11,300 Years," *Science* 339, 6124 (2013): 1199–1200.

3. The Pleistocene epoch brought relatively regular cycles in which a glacial maximum, with deep ice sheets covering polar areas, was followed by rapid warming and melting and then by gradual cooling. The Last Glacial Maximum, some 20,000 years ago, was preceded by other glacial maxima at 140,000, 240,000, and 335,000 years ago. Lovelock, *Gaia*; Brooke, *Climate Change*.
4. Remains of a shelter at Terra Amata, in Mediterranean France, have been dated to some 400,000 years ago. Henry de Lumley, *La grande histoire des premiers hommes européens* (Paris : O. Jacob, 2007).
5. Dwellings at Mezherich, south of Kiev and dated 18,000–14,000 years ago, were some 20 feet across, with roofs of mammoth hide and walls made of mammoth bones. Flannery and Marcus, *Creation of Inequality*, 13.
6. The remains of a boy found in the original 1920s archaeological dig, now analyzed, are known as MA-1. https://en.wikipedia.org/w/index.php?title=Mal%27ta-Buret%27_culture&oldid=867852592. Other examples of early dwellings are subterranean houses in Ukraine, 28,000–24,000 years ago, and a great longhouse at Kostienki in the Russian plain, with 10 hearths for families in a building 119 feet long and 50 feet wide, dated between 42,000 and 30,000 years ago. Flannery and Marcus, *Creation of Inequality*, 12–13.
7. See www.youtube.com/watch?v=wbsURVgoRD0. After Arthur S. Dyke, "An Outline of North American Deglaciation," 2004. Cited hereafter as "Deglaciation."
8. Ross D. E. MacPhee, *End of the Megafauna: The Fate of the World's Hugest, Fiercest, and Strangest Animals* (New York: W. W. Norton, 2018), 41; "Deglaciation," www.youtube.com/watch?v=wbsURVgoRD0.
9. In 52 short chapters dated from 20,000 BCE to 4000 BCE, Mithen takes the reader on a tour of archaeological sites in every region of the world, tracing the challenges and changes faced by people in a mosaic of conditions. Steven Mithen, *After the Ice: A Global Human History, 20,000–5000 BC* (Cambridge, MA: Harvard University Press, 2003).
10. Wu Xiaohong et al., "Early Pottery at 20,000 Years Ago in Xianrendong Cave, China," *Science* 336 (2012): 1696–1700; Suzanne Staubach, *Clay: The History and Evolution of Humankind's Relationship with Earth's Most Primal Element* (New York: Berkley Books, 2005).
11. Douglas Kenrick, *Jomon of Japan: The World's Oldest Pottery* (London: Kegan Paul International, 1995); Friederike Jesse, "Early Pottery in Northern Africa – An Overview," *Journal of African Archaeology* 8 (2010): 219–38.
12. Staubach, *Clay*.
13. Harvesting of wild grains by speakers of Afroasiatic languages is projected as far back as 20,000 years. Christopher Ehret, personal communication.
14. By a "confederation" I mean the amalgamation or confederation of language communities (with 150 individuals or more) into groups roughly three or four times as large. I avoid using the term "society" at this stage, desiring to save it for a more general use in the many types of societies that formed at later times: see Chapter 6.

15. Further, neither of these changes took place suddenly or evenly, and local variations were considerable. For instance, many foraging communities may have declined to make the shift to creating aggregated confederations or to emphasizing production rather than foraging to meet their needs. Verifying this thesis of significant expansion in late Pleistocene size of human groups will require considerable research. Nevertheless, "The archaeological evidence suggests that some [Ice-Age foragers] created large, permanent groups of people who considered themselves related, whether it was true or not." Flannery and Marcus, *Creation of Inequality*, 549.
16. As Flannery and Marcus argue, for clanless foragers who, even if they participated in initiation and bride service, their "us" versus "them" mentality justified raiding of other groups. Clans, further, were not necessarily a second stage of foraging. Flannery and Marcus, *Creation of Inequality*, 14, 42–49, 53, 550.
17. As a reminder, Miller's subsystems for reproduction are the reproducer and boundary; the subsystems for processing information are the input transducer, internal transducer, channel and net, decoder, associator, memory, decider, encoder, and output transducer. Miller presents a brief discussion of cultural evolution at the level of society, drawing on the work of Lomax and Berkowitz. That work, however, is limited to a factor analysis of the presence of cultural attributes, with no attention to the process of selection. Miller, *Living Systems*, 756–60; A. Lomax and N. Berkowitz, "The Evolutionary Taxonomy of Culture," *Science* 177 (21 July, 1972): 230, 233.
18. Individuals and families were governed by processes of biological and cultural evolution; communities and confederations by social evolution.
19. While I wish to raise the possibility of such a generalization of emotions, I acknowledge that Tuomela identifies limits to the passage of group emotions on to changed groups or larger groups. "For emotion transferal, the invariance pertaining to the existence of the same cognitive elements of conditions, and hence action based on these elements, must be there." Tuomela, *Social Ontology*, 261–62.
20. The most distant Eurasiatic groups were Eskimo-Aleut, to the east, and Indo-European and Etruscan, to the west. Since roughly 1700 CE, Indo-European languages have spread to every region of the world.
21. Ruhlen, *Origin of Language*, 49, 164–66.
22. Johanna Nichols, in earlier work, has suggested that Indo-Pacific-speakers may have participated in the settlement of northeastern Asia and the Americas. Fu, "Tianyuan Cave"; Johanna Nichols, "The Origin and Dispersal of Languages: Linguistic Evidence," in Jablonski and Aiello (eds.), *The Origin and Diversification of Language*, 127–70 (San Francisco: California Academy of Sciences, 1998). On maritime travel via Okinawa 35,000 years ago, see Erlandson, "Ancient Immigrants," 201–02.
23. Johnstone, *Sea-Craft of Prehistory*, 26–43.
24. On language distribution, see Ehret *Proto-Afroasiatic*. On parallel genetic distribution, Keita and Boyce conducted Y-chromosome analyses of selected haplotypes (p49 a,f *Taq*I RFLP) in Egypt, showing that the most prominent of these haplotypes were established early (before the Middle Kingdom) and from the south. Later conquests of Egypt in Greco-Roman and Islamic eras

added other haplotypes. See S. O. Y. Keita and A. J. Boyce, "Genetics, Egypt, and History: Interpreting Geographical Patterns of Y Chromosome Variation," *History in Africa* 32 (2005): 221–46.

25. The indirect evidence supporting this argument is a genome from Mal'ta in Siberia, dated 24,000 years ago, and a genome from Upward Sun River in the middle Yukon Valley, dated 11,500 years ago. Maanasa Raghavan et al., "Upper Palaeolithic Siberian Genome Reveals Dual Ancestry of Native Americans," *Nature* 505 (2014): 87–91; J. Victor Moreno-Mayar et al., "Terminal Pleistocene Alaskan Genome Reveals Founding Population of Native Americans," *Nature* 553 (2018): 203–08.
26. "Deglaciation," www.youtube.com/watch?v=wbsURVgoRD0.
27. The genomic analysis of Scheib et al. shows, in the inset, that populations of northern and eastern North America were separate from other American populations; this parallels quite closely the linguistic analysis separating the Almosan–Keresiouan branches of Northern Amerind from the rest of Amerind. C. L. Scheib et al., "Ancient Human Parallel Lineages within North America Contributed to a Coastal Expansion," *Science* 360 (2018): 1024–27; Moreno-Mayar et al., "Terminal Pleistocene Alaskan Genome."
28. Jon M. Erlandson, Michael H. Graham, Bruce J. Bourque, Debra Corbett, James A. Estes, and Robert S. Steneck, "The Kelp Highway Hypothesis: Marine Ecology, the Coastal Migration Theory, and the Peopling of the Americas," *Journal of Island and Coastal Archaeology* 2 (2007): 161–74.
29. For the initial multidisciplinary statement on settlement of the Americas, see Greenberg et al., "Settlement of the Americas"; see also Greenberg, *Language in the Americas*.
30. Tom D. Dillehay et al., "New Archaeological Evidence for an Early Human Presence at Monte Verde, Chile," *PLoS ONE* (November 18, 2015): 1–27, DOI: 10.137.
31. Scheib et al., "Ancient Human Parallel Lineages."
32. Studies of additional detail by Scheib et al. centered on Algonkian speakers of Ontario, within the northern group, and on data from California's coastal islands, within the southern group.
33. The main exceptions to this fit of language and genomic data are that Scheib's northern genomic group includes, in addition to Almosan–Keresiouan speakers, the speakers of Tsimshian languages, part of the Penutian (otherwise southern) group, and the Na–Dene languages, which Greenberg classifies as separate from Amerind languages. I believe that these discrepancies indicate that the main admixture and perhaps even the preceding separation of northern and southern genetic groups took place in what we now call the Pacific Northwest.
34. Scheib et al., "Ancient Human Parallel Lineages," 1024; P. Skoglund et al., "Genetic Evidence for Two Founding Populations of the Americas," *Nature* 525 (2018), 104–08 (2015).
35. Dene–Yeniseian languages, in turn, are arguably related to Trans-Himalayan languages.

36. That Haida Gwaii was ice free, perhaps even as early as 18,000 years ago is shown clearly in "Deglaciation," www.youtube.com/watch?v=wbsURVgoRD0.
37. As the Beringians are known only through genomic and archaeological evidence, there is no indication as to what language they spoke.
38. Moreno-Mayar et al., "Terminal Pleistocene Alaskan Genome"; Scheib et al., "Ancient Human Parallel Lineages," 1024.
39. Clovis points, though most densely located within the lands of Keresiouan-speaking peoples, have also been collected and mapped in almost every region of North America that lay south of the glaciated region.
40. The genome associated with Clovis is of infant boy Anzick-1. Scheib et al., "Ancient Human Parallel Lineages," 1024.
41. The flowering *Dryas octopetala* expands in times of sudden cold. The three periods of cooling were the Oldest Dryas (15,070–14,670 years ago), Older Dryas (ca. 14,000–ca. 13,800 years ago), and Younger Dryas (12,900–11,500 years ago). Specific dates were confirmed when radiocarbon dating became available.
42. Wendy S. Wolbach et al., "Extraordinary Biomass-Burning Episode and Impact Winter Triggered by the Younger Dryas Cosmic Impact ~12,800 Years Ago. 1. Ice Cores and Glaciers," *Journal of Geology* 126, 2 (2018): 165–84. Bibcode: 2018JG .... 126..165W. DOI: 10.1086/695703.
43. Kurt H. Kjaer, "A Large Impact Crater beneath Hiawatha Glacier in Northwest Greenland," *Science Advances* 4, 11 (2018), DOI: 10.1126/sciadv.aar8173.
44. Anna Groves, "There's a second impact crater under Greenland," *Astronomy Magazine*, February 13, 2019, www.astronomy.com. In addition, William Ruddiman has emphasized that, during the Younger Dryas event, levels of methane collapsed and then rose again while levels of carbon dioxide halted their growth; this might be related to populations of large animals. William F. Ruddiman, *Earth Transformed* (New York: W. H. Freeman, 2014), 27, 39.
45. MacPhee, *End of the Megafauna.*
46. The distribution of language groups, as described primarily through lexical analysis, appears in the work shown here to be consistent with some very early migratory patterns. Johanna Nichols reaffirms a common view that "the comparative method is valid only for time depths up to about 10,000 years," but suggests that lexical studies in association with other linguistic patterns can provide information on earlier distribution of languages. Johanna Nichols, *Linguistic Diversity in Space and Time* (Chicago: University of Chicago Press, 1992), 184.
47. L. L. Cavalli-Sforza, *Genes, Peoples, and Language* (New York: North Point Press, 2000).
48. On skin color, I rely tentatively on a world map of its global distribution, as published in various scientific studies. The original version of this map, however, was researched and drawn by Biasutti in 1941, drawing on (but not replicating) the 36-tone skin-color scheme published by Luschan in 1922. Research on the biology of changes in skin color has advanced impressively, but the representation of global distribution in skin color appears

outdated. Cavalli-Sforza et al. published, in 1994, a slightly simplified version of Biasutti's Pacific-centered Robinson projection; the map shown on Wikipedia is an Atlantic-centered Mercator projection, redrawn from Biasutti's Pacific-centered Robinson projection. N. G. Jablonski and G. Chaplin, "The Evolution of Skin Coloration," *Journal of Human Evolution* 39 (2000): 57–106; L. Luca Cavalli-Sforza, Paolo Menozzi, and Alberto Piazza, *The History and Geography of Human Genes*, abridged (Princeton, NJ: Princeton University Press, 1994), 145; "Unlabeled Renatto Luschan Skin color map.svg," Wikipedia Commons; Felix von Luschan, *Völker, Rassen, Sprachen* (Berlin: Welt-Verlag,1922); "Tavole VI. Distribuzione della varia intensità del colore della pelle," in Renato Biasutti (ed.), *Le Razze e I popoli della terra*, 4 vols. (Turin: Unione Tipografico – Editrice Torinese, 1967 [1941]), 1: 224–29.

49. Learning a language, for adults as for children, requires joining a we-group because of the required commitment to group norms.
50. Flannery and Marcus, *Creation of Inequality*, 189.
51. Patrick Manning, *The African Diaspora: A History through Culture* (New York: Columbia University Press, 2009), 40, 51.
52. The instances included settlements of Beringians, Amerind-speakers (including those of Indo-Pacific ancestry), Na–Dene speakers, and later Eskimo–Aleut speakers, plus those argued to have returned to Asia.
53. Indian Ocean navigation is documented by settlement of the Andaman Islands and Sahul; western Pacific navigation is documented by archaeology on Okinawa; eastern Pacific navigation is documented by archaeology in the Salish Sea and at Monte Verde.

SOCIETY: NETWORK VERSUS HIERARCHY

1. For instance, on an early Holocene burial ground in northwestern Russia, see Mithen, *After the Ice*, 168–77.
2. As Flannery and Marcus argue, the historical and anthropological literatures have tended to conclude that the transition to hierarchical chiefdoms was more rapid and more automatic than can be documented. For the rise of states, similarly, James Scott has recently argued that state formation was commonly delayed to a later time than is often thought. In Africa, for instance, numerous societies that had long used iron and had sophisticated agricultural systems did not form states until the fifteenth century or later. Flannery and Marcus, *Creation of Inequality*; Scott, *Against the Grain*; Peter Turchin et al., "Quantitative Historical Analysis Uncovers a Single Dimension of Complexity that Structures Global Variation in Human Social Organization," *PNAS* (December 21, 2017), E144–51; www.pnas.org/cgi/doi/10.1073/pnas.1708800115 .
3. My definition of "society" is parallel to that of James G. Miller: "A society is a large, living, concrete system with organizations and lower levels of living systems as subsystems and components." His definition includes modern nations, ancient city-states and kingdoms, and small societies. Miller discusses cultural evolution of societies, but only by listing their characteristics rather than

proposing a mechanism or process of cultural evolution. Tylor, Morgan, and Engels proposed patterns of societal evolution through stages of development; the only mechanism proposed for these stages was gradual adoption of hierarchy. Carneiro proposed a much more elaborate set of social characteristics and analyzed numerous societies through factor analysis, but again classified by phenotypical distinctions rather than proposing a mechanism of social evolution. Miller, *Living Systems*, 747, 756–60; R. L. Carneiro, "Scale Analysis, Evolutionary Sequences, and the Rating of Cultures," in R. Naroll and R. Cohen (eds.), *A Handbook of Method in Cultural Anthropology* (Garden City, NY: Doubleday, 1970), 846.

4. The notion that humanity has sought to master nature is oft repeated, for instance by Rousseau; this is a formulation of the idea in the context of social evolution.
5. These proposed group sizes are similar to the group sizes identified by Dunbar and Solis in their study of contemporary humans. (See notes in Chapters 2 and 3.) The important point is that, based on the work of Dunbar and Solis, one must assume that speaking humans at any time organized groups of several sizes, though the group of 150 appears to have persisted. Dunbar and Sosis, "Optimising Human Community Sizes."
6. In addition, comparative linguistics has great promise in providing information on the Holocene epoch, though the precision of historical linguistics is currently limited by unresolved inconsistencies in method.
7. See the sections "Selection among Institutions" and "Social Fitness" in Chapter 3.
8. In addition to this question of the *standard* for assessing the fitness of an institution, one could measure the *level* of fitness in terms of the resulting population size, quantity of material gain, or overall level of social welfare.
9. Diamond, *Guns, Germs, and Steel*, 83–130.
10. I have described these institutions in terms of social evolution. Richerson and Boyd explain agriculture and animal husbandry in terms of cultural evolution, as follows: "We hypothesize that two sets of social 'instincts' underpin and constrain the evolution of complex societies. One set is ancient and shared with other social primate species, and one is derived and unique to our lineage. The latter evolved by the late Pleistocene, and led to the evolution of institutions of intermediate complexity in acephalous societies. The institutions of complex societies often conflict with our social instincts. The complex societies of the past few thousand years can function only because cultural evolution has created effective 'work-arounds' to manage such instincts." Richerson and Boyd, "Complex Societies: The Evolutionary Origins of a Crude Superorganism," *Human Nature* 10 (1999), 253.
11. Dmitry Yakolevich Telegin, trans. V. A. Pyatkovskiy, ed. J. P. Mallory, *Dereivka. A Settlement and Cemetery of Copper Age Horse Keepers on the Middle Dnieper* (Oxford: BAR International Series, 1986); Anon., "Ancient Krasnyi Yar," www.amnh.org/exhibitions/horse/domesticating-horses/ancient-krasnyi-yar; American Museum of Natural History.
12. Great and unresolved debates continue on the region from the Pontic steppes south to the Levant, on the timing, location and nature of the rise of

agriculture, animal husbandry, chariot warfare, cultural practices, and waves of migration, as related to speakers of Indo-European, Altaic, Semitic, and North Caucasian languages. J. P. Mallory, *In Search of the Indo-Europeans: Language, Archaeology, and Myth* (London: Thames and Hudson, 1989); David W. Anthony, *The Horse, the Wheel, and Language: How Bronze-age Riders from the Eurasian Steppes Changed the World* (Princeton, NJ: Princeton University Press, 2007); Marija Gimbutas, *The Goddesses and Gods of Old Europe, 6500–3500 BC, Myths and Cult Images*, new and updated ed. (Berkeley: University of California Press, 1982).

13. "Neolithic Site of Çatalhöyük," UNESCO World Heritage Centre, https://whc.unesco.org/en/list/1527.
14. Erlandson, "Ancient Immigrants."
15. Paul C. Sereno et al., "Lakeside Cemeteries in the Sahara: 5000 Years of Holocene Population and Environmental Change," *PLoS ONE* 3(8): e2995. https://doi.org/10.1371/journal.pone.0002995.
16. "Among rank societies, war became a tool for chiefly aggrandizement." Flannery and Marcus, *Creation of Inequality,* 552, 555.
17. Flannery and Marcus, *Creation of Inequality*, 153–83.
18. Ibid., 79, 105–09, 551.
19. This issue can be linked to the question of violence and oppression by social scale. Does larger-scale society mean more violence at the societal level and less at the family level? See Turchin *Ultra Society*; Walter Scheidel, *The Great Leveler: Violence and the History of Inequality from the Stone Age to the Twenty-first Century* (Princeton, NJ: Princeton University Press, 2017); Ian Morris, *The Measure of Civilization: How Social Development Decides the Fate of Nations* (Princeton, NJ: Princeton University Press, 2013).
20. Further, hereditary rank society arose by manipulation of "cosmologies, reciprocal exchange, social obligations, wealth transfer, and the subservience of junior lineages." Flannery and Marcus, *Creation of Inequality*, 65, 87.
21. For details of the Price Equation, see Chapter 2. Successful groups are assumed to be homogeneous in this model. Further, this analysis implicitly assumes a gradual rather than rapid emergence of spoken language.
22. Turchin, *Ultra Society,* 41.
23. Turchin, *Ultra Society*, 168–78, 213–25; Steven Pinker, *The Better Angels of Our Nature: Why Violence Has Declined* (New York: Viking, 2011). For a modified approach to theory in cultural evolution, see Paul E. Smaldino, "The Cultural Evolution of Emergent Group-Level Traits," *Behavioral and Brain Sciences* 37 (2014): 243–54.
24. Turchin, *Ultra Society*; see also the special issue on "Evolutionary Biology Arguments in Political Economy" in the *Journal of Bioeconomics* 17, 1 (2015). For a discussion of Turchin's earlier analysis of warfare, see Timothy A. Kohler, Sarah Cole, and Stanca Ciupe, "Population and Warfare: A Test of the Turchin Model in Pueblo Societies," in Shennan (ed.), *Pattern and Process in Cultural Evolution*, 277–96; and Peter Turchin, *Historical Dynamics: Why States Rise and Fall* (Princeton, NJ: Princeton University Press, 2003).
25. This model of cultural evolution relies on multilevel selection; in contrast, the social evolution model relies on institutions, social selection, networks linking

societies, and the distinction of achievement and hierarchical modes. On multilevel selection, see the appendix on Frameworks for Analysis, part 2; and Turchin, *Ultra Society*, 19, 81–82, 92.

26. Ruddiman, *Earth Transformed*, 19–42.
27. Ibid., 231–44, 346–47.
28. Of Oceania's total area, Australia comprises 90%. In addition, one could add the Malay Archipelago, with 2 million km^2 in area, to Oceania.
29. One difficulty with the term "Afro-Eurasia" is that, while its meaning includes all of Eurasia, its African reference is ambiguous: usually limited to North Africa and the Sahara, sometimes including all of Africa.
30. State formation and consolidation often took some time, even thousands of years. Flannery and Marcus, *Creation of Inequality*, 555. See also Scott, *Against the Grain*; and Ronald Cohen and Elman R. Service, *Origins of the State: The Anthropology of Political Evolution* (Philadelphia: Institute for the Study of Human Issues, 1978).
31. As a rough guide, earthenwares are normally fired at temperatures in the range of about 1000°C (1830°F) to 1200°C (2190°F); stonewares at between about 1100°C (2010°F) and 1300°C (2370°F); and porcelains at between about 1200°C (2190°F) and 1400°C (2550°F).
32. Richard W. Bulliet, *The Wheel: Invention and Reinvention* (New York: Columbia University Press, 2016), 51–70, 79.
33. Ibid., 71–91.
34. For an updated global accounting of urbanism, see Meredith Reba, Femke Reitsma, and Karen C. Seto, "Spatializing 6,000 Years of Global Urbanization from 3700 BC to AD 2000," *Scientific Data* 3 (2016), article no. 160034.
35. Peter Bellwood, James J. Fox, and Darrell Tryon, *The Austronesians: Historical and Comparative Perspectives* (Canberra: Australian National University, 1995); and Catherine Cymone Fourshey, Rhonda M. Gonzales, and Christine Saidi, *Bantu Africa: 3500 BCE to Present* (New York: Oxford University Press, 2017).
36. It is likely that Indo-Pacific-speakers occupied Taiwan before the settlement of Austronesian speakers.
37. Bellwood et al., *Austronesians*, 102.
38. Fourshey et al., *Bantu Africa.*
39. Bantu expansion led ultimately to the Zimbabwe state; Austronesian expansion brought formation of hierarchical Pacific states; and Indo-Aryan migration led to North Indian states.
40. Scott DeLancy and Victor Golla, "The Penutian Hypothesis: Retrospect and Prospect," *International Journal of American Linguistics* 63 (1997), 171–202; Jason A. Eshelman et al., "Mitochondrial DNA and Prehistoric Settlements: Native Migrations on the Western Edge of North America," *Human Biology* 76 (2004), 55–75.
41. One may assume that previous inhabitants of Yucatán adopted the language of Penutian-speaking immigrants.
42. Other migrations of this era (yet to be dated with precision) included, in Eurasia, westward movements of Uralic speakers in the Arctic zone, Indo-

European movement into Europe, Austric migrations west to India; in Africa, Adamawa–Eastern languages spread eastward from Cameroon; in the Americas, migration of Navajo and other Na–Dene speakers from Athabasca to Arizona and New Mexico, and migration of Algonkian speakers from the Pacific to occupy most of the Great Lakes region. For a valuable overview of the contribution of linguistic evidence to historical analysis, see Christopher Ehret, *History and the Testimony of Language* (Berkeley: University of California Press, 2011).

43. Eric H. Cline, *1177 BC: The Year Civilization Collapsed* (Princeton, NJ: Princeton University Press, 2014); on the Shang dynasty, see Robert L. Thorp, *China in the Early Bronze Age: Shang Civilization* (Philadelphia: University of Pennsylvania Press, 2006).
44. Iron use did not, however, spread into the Pacific islands. And while metallurgy expanded in many parts of the Americas, the furnaces were not hot enough to move beyond smelting of gold, silver, and copper.
45. Flannery and Marcus clearly distinguish achievement and inheritance of leadership; treating this as a difference between African and Eurasian societies is my own suggestion. Flannery and Marcus, *Creation of Inequality*.
46. Bulliet, *The Wheel*.
47. Philip D. Curtin, *Cross-Cultural Trade in World History* (New York: Cambridge University Press, 1984).
48. Bin Yang, *Cowrie Shells and Cowrie Money: A Global History* (London: Routledge, 2018), 94–110, 161–71.
49. Further, van Bavel advances a theory of the internal dynamics of commercial institutions, generating stages from social revolt and economic equality, to institution-building, economic growth, inequality, and decline over a period of three to seven centuries. For further discussion, see Chapter 8. Bas van Bavel, *The Invisible Hand? How Market Economies Have Emerged and Declined since AD 500* (Oxford: Oxford University Press, 2016).
50. Herodotus (ca. 484–ca. 425 BCE); Xuanzhang (ca. 602–664 CE); Ibn Fadlan (877–960 CE). Richard N. Frye, ed. and trans., *Ibn Fadlan's Journey to Russia: A Tenth-Century Traveler from Baghdad to the Volga River* (Princeton, NJ: Markus Wiener, 2005).
51. In an early contribution to large-scale religion, Zarathustra preached in northeast Persia as much as a thousand years earlier. Michael Stausberg, Yuhan Sohrab-Dinshaw Vevaina, and Anna Tessmann, *The Wiley-Blackwell Companion to Zoroastrianism* (Chichester, UK: Wiley, 2015), 57–59.
52. The parallels of Roman and Greek pantheons are noted; equally significant is the ancestry of both in the broader Indo-European pantheon.
53. Richard J. Smith, "Buddhism and the 'Great Persecution' in China," in *Critical Moments in Religious History*, ed. Kenneth Keulman (1993); E. O. Reischauer, *Ennin's Travels in T'ang China* (New York: The Ronald Press, 1955).
54. On the Abbasid economy, see van Bavel, *Invisible Hand?*, 41–96; on the modification of Islamic tradition, see Chouki El Hamel, *Black Morocco: A History of Slavery, Race, and Islam* (Cambridge: Cambridge University Press, 2013), 15–59.

55. Flannery and Marcus speak of the deep ancestry of colonial rule, arguing that it is "4300 years old, the product of kings who sought to add land and tribute to their realms." I acknowledge their point but find it more practical to introduce the notion of colonial rule with the realms of the Achaemenids, Mauryans, and Carthage. Flannery and Marcus, *Creation of Inequality*, 557.
56. Jane Burbank and Frederick Cooper, *Empires in World History: Power and the Politics of Difference* (Princeton, NJ: Princeton University Press, 2010).
57. Translations into Arabic included the works of Aristotle and Claudio Ptolemy from Greek and Syriac, an initial decoding of Egyptian hieroglyphics, and translation of Sanskrit texts on mathematics. Al-Khwarizmi (ca. 780–850), in addition to translation work, developed the first stages of algebra and did much to spread use of the zero in arithmetic notation and calculations.
58. Sheldon Pollock, *The Language of the Gods in the World of Men: Sanskrit, Culture, and Power in Premodern India* (Berkeley: University of California Press, 2006); for generalization of the term, see Alexander Beecroft, *An Ecology of World Literature: From Antiquity to the Present Day* (London: Verso, 2015). Of major 2 cosmopolis, it is remarkable how many were centered on Indo-European languages: Greek, Latin, Sanskrit, Persian, Russian, and some languages of Western Europe.
59. McNeill and McNeill, *The Human Web.*
60. Xinru Liu, *The Silk Road in World History* (New York: Oxford University Press, 2010).
61. Edmund Burke III, "Islam at the Center: Technological Complexes and the Roots of Modernity," *Journal of World History* 20 (2009): 165–86.

COLLISIONS AND CONTRACTION

1. Shaun Marcott et al., "Regional and Global Temperature," 1198–1201.
2. "Climate and society, ecology and biology, and microbes and humans comprise the six core components of such a dynamical system . . . Each component should be viewed as a semi-autonomous sub-system in its own right, made up of sub-elements linked by direct and indirect feedbacks, so that each possesses its own independent dynamic. None, however, exists in isolation, hence change in any one of the six core components will elicit change in one or more of the others, with the character of that change in part mediated by the prevailing state of those other components. . . . To privilege endogenous human processes over ostensibly exogenous environmental events is also to create a false dichotomy, since there is nothing in this model that is not endogenous." Bruce M. S. Campbell, *The Great Transition: Climate, Disease and Society in the Late-Medieval World* (Cambridge: Cambridge University Press, 2016), 21–22.
3. Campbell, *Great Transition*, 1.
4. Monica Green, "Taking 'Pandemic' Seriously: Making the Black Death Global," *The Medieval Globe* 1, 1 (2014), https://scholarworks.wmich.edu/tmg/vol1/iss1/4.
5. Ruddiman, *Earth Transformed*, 291–92.
6. Ibid., 312–13.

7. Gérard Chouin, "Reflections on Plague in African History (14th–19th c.)," *Afriques: débats, méthodes, et terrains d'histoire* 9 (2018), https://journals.openedition.org/afriques/2084.
8. Monica Green, "Putting Africa on the Black Death Map: Narratives from Genetics and History," *Afriques: débats, méthodes, et terrains d'histoire* 9 (2018), https://journals.openedition.org/afriques/2084.
9. P. D. Jones and M. E. Mann, "Climate over Past Millennia," *Reviews of Geophysics* 42, 2 (2004); Alfred W. Crosby, *The Columbian Exchange: Biological Consequences of 1492* (Westport, CT: Greenwood Publishers, 1972).
10. This research projects overall decline of Amerindian population by 50% to 90% of the total. Noble David Cook, *Born to Die: Disease and New World Conquest, 1492–1650* (Cambridge: Cambridge University Press, 1998).
11. My appreciation to Ann Jannetta, who documented this set of events through Japanese records.
12. The genetic study is based on mitochondrial DNA from 137 modern samples and 67 ancient samples from regions throughout the Americas. Brendan O'Fallon and Lars Fehren-Schmitz, "Native Americans Experienced a Strong Population Bottleneck Coincident with European Contact," *PNAS* 108, 51 (2011): 20444–48.
13. Ruddiman, *Earth Transformed*, 312–28. Most estimates of long-term population change tend to assume steady rates of growth, but more detailed and multidisciplinary information makes it possible to identify fluctuations. Cf. Colin McEvedy and Richard Jones, *Atlas of World Population History* (Harmondsworth: Penguin, 1978).
14. "The use of the concept of bellicosity in part overcomes the unhelpful distinction between rationality and irrationality in discussing motivation and conduct. Bellicosity can be regarded as crucial to both motivation and conduct, as well as both (or either) a necessary condition for and even, definition of war.... Bellicosity also helps explain the continuation of wars once begun." My question is whether this model can reasonably be linked to biological and social expressions of emotions, and their variations. Jeremy Black, *War and Its Causes* (London: Rowman and Littlefield, 2019), 8–9.
15. Barrett, *How Emotions Are Made*; Adolphs and Anderson, *The Neuroscience of Emotion.* See also Christian von Scheve and Mikko Salmela, *Collective Emotions: Perspectives from Psychology, Philosophy, and Sociology* (Oxford: Oxford University Press, 2014).
16. In 1420, however, the Yongle Emperor moved the capital back to Beijing, where it stayed, although the capital went through one more episode of conquest and displaced archives as the Qing dynasty took hold in 1644.
17. At the eastern end of the Old World, in a region of Hindu and Buddhist faith, the maritime state of Srivijaya had carried and protected centuries of trade in the southern seas through two crucial passages, the straits of Melaka to the north and the straits of Sunda to the south of Srivijaya's base in Sumatra. But when Srivijaya elevated its taxes on trade, the south Indian state of Chola responded vigorously, mounting naval attacks across the Bay of Bengal, including a huge expedition in 1025, which weakened Srivijaya.

18. The era of warming brought expansion of states in sub-Saharan Africa, surely linked to warfare. The kingdom of Mapungubwe rose in the twelfth century, only to be succeeded by Zimbabwe in the thirteenth century. The same era also saw the emergence of the Bunyoro state in the East African highlands and, in West Africa, the rise of Benin on the coast and the Mossi states in the interior.
19. R. J. Barendse, "The Feudal Mutation: Military and Economic Transformations of the Ethnosphere in the Tenth to Thirteenth Centuries," *Journal of World History* 14 (2003): 503–29; Stephen Morillo, "A 'Feudal Mutation'? Conceptual Tools and Historical Patterns in World History," *Journal of World History* 14, 4 (2003), 549–50.
20. Nicola Di Cosmo, "Climate Change and the Rise of an Empire." *The Institute Letter* (Institute for Advanced Study), Spring 2014, 1, 15.
21. Janet Abu-Lughod, *Before European Hegemony: The World System AD 1250–1350* (New York: Oxford University Press, 1989).
22. Giorgio Riello, in his analysis of cotton in the Indian Ocean, suggests somewhat greater continuity in Indian Ocean prosperity. See also Bin Yang's analysis of the cowrie currency zone, centering on the Indian Ocean trading zone but extending beyond that realm, which also suggests greater continuity in prosperity. Giorgio Riello, *Cotton: The Fabric that Made the Modern World* (Cambridge: Cambridge University Press, 2013); Yang, *Cowrie Shells and Cowrie Money*.
23. Abu-Lughod, *Before European Hegemony*; Immanuel Wallerstein, *The Modern World-System*, vol. 1 (New York: Academic Press, 1974). For a long-term interpretation of economic and political cycles, see Barry K. Gills and Andre Gunder Frank, "World System Cycles, Crises, and Hegemonic Shifts, 1700 BC to 1700 AD," in Frank and Gills (eds.), *The World System: Five Hundred Years or Five Thousand?* (London: Routledge, 1992), 143–99.
24. Campbell, *Great Transition;* Maarten Prak, *Citizens without Nations: Urban Citizenship in Europe and the World, c. 1000–1789* (Cambridge: Cambridge University Press, 2018); van Bavel, *Invisible Hand?*
25. Barendse, "Feudal Mutation"; Morillo, "A 'Feudal Mutation'?"; Abu-Lughod; *Before European Hegemony*; Gills and Frank, "World System Cycles."
26. Roxann Prazniak, "Marāgha Observatory: A Star in the Constellation of Eurasian Scientific Translation," in Patrick Manning and Abigail Owen (eds.), *Knowledge in Translation: Global Patterns of Scientific Exchange, 1000–1800 CE* (Pittsburgh: University of Pittsburgh Press, 2018), 227–43.
27. John Darwin, *After Tamerlane: The Global History of Empire after 1405* (London: Bloomsbury Press, 2008). Arabia and North Africa became major centers of horse breeding. Still further, horses traded across the Sahara had become essential in the warfare of the African savanna, and horses as ceremonial symbols of state reached the kingdom of Benin in the forest of the lower Niger Valley.
28. Ming, Korean, and Japanese regimes each experienced a remarkable cultural flowering in the post-Mongol era. One may ask whether these patterns fit with Ibn Khaldun's vision of political dynamics in the world of Islam.
29. "A bellicose culture, society, and politics can explain the frequency of wars. That, however, is not the same as accounting for the outbreak of particular wars. There is a link in that the bellicosity encouraged the vigorous and

violent pursuit of interests, claims, and disputes, whatever their origins and nature." Black, *War and Its Causes*, 42.

30. See also Tuomela, *Social Ontology*, 260–62.
31. In the view of Donald Campbell, biological evolution optimizes the individual person and the gene-frequency system, while social evolution optimizes the functioning of the social system. "For many behavioral dispositions, the two systems redundantly support each other. For others, the two are in conflict, and curb each other." For the areas of conflict between motivations, "one would expect the modal average socialized adult to be optimally inhibited and repressed in the areas where the social and biological systems are in conflict." Campbell, "On the Conflicts between Biological and Social Evolution," 1116. On emotions, see also Adolphs and Anderson, *Neuroscience*, 40–41; and Barrett, *How Emotions Are Made*, 169–70, as these are discussed in Chapter 2.
32. Disciplined armies and monarchies were we-groups, determined by a common logic to conquer and rule. A zeitgeist emphasizing bellicosity was not necessarily an inherent dynamic in such groups, yet there was a common likelihood that it would develop.
33. The 1381 anti-tax revolt of Wat Tyler in England is a well-known example.
34. Robert Irwin, *Ibn Khaldun: An Intellectual Biography* (Princeton, NJ: Princeton University Press, 2018), 108–17.
35. Ibn Khaldun developed the term "asabiyya" to refer to the tight alliances within the nomadic tribes that were central to North African wars; this was in effect the we-mode agreement that made the tribes into effective military institutions. Ibn Khaldun, *The Muqaddimah: An Introduction to History*, 3 vols., translated from the Arabic by Franz Rosenthal (Princeton, NJ: Princeton University Press, 1958). His critique referred especially to the Almoravids and Almohads.
36. Irwin, *Ibn Khaldun;* Mohamad Ballan, "The Scholar and the Sultan: A Translation of the Historic Encounter between Ibn Khaldun and Timur." See also N. Levtzion and J. F. P. Hopkins (eds.), *Corpus of Early Arabic Sources for West African History* (Cambridge: Cambridge University Press), 332–35, for Ibn Khaldun's description of the kings of the Sudan.
37. John Darwin, *After Tamerlane.*
38. Guido Alfani, trans. Christine Calvert, *Calamaties and the Economy in Renaissance Italy: The Grand Tour of the Four Horsemen of the Apocalypse* (New York: Palgrave Macmillan, 2013).
39. African states that arose in the fourteenth and fifteenth centuries included Kongo, Luba, Kuba (Central Africa), Hausa, Gbe, Akan (West Africa), Rwanda, Maravi, and Funj (East Africa). In the Americas, Aztec and Inca empires arose in the mid-fifteenth century.
40. Lincoln Paine, *Sea and Civilization: A Maritime History of the World* (New York: Alfred A. Knopf, 2013). For an analytical comparison of fields, see Patrick Manning, "Global History and Maritime History," *International Journal of Maritime History* 15 (2013): 1–22.
41. The complement of a ship was therefore a we-group – if not at the moment of recruitment, it had a common purpose by the time the captain had made his will clear. Where several vessels were combined into a fleet, they could form

a larger we-group. The commercial network within which these tightly managed vessels sailed, however, was an I-network, in which most participating groups were there of their own volition.

42. Robert Batchelor, "The Global and the Maritime: Divergent Paradigms for Understanding the Role of Translation in the Emergence of Early Modern Science," Patrick Manning and Abigail Owen (eds.), *Knowledge in Translation: Global Patterns of Scientific Exchange, 1000–1800 CE* (Pittsburgh: University of Pittsburgh Press, 2018), 77–87.
43. These alternations of war and commerce, as hinted by Abu-Lughod, appear to have been inherent in the Old World network of commerce. Abu-Lughod, *Before European Hegemony*.
44. Regina Grafe and Oscar Gelderblom, "The Rise and Fall of the Merchant Guilds: Re-thinking the Comparative Study of Commercial Institutions in Premodern Europe," *Journal of Interdisciplinary History* 40 (2010): 477–79.
45. Van Bavel, *Invisible Hand?*
46. On the expansion of cotton production and trade in Persia, see Richard Bulliet, *Cotton, Climate, and Camels in Early Islamic Iran: A Moment in World History* (New York: Columbia University Press, 2011).
47. Lucassen, Jan, and Leo Lucassen. "The Mobility Transition Revisited, 1500–1900: What the Case of Europe Can Offer to Global History," *Journal of Global History* 4 (2009): 355.
48. Ahmad Ibn Mājid al-Sa'dī, *Kitāb al-fawā'id fī uṣūl 'ilm al-baḥr wa-al-qawā'id* (Damascus: al-Maṭba'ah al-Ta'āwunīyah, 1971).
49. Ibn Majid contrasted European navigation, with its observations of the sun, with Arab navigation, focusing on observation of the stars; he may have learned of innovations in European navigation from Italian merchants in the Indian Ocean. Pat Seed, "Celestial Navigation," in Manning and Owen (eds.), *Knowledge in Translation*, 275–91.
50. During the era of the voluntary and fluctuating Old World commercial network, there were various efforts to build institutions within it: the Hanseatic League of the Baltic; the Fugger banking family of Augsburg; the maritime realm of Majapahit; and the commercial empires of Venice, Genoa, and Portugal.
51. The latitude Columbus selected for his westward voyage was that of Shanghai.
52. Robert Batchelor, "The Global and the Maritime: Divergent Paradigms for Understanding the Role of Translation in the Emergence of Early Modern Science," in Manning and Owen (eds.), *Knowledge in Translation*, 75–90.
53. The date of 1580 was chosen for Map 7.1 because that is a date by which the Manila galleon trade across the Pacific had been added to the other new routes.
54. Volumes of Indian Ocean slave trade tended to fluctuate, while volumes of Atlantic slave trade tended to increase until the nineteenth century.
55. Batchelor, "The Global and the Maritime."
56. Crosby, *Columbian Exchange.*
57. In another tale, still told in Southeast Asia, a Malay man had sailed as crew with the Portuguese to Lisbon, joined the voyage of Magellan, and left the

expedition as it passed Melaka, to return home. The late Adam McKeown related this story to me.

58. Helen M. Wallis and E. D. Grinstead, "A Chinese Terrestrial Globe," *The British Museum Quarterly* 25 (1962): 83–91.
59. As described in the *Yuan shih*, a Ming-era compilation of Yuan-era documents, Jamal al-Din contributed a terrestrial globe of wood, fashioned into "a sphere, seven parts [of which] are water, its color being green, and three parts [of which] are land, its color being white. They drew the streams, rivers, lake, and seas [like] interconnected veins and arteries over the whole [of the sphere]. They also marked off small squares in order to calculate the breadth and length of countries and distance . . . in miles [li] of the roads." The creator of the 1623 globe was aware of this 1267 globe. Thomas T. Allsen, *Culture and Conquest in Mongol Eurasia* (Cambridge: Cambridge University Press, 2011), 107–8; Wallis and Grinstead, "Terrestrial Globe," 83.
60. As a contrast to Guicciardini's views of the sixteenth-century world, Sanjay Subrahmanyam offers summaries of the views of four other sixteenth-century historians of the world. Kenneth R. Bartlett, "Burckhardt's Humanist Myopia: Machiavelli, Guicciardini and the Wider World," *Scripta Mediterranea* 16–17 (1995–96): 17–30; Sanjay Subrahmanyam, "On World Historians in the Sixteenth Century," *Representations* 91, 1 (2005): 26–57.
61. Katharine Park and Lorraine Daston, "Introduction: The Age of the New," *The Cambridge History of Science, vol. 3: Early Modern Science*, ed. Park and Daston (Cambridge: Cambridge University Press, 2008), 15–16.
62. Patrick Manning, "Introduction," in Manning and Owen (eds.), *Knowledge in Translation*, 9–11, 15–16.
63. In addition, the Abbasids sent an archaeological expedition to Egypt, including an early deciphering of hieroglyphics. Hayrettin Yücesoy, "Translation as Self-Consciousness: Ancient Sciences, Antediluvian Wisdom, and the 'Abbāsid Translation Movement," *Journal of World History* 20 (2009): 523–57.

FROM GLOBAL NETWORKS TO CAPITALISM

1. Sebouh David Aslanian, *From the Indian Ocean to the Mediterranean: The Global Trade Networks of Armenian Merchants from New Julfa* (Berkeley: University of California Press, 2011); Claude Markovits, *Global World of Indian Merchants, 1750–1947: Traders of Sind from Bukhara to Panama* (Cambridge: Cambridge University Press, 2000); Curtin, *Cross-Cultural Trade*.
2. On craft and industrial work in various markets, see Maxine Berg, "Skill, Craft, and Histories of Industrialisation in Europe and Asia," *Transactions of the Royal Historical Society* 24 (2014): 127–48.
3. Current research at the International Institute of Social History in Amsterdam is documenting the many categories of labor worldwide as they have changed since 1500. Karin Hofmeester, Jan Lucassen, Leo Lucassen, Rombert Stapel, and Richard Zijdeman, "The Global Collaboratory on the History of Labour Relations, 1500–2000: Background, Set-up, Taxonomy,

and Applications," Working paper, October 2013, http://hdl.handle.net/10622/40GRAD; see also Marcel van der Linden, *Workers of the World: Essays Toward a Global Labor History* (Leiden: Brill, 2008).

4. Karin Hofmeester and Marcel van der Linden (eds.), *Handbook Global History of Work* (Berlin: de Gruyter, 2018); Merry Wiesner-Hanks, *Gender in History: Global Perspectives*, 2nd ed. (Malden, MA: Wiley-Blackwell, 2011).
5. Marcott et al., "Regional and Global Temperature."
6. Geoffrey Parker, *Global Crisis: War, Climate Change, & Catastrophe in the Seventeenth Century* (New Haven, CT: Yale University Press, 2013).
7. As noted in Chapter 7, Ruddiman argues that pandemic-generated population declines in the Old World and then the New World contributed to temperature decline. Ruddiman, *Earth Transformed*, 312–28.
8. Anne Gerritsen, "Ceramics for Local and Global Markets: Jingdezhen's Agora of Technologies," in Dagmar Schäfer (ed.), *Cultures of Knowledge: Technology in Chinese History* (Leiden: Brill, 2011), 161–84.
9. Innovative miners at Potosí developed the technique for extracting silver from ores with the use of mercury in the mid-sixteenth century. Saul Guerrero, "The Environmental History of Silver Mining in New Spain and Mexico, 16c to 19c: A Shift of Paradigm," PhD dissertation, McGill University, 2015. For earlier silver trade, see Akinobu Kuroda, "The Eurasian Silver Century, 1276–1359: Commensurability and Multiplicity," *Journal of Global History* 4 (2009), 245–69.
10. Dennis O. Flynn and Arturo Giráldez. "Born with a 'Silver Spoon': World Trade's Origin in 1571," *Journal of World History* 6 (1995): 201–21.
11. A. G. Frank, *ReOrient: Global Economy in the Asian Age* (Berkeley: University of California Press, 1998); Prasannan Parthasarathi, *Why Europe Grew Rich and Asia Did Not: Global Economic Divergence, 1600–1850* (Cambridge: Cambridge University Press, 2011).
12. As early as 1517 a Portuguese vessel brought cowrie shells from the Indian Ocean, where they were used as money, to Forcados at the Niger River mouth, where they were also used as money. This trade was later dominated by the Dutch. Yang, *Cowrie Shells and Cowrie Money*, 171–73. On pearls, see Molly Warsh, *American Baroque: Pearls and the Nature of Empire, 1492–1700* (Chapel Hill: University of North Carolina Press, 2018).
13. "Population grew at a rate of 0.45% per year while the silver stock grew at a rate of 0.7% per year [from 1700 to 1900] – close to double the rate of population growth and an impressive rate of average global silver holdings, which may mean expanding monetization. In the sixteenth century, population grew at a slower rate (perhaps declining overall), while silver stocks grew at a greater rate." These estimates assume a 1% annual loss in silver stock. Patrick Manning, Dennis O. Flynn, and Qiyao Wang, "Silver Circulation Worldwide: Initial Steps in Comprehensive Research," *Journal of World-Historical Information* 3–4 (2016–17): 10.
14. Slave imports, expanding in the seventeenth century, were employed in the military, as galley slaves, in agriculture, and as domestics. Lucassen and Lucassen, "Mobility Transition," 355.

15. Coffee, a highland crop, was consumed locally in Ethiopia and Yemen until the technique of roasting permitted coffee beans to be transported a distance. The expanding Ottoman and Safavid regimes each facilitated trade in coffee, and coffee houses opened in Istanbul, Isfahan, and elsewhere. Europeans began to purchase coffee in Yemen and were able from the seventeenth century to transplant coffee seedlings to Java and to the Caribbean. Michel Tunscherer, "Coffee in the Red Sea Area from the Sixteenth to the Nineteenth Century," in William Gervase Clarence-Smith and Steven Topik (eds.), *The Global Coffee Economy in Africa, Asia, and Latin America* (Cambridge: Cambridge University Press, 2003), 50–66.
16. Om Prakash, *The Dutch East India Company in Bengal* (Princeton, NJ: Princeton University Press, 1985); Titas Chakraborty, "Work and Society in the East India Company Settlements in Bengal, 1650–1757," PhD dissertation, University of Pittsburgh, 2016.
17. Lauren A. Benton, *Law and Colonial Cultures: Legal Regimes in World History, 1400–1900* (Cambridge: Cambridge University Press, 2002).
18. Ray A. Kea, *Settlements, Trade, and Polities in the Seventeenth-Century Gold Coast* (Baltimore: Johns Hopkins University Press, 1982).
19. The *Siete Partidas*, drawn up to govern slavery in thirteenth-century Castile, did not identify slaves by race, but its enforcement gradually took on a racial tinge. Jacques Gillot and Jean-François Niort, *Le Code Noir* (Paris: Editions le Cavalier Bleu, 2015); Elsa Goveia, "The West Indian Slave Laws of the Eighteenth Century," Revista de Ciencias Sociales 4 (1960): 75–105.
20. El Hamel, *Black Morocco*, 95. In another form of slavery, impoverished Russian peasants sold themselves or their children into slavery, especially during the seventeenth century. With time, their condition was transferred to that of serfdom. Richard Hellie, *Slavery in Russia, 1450–1725* (Chicago: University of Chicago Press, 1982).
21. Even in the Atlantic, where European merchants faced almost no competition, canoe-based African shipping expanded along Africa's west coast in association with long-distance trade carried by Europeans. A. F. C. Ryder, *Benin and the Europeans, 1485–1897* (New York: Humanities Press, 1969), 73–74.
22. Pim de Zwart has kindly pointed out to me citations indicating that the bulk of trade in the Indian Ocean was still dominated by Asian merchants in the late eighteenth century. For Indonesia, see Gerrit Knaap, "Shipping and Trade in Java, c. 1775: A Quantitative Analysis," *Modern Asian Studies* 33 (1999): 405–20; Knaap, "All About Money: Maritime Trade in Makassar and West Java around 1775," *Journal of the Economic and Social History of the Orient* 49 (2006): 482–502; Gerrit Knaap and Heather Sutherland, *Monsoon Traders. Ships, Skippers and Commodities in Eighteenth-Century Makassar* (Leiden: Brill, 2004). For India, see Sushil Chaudhury, *From Prosperity to Decline, Eighteenth Century Bengal* (New Delhi: Manohar, 1995), and Sushil Chaudhury, "The Asian Merchants and Companies in Bengal's export trade ca. Mid-eighteenth century," in Chaudhury and Morineau, *Merchants, Companies and Trade: Europe and Asia in the Early Modern Era* (Cambridge: Cambridge University Press, 1999).

23. The figures in Table 8.1 are somewhat speculative reflections of the two sources; I have interpolated to fill in missing data for certain periods. Matthias van Rossum, *Werkers van de wereld: Globalisering, arbeid en interculturele ontmoetingen tussen Aziatische en Europese zeelieden in dienst van de VOC, 1600–1800* (Hilversum: Verloren, 2014), 63, 65, 70–71; Richard W. Unger, "The Tonnage of Europe's Merchant Fleets 1300–1800," *The American Neptune* 52 (1992): 256–61. See also Jan de Vries for estimates on European shipping to Asia: de Vries, "Connecting Europe and Asia: A Quantitative Analysis of the Cape Route Trade, 1497–1795," in Dennis Flynn, Arturo Giráldez, and Richard von Glahn (eds.), *Global Connections and Monetary History, 1470–1800* (London: Ashgate, 2003), 35–106.
24. Matthias van Rossum notes that the estimates of "Asian shipping" are limited to shore-to-shore voyages within the Indian Ocean; they exclude the significant coastal trades along each shore. Van Rossum, personal communication; Knaap, "All about Money"; Chaudhury, "Asian Merchants."
25. Van Bavel, Invisible Hand?; Gills and Frank, "World System Cycles."
26. Regina Grafe and Oscar Gelderblom, "The Rise and Fall of Merchant Guilds: Re-thinking the Comparative Study of Commercial Institutions in Premodern Europe," *Journal of Interdisciplinary History* 40 (2010), 477–511.
27. In the terms I use here, the enterprise, the merchant guild, and the national proprietors' association all qualify as we-groups and social institutions, with members (organized into a hierarchy) and with common objectives, seeking to establish influence within the global economic network.
28. Kocka notes that enterprises gradually became more focused on profitability. Jürgen Kocka, trans. Jeremiah Riemer, *A Short History of Capitalism* (Princeton, NJ: Princeton University Press, 2016), 21–22, 40–46, 58–64, 69. For other descriptions of enterprises, see Gerritsen, "Jingdezhen's Agora"; Grafe and Gelderblom, "Merchant Guilds"; and Prak, *Citizens without Nations*.
29. "Only in the late eighteenth century did merchant guilds disappear from most parts of Europe . . . . Protection of traders and their goods, contract enforcement, and mitigation of risks largely became functions of either state activity or vertically integrated businesses." Grafe and Gelderblom, "Merchant Guilds," 478–79.
30. In practice, organizations get little detailed attention in this framework. Douglass C. North, John Joseph Wallis, and Barry R. Weingast, *Violence and Social Orders: A Conceptual Framework for Interpreting Recorded History* (New York: Cambridge University Press, 2009). See also Udehn, *Methodological Individualism*; and Daron Acemoglu and James A. Robinson, *Why Nations Fail: The Origins of Power, Prosperity, and Poverty* (London: Profile, 2012), 209–12.
31. Further, institutions must ultimately demonstrate their institutional fitness. Searle, *Construction of Social Reality*; and Tuomela, *Social Ontology*. For further discussion of Tuomela's insights, see Gerhard Preyer and Georg Peter (eds.), *Social Ontology and Collective Intentionality: Critical Essays on*

the Philosophy of Raimo Tuomela with His Responses (Cham, Switzerland: Springer, 2017).

32. Up to this point, of course, my analysis of the causes of group formation and institutional change is qualitative rather than quantitative, though it does allow for comparison with the rise of other institutions.
33. Alston et al. define institutions as "rules that recognized authorities create and enforce." Eric Alston, Lee J. Alston, Bernardo Mueller, and Tomas Nonnenmacher, *Institutional and Organizational Analysis: Concepts and Applications* (Cambridge: Cambridge University Press, 2018), 1, 26–27, 276–82. See also Leo Lucassen, "Working Together: New Directions in Global Labour History," *Journal of Global History* 11 (2016): 66–87.
34. Of the resources on which I rely in this section, others use the term "institution," sometimes with descriptions but without definitions: see Kocka, *Capitalism*; Van Bavel, *Invisible Hand?*; Prak, *Citizens without Nations*; and Grafe and Gelderblom, "Merchant Guilds."
35. Charles, born 1500, became Duke of Burgundy in 1506, became king of Castile in 1516, and Holy Roman Emperor in 1519. He abdicated in 1556 and died in 1558. Later English voyager Francis Drake had circumnavigated the world 1577–80, seizing Spanish treasure from two vessels near Lima.
36. Pepijn Brandon, *War, Capital, and the Dutch State (1588–1795)* (Leiden: Brill, 2015), 2.
37. The English settlement at Jamestown was nearly eliminated by the cold. Sam White, *A Cold Welcome: The Little Ice Age and Europe's Encounter with North America* (Cambridge, MA: Harvard University Press, 2017).
38. Brandon adopts the term "historic bloc" from Gramsci and draws on Tilly's notion of brokerage. Brandon, *War, Capital*, 22–24, 33–34.
39. Brandon, *War, Capital*, 288–89.
40. Arguably, this proposal was rather close to the relationship achieved, if briefly, in 1688. Brandon, *War, Capital*, 50–51.
41. Brandon, *War, Capital*, 86, 100–01.
42. Regarding Louis XIV of France, Black argues that, "A key element of Louis's bellicosity was that he had little realistic sense of how the wars he began would develop diplomatically, militarily, or politically." Black, *War and Its Causes*, 50. See also Brandon, *War, Capital*, 86–93.
43. The English, after their surrender to the Mughals, were allowed to stay in Bombay rather than be expelled.
44. On the role of French merchants, see Abdoulaye Ly, *La Compagnie du Sénégal* (Paris: Présence africaine, 1958).
45. Followers of Wallerstein have applied the term "world-system" to earlier times and smaller regions.
46. This point was developed more fully in Paul Kennedy, *The Rise and Fall of the Great Powers* (New York: Random House, 1988).
47. Wallerstein does not mention the VOC in volume 1 of *The Modern World-System* but argues explicitly in volume 2 that it should be excluded from analysis. Wallerstein, *Modern World-System*, 2: 46–51.

48. Similarly, Wallerstein excluded Africa, the source of essential slave labor. "But why Africans as the new slaves? . . . it had to be from a region that was outside its world-economy so that Europe could feel unconcerned about the economic consequences for the breeding region of wide-scale removal of manpower as slaves." Wallerstein, *Modern World-System,* 1: 89. He affirms in volume 2 that Africa was external to the system: Wallerstein, *World-System,* 2: 9, 17. Analyzing a system does require attributing borders to the system, but Wallerstein had the option of tracing the movement of goods across borders rather than assuming an isolated system. Other examples of cross-regional links include Russian exchange of furs with China and Iran and the substantial commerce linking India, Southeast Asia, and East Asia.
49. Kocka, *Short History of Capitalism,* 6–24.
50. Ibid., 63–64.
51. Ibid., 54–84, plus later chapters for the era after 1800. For a survey of social theories relevant to citizenship, see Prak, *Citizens without Nations,* 1–17; on capitalism, see van Bavel, *Invisible Hand?,* 270–76.
52. Nuala Zahedieh, *The Capital and the Colonies: London and the Atlantic Economy 1660–1700* (Cambridge: Cambridge University Press, 2010). Her key individual example was Sir Gilbert Heathcote (1652–1733), a London merchant whose trade connected him to the Baltic, Africa, Spain, Newfoundland, New York, and Jamaica. His fortune reached £ 750,000 by the end of his life.
53. Zahedieh, *The Capital and the Colonies,* 126–36.
54. Zahedieh amply demonstrates the commercial success of the leading London merchants but gives little detail on their influence on national policies. Brandon, providing less detail than Zahedieh on Dutch merchant communities, gives a specific picture of their political influence, though emphasizing that it had to be accomplished in coalition with other factions. Zahedieh, *The Capital and the Colonies*; Brandon, *War, Capital.*
55. Brandon, *War, Capital,* 13; Giovanni Arrighi, *The Long Twentieth Century: Money, Power, and the Origins of Our Times* (London: Verso, 1994).
56. Jonathan I. Israel, *Dutch Primacy in World Trade, 1585–1740* (Oxford: Clarendon Press, 1989).
57. Van der Linden, *Workers of the World.* For other leading contributions drawn from a large literature, see E. P. Thompson, *The Making of the English Working Class* (New York: Pantheon, 1964); and Peter Linebaugh and Marcus Rediker, *The Many-Headed Hydra: Sailors, Slaves, Commoners, and the Hidden History of the Revolutionary Atlantic* (Boston: Beacon, 2000).
58. For a discussion on institutional change and growth, see Pim de Zwart and Jan Luiten van Zanden, *The Origins of Globalization: World Trade in the Making of the Global Economy, 1500–1800* (Cambridge: Cambridge University Press, 2018), 252–57.
59. De Zwart and van Zanden have taken a step forward by integrating global summaries and detailed chapters on most world regions, tracing growth and connection in trade from the early second millennium to 1800: de Zwart and van Zanden, *Origins of Globalization.* For major earlier studies, focusing

principally on European economic expansion, see Douglass C. North and Robert Paul Thomas, *The Rise of the Western World* (Cambridge: Cambridge University Press, 1973); Eric L. Jones, *The European Miracle*; David S. Landes, *The Wealth and Poverty of Nations* (Boston: Little, Brown, 1998); and Nathan Rosenberg and L. E. Birdzell, Jr., *How the West Grew Rich: The Economic Transformation of the Industrial World* (New York: Basic Books, 1986). Global explorations included Wallerstein, *Modern World-System*; Samir Amin, *L'accumulation à l'échelle mondiale: critique de la théorie du sous-développement* (Dakar : IFAN, 1970); and Andre Gunder Frank, *World Accumulation 1492–1789* (New York: Monthly Review Press, 1978. See also Paul Bairoch, *Le Tiers-Monde dans l'impasse. De démarrage économique du XVIIIe au XXe siècle* (Paris: Gallimard, 1971).

60. De Zwart calculated three measures of price convergence for 13 commodities. Convergence was in fact strongest for those products bought in competitive markets in China and India, like tea, silk, textiles, and saltpeter. Pim de Zwart, "Globalization in the Early Modern Era: New Evidence from the Dutch-Asiatic Trade, c. 1600–1800," *Journal of Economic History* 76 (2016): 526–31; de Zwart, personal communication. For parallel work on Atlantic price convergence see, for instance, Paul Sharp and Jacob Weisdorf, "Globalization Revisited: Market Integration and the Wheat Trade between North America and Britain from the Eighteenth Century," *Explorations in Economic History* 50 (2013): 88–98. See also de Zwart and van Zanden, *The Origins of Globalization.*
61. De Zwart, "Globalization," 541–43. Overall, de Zwart's estimates provide the profit margins of European merchants as a contribution to the study of globalization: one may hope for future studies of the profits of Asian merchants.
62. Kevin H. O'Rourke and Jeffrey G. Williamson, *Globalization and History: The Evolution of a Nineteenth-Century Atlantic Economy* (Cambridge, MA: MIT Press, 1999).
63. Colin Palmer, *Human Cargoes: The British Slave Trade to Spanish America, 1700–1739* (Urbana: University of Illinois Press, 1981).
64. Gregory Clark and David Jacks, "Coal and the Industrial Revolution, 1700–1869," *European Review of Economic History* 11 (2007): 39–72.
65. While Indian armies too relied on this saltpeter, they never managed to deny its supply to European purchasers.
66. Frank. *ReOrient*; R. Bin Wong, *China Transformed: Historical Change and the Limits of European Experience* (Ithaca, NY: Cornell University Press, 1997); Kenneth Pomeranz, The Great Divergence: China, Europe, and the Making of the Modern World Economy (Princeton, NJ: Princeton University Press, 2000).
67. Population growth enabled aggregate Chinese output to grow even as per capita output declined. Stephen Broadberry, Hanhui Guan, and David Daokui Li, "China, Europe and the Great Divergence: A Study in Historical National Accounting, 980–1850," *Journal of Economic History* 78 (2018): 955–1000.
68. He adds, " . . . if culture does evolve, it is logically impossible for it to do so on Darwinian lines." Fernández-Armesto, *A Foot in the River*, 148, 163.

69. These quotes appear to apply especially to group-level culture, more so than to individual-level culture. Fernández-Armesto, *A Foot in the River*, 178, 180.
70. Various meanings of the terms "culture" and "cultural" are discussed in this volume. Here I consider categories of group culture that refer to practices from the small scale of shared daily meals to the massive scale of civilizational identity.
71. One must make an exception for China, where tobacco and sweet potatoes seem to have arrived rapidly. The Chinese administration appears to have speeded the spread of several American crops.
72. The North American banjo, developed out of African models, was better able to survive and expand because of its similarity to the guitar. Similarly, the balalaika, a stringed instrument shared in Russian and Kazakh traditions, was reinforced by guitars played in Black Sea ports.
73. Ning Ma, *The Age of Silver: The Rise of the Novel East and West* (New York: Oxford University Press, 2018). For a valuable compendium of new thinking that links world literature and world history, see May Hawas (ed.), *The Routledge Companion to World Literature and World History* (London: Routledge, 2018).
74. In an example of global vision, Jahangir, the seventeenth-century Mughal sultan, took a throne name meaning "world crusher," indicating something about his ambition if not his reach.
75. Dena Goodman, *The Republic of Letters: A Cultural History of the French Enlightenment* (Ithaca, NY: Cornell University Press, 1994).
76. Sir William Jones is thought to have relied significantly on Indian scholarship in his study of Sanskrit; on the other hand, Montesquieu, in his *Lettres persanes*, relied on imaginary rather than actual Persian scholars. Sebastian Conrad, "Enlightenment in Global History: A Historiographical Critique," *American Historical Review* 117 (2012): 999–1027. For a visual representation of the European Republic of Letters, see http://republicofletters.stanford.edu; see also Muhsin J. al-Musawi, *The Medieval Islamic Republic of Letters: Arabic Knowledge Construction*; and Subrahmanyam, "On World Historians in the Sixteenth Century."
77. These cycles of exchange in global and scientific analysis might best be studied over the long term, including analysis and voyages of Ptolemy (Roman times), Xuanzhang (Tang era), al-Khwarizmi (Abbasid), Ibn Sina (late Abbasid), al-Tusi and Rashid al-Din (Mongol era), Hernández (Spanish empire), Linnaeus and Cook (eighteenth century), and Humboldt (nineteenth century).
78. James Kloppenberg, *Toward Democracy: The Struggle for Self-Rule in European and American Thought* (New York: Oxford University Press, 2016).
79. Ibid., 6.
80. Ibid., 13.
81. Benedict Anderson, *Imagined Communities: Reflections on the Origin and Spread of Nationalism*, rev. ed. (London: Verso, 2006).
82. On Cherokee, see Robert J. Conley, *The Cherokee Nation: A History* (Albuquerque: University of New Mexico Press, 2008); on Tipu Sultan, see Parthasarathi, *Why Europe*, 198–200, 207–08; on Muhammad Ali

Pasha, see Khaled Fahmy, *All the Pasha's Men: Mehmed Ali, His Army and the Making of Modern Egypt* (Cambridge: Cambridge University Press, 1997).

SYSTEMIC THREATS

1. My choice for the turning point is based on what I think was the biggest shift in the Human System, 1800–50, rather than the biggest shift in the natural world. For the latter, observers have proposed dates as early as 1600 CE and as late as 1960 CE. See J. R. McNeill, *Something New under the Sun: An Environmental History of the Twentieth-Century World* (New York: W. W. Norton, 2000); and Ruddiman, *Earth Transformed*. An early concept for the Anthropocene was the Noosphere of Vladimir Vernadsky, who in 1938 wrote of "scientific thought as a geological force." Scientists in the Soviet Union appear to have used the term "Anthropocene" as early as the 1960s to refer to the Quaternary, the most recent geological period. Ecologist Eugene F. Stoermer subsequently used "anthropocene" with a different sense in the 1980s; the term was widely popularized in 2000 by atmospheric chemist Paul J. Crutzen, who regarded the influence of human behavior on Earth's atmosphere in recent centuries as so significant as to constitute a new geological epoch.
2. In describing this natural world, I persist in labeling it as Gaia, to pair it with the Human System and focus on the balance or imbalance of the two. Gaia is not the whole of the natural world – it is the living creatures and organic material of the Earth along with adjacent inorganic materials, whose interactions preserve a temperature and chemical composition that enables life to survive – up to a point.
3. In this institutional framework, is nineteenth-century capitalism to be identified as an I-group network? Or perhaps as a we-group network or a we-group hierarchy? Hypotheses and analysis on these points can be developed if the framework is seen to be of interest.
4. On evolutionary dimensions of cotton fabrics and the technological change of breeding, see Russell, *Evolutionary History*, 103–44.
5. See Osterhammel's excellent chapter on "Civilization and Exclusion," especially pp. 834–36." Jürgen Osterhammel, *The Transformation of the World: A Global History of the Nineteenth Century* (Princeton, NJ: Princeton University Press, 2014), 826–72.
6. Kenneth Pomeranz, *The Great Divergence: China, Europe, and the Making of the Modern World Economy* (Princeton, NJ: Princeton University Press, 2000); Andre Gunder Frank, ed. Robert A. Denemark, *ReOrienting the Nineteenth Century: Global Economy in the Continuing Asian Age* (Boulder, CO: Paradigm Publishers, 2014).
7. Gran identifies the period after 1850 as "a period during which one sees the banding together of a growing number of the world's dominant elements in an increasing interdependence as they seek to promote market capitalism." Peter Gran, *The Rise of the Rich: A New View of Modern World History* (Syracuse, NY: Syracuse University Press, 2009), 100.

8. Texas had broken away from Mexico in 1836 to form a slave-holding republic led by American settlers; Texas was annexed by the United States in 1845, and war between the United States and Mexico broke out in 1846.
9. In the Second Opium War, 1856–60, also known as the Arrow War, Britain pressed on China to expand the scale of treaty ports, using an incident with the vessel "Arrow" as a pretext for the first of two occupations of Guangzhou by British and French troops. The 1860 Convention of Beijing ended the war, expanding treaty ports further and allowing opium trade to continue.
10. For a broad and effective survey of cotton and empire that sets the US Civil War in global context, see Sven Beckert, *Empire of Cotton: A Global History* (New York: Alfred A. Knopf, 2014).
11. Further, in a conflict of great powers, the Crimean war of 1853–56 pitted Britain, France, and the Ottoman Empire against the expanding Russian empire, with a defeat for Russia, and with casualties of nearly a million dead.
12. By 2020, the number of monarchies had declined to 29 out of 193 UN members; all but 10 of them had constitutional governments.
13. Haiti – more precisely the northern half of Haiti – was a monarchy for most of the time from 1804 to 1818.
14. A. G. Hopkins,"Globalization with and without Empires: From Bali to Labrador," in A. G. Hopkins (ed.), *Globalization in World History*, 222–30 (New York: W. W. Norton, 2002). See also Hopkins, *American Empire: A Global History* (Princeton, NJ: Princeton University Press, 2018).
15. The strangest case of all was that of Belgium, where the constitutional monarch had almost no power at home, but where King Leopold II managed in the course of the 1880s to gain personal control of an immense African empire – virtually the whole of the Congo basin – for more than 20 years through adept playing of diplomatic games. Adam Hochschild, *King Leopold's Ghost: A Story of Greed, Terror, and Heroism in Colonial Africa* (Boston: Houghton Mifflin, 1998).
16. The list included Canada, Brazil, Argentina, Mexico, Chile, and even others.
17. Decolonization also meant Balkanization. Efforts to create regional federations, as of Arab states or African states, failed in part because of big-power opposition. European countries, however, were able to move step by step toward creating a regional union.
18. Arabic and Mandarin each spread widely, without direct imperial support.
19. Hopkins shows that, with the opening of this phase, the United States sought to maintain its Cold War–era economic and military hegemony while competing nations grew in influence. Hopkins, *American Empire*, 821–46.
20. Hannah Ritchie and Max Roser, "Our World in Data," "Global Primary Energy by Source." https://ourworldindata.org, citing "BP Statistical Review of World Energy" (online) and Vaclav C. Smil, *Energy Transitions: Global and National Perspectives*, 2nd ed. (Santa Barbara, CA: Praeger, 2017).
21. From its 1859 start to 1910, US crude oil production reached 0.6 million barrels per day and rose to a peak of 10 million barrels per day in 1970. World crude oil production rose from 10 million barrels per day in 1950 to 75 million barrels per day in 2000.

22. United Nations Conference on Trade and Development, *Review of Maritime Transport, 2017.* https://unctad.org/en/PublicationsLibrary/rmt2017_en.pdf.
23. World population had grown from roughly 570 million in 1600, at a rate of less than 0.3% per year.
24. That is, roughly 4 billion people live in cities.
25. Patrick Manning with Tiffany Trimmer, *Migration in World History*, 2nd ed. (London: Routledge, 2012), 148–65.
26. Jan Luiten van Zanden et al., *How Was Life? Global Well-Being since 1820* (Amsterdam: OECD, 2014), 42. I have made corrections to the African population for 1820.
27. Hopkins, *American Empire*, 705–06.
28. Jean-Baptiste Fressoz and Christophe Bonneuil,"Growth Unlimited: The Idea of Infinite Growth from Fossil Capitalism to Green Capitalism," in Iris Borowy and Matthias Schmelzer (eds.), *History of the Future of Economic Growth: Historical Roots of Current Debates on Sustainable Growth* (London: Routledge, 2017), 52–68.
29. Marcus Tullius Cicero (106–43 BCE), the Roman lawyer, statesman, and writer, attributed this term to Cassius.
30. Biological species are made up of modular organisms, each composed of modular organs and cells: fitness resulting from genomic change is most obviously measured through offspring of the organism and the species. Human society is made up of more flexible societies, communities, and subgroups. Fitness resulting from institutional change must be measured by welfare of a society or the species, but in practice the measurement may arguably be made based on various subgroups. Measuring fitness based on a beneficiary population allows desirable flexibility but introduces ambivalence and debate into the results.
31. As I grew up in Los Angeles, the police cars all bore the slogan, "To protect and serve." That slogan implicitly identified the inhabitants of the city as the beneficiary population, the slogan as institutional objective of the police force, and the police force as members of the institution; in addition, the police institution clearly had its directors.
32. The literature on trade unionism is huge and disaggregated. For the most wide-ranging narrative of trade unions, 1760–1950, see William Z. Foster, *Outline History of the World Trade Union Movement* (New York: International Publishers, 1956).
33. Samuel Gompers, president of the craft-oriented American Federation of Labor, was enabled by his association with US President Woodrow Wilson to gain a leading role in drafting its constitution. The United States, however, did not join the ILO until 1934, the same year that the Soviet Union joined.
34. Since the ILO, the Soviet Union, and the RILU were all new institutions, one could subject them to a parallel analysis of the processes of innovation, reproduction, selection, and fitness. The Soviet Union was a member of ILO 1933–37 and again beginning 1954. The RILU was dismantled in 1937.
35. Patrick Manning, "The Life-Sciences, 1900–2000: Analysis and Social Welfare from Mendel and Koch to Biotech and Conservation," *Asian Review of World Histories* 6 (2018), 185–208.

36. These included arguments that unions and their members represent a special interest, while corporations represent a more general interest. that unions oppose economic freedom, supporting socialism, that they are socially prejudiced (racist and sexist), that they support a labor aristocracy, undeserving of their wages and benefits, that union leaders are corrupt, that workers should rely simply on employers and not pay to support an employee organization.
37. I have chosen to give minimal attention to the place of socialism in this survey of trade unionism. I simplified the argument in order to illustrate the application of principles of social evolution to the dynamics of trade unions. With more time and space, one could prepare a similar analysis of the social evolution of socialism and also a broader analysis tracing the overlaps and oppositions of trade unionism and socialism.
38. For each of these social movements, one can pose the full set of questions posed here for trade unions, regarding their objectives, their membership, their activities, the changes they achieved, and the opposition they faced.
39. The Detroit Chaldeans are migrants from an Iraqi Christian community who maintain a language descended from ancient Aramaic. Natalie Henrich and Joseph Henrich, *Why Humans Cooperate: A Cultural and Evolutionary Explanation* (Oxford: Oxford University Press, 2007).
40. In the terms I have used here, a group of cooperating people is an "I-group." Henrich and Henrich, *Why Humans Cooperate*, 38–39, 57.
41. Henrich and Henrich, *Why Humans Cooperate*, 38–39.
42. Ibid., 51.
43. Ibid., 55.
44. Ibid., 180–84.
45. The 1971 collapse of the postwar Bretton Woods system of currency rates, associated with American debt in the Vietnam War, brought another dimension to economic change.
46. Iris Borowy, "Science and Technology for Development in a Postcolonial World. Negotiations at the United Nations, 1960–1980," *NTM Journal of the History of Science, Technology and Medicine* 26 (2018): 31–62.
47. Donnella H. Meadows et al., *The Limits to Growth: A Report for the Club of Rome's Project on the Predicament of Mankind* (New York: Universe Books, 1972); Jay W. Forrester, *World Dynamics* (Cambridge, MA: Wright-Allen Press, 1971).
48. Rachel Carson, *Silent Spring* (New York: Fawcett Crest, 1962). Environmental concerns arose not only in Europe and North America, but also in industrializing Asian countries and wherever waste was dumped.
49. In 1978 the United States banned use of Freon, a chlorofluorocarbon, in aerosol cans; the Vienna Protocol (1985), and Montreal Protocol (1987) completed international agreement on restricting these chemicals.
50. Hayek, born in Austria in 1899, moved to Britain in 1931, to the United States in 1950, and to Germany in 1962.
51. For an outstanding statement of the wide-ranging outlook in biological studies, see François Jacob, "Evolution and Tinkering," *Science* 196 (1977): 1161–66. For social evolution I assume, in contrast to Hayek, the conscious development of institutions. I therefore assume an important process of regulating and reforming existing institutions rather than a take-it-or-leave-it approach to

existing institutions. F. A. Hayek, *Studies in Philosophy, Politics and Economics* (Chicago: University of Chicago Press, 1967); D. Schmidtz, "Friedrich Hayek," *Stanford Encyclopedia of Philosophy* (updated December 14, 2016), h ttps://plato.stanford.edu/entries/friedrich-hayek/. For a vision of neoliberalism that locates its origins in the 1920s with Ludwig von Mises, see Philip W. Magness, "The Pejorative Origins of the Term 'Neoliberalism'," *AIER*, www.aier.org/article,pejorative-origins-term-"neoliberalism." See also Manning, *Methods*.

52. Population thinking was an important advance in biological thinking, led by R. A. Fisher in his development of the field of population genetics: it treats the elements of a population as equal individuals. Yet population thinking neglects possible constituents within population elements and minimizes possible groupings of population elements.
53. In this reasoning, Hayek left out history. The Human System and its networks dealt with the allocation of scarce resources from the beginning but did not gain a formal price network until a few thousand years ago. Other aspects of human networks deal with social and cultural issues, not necessarily by way of profit maximization.
54. Rational choice theory, emphasizing individual-level decisions, also gained wide attention at this time in economics, sociology, and other fields.
55. Irving Janis's 1982 critique of "groupthink" shows how desire for conformity in a group can lead to irrational decisions. While the 1972 first edition focused on US foreign policy disasters, the second edition reinforced aspects of neoliberal thinking by expanding skepticism about group behavior. Irving L. Janis, *Groupthink: Psychological Studies of Policy Decisions and Fiascoes*, 2nd ed. Boston: Houghton Mifflin Co., 1982.
56. Thatcher thus rejected the idea of multiple perspectives at the very moment when multiplicity was being recognized increasingly in all fields of knowledge; yet the device served her party well in marginalizing other perspectives.
57. The US left UNESCO in 1984, returning in 2003 and leaving again in 2017. The UK left in 1985 and returned in 1997. Singapore left UNESCO in 1984, returning in 2007. Nevertheless, local groups in the USA and UK gave active support to UNESCO after 1985; academic organizations in those countries continued to participate in UN-affiliated international academic organizations.
58. Naomi Oreskes and Erik M. Conway, *Merchants of Doubt: How a Handful of Scientists Obscured the Truth on Issues from Tobacco Smoke to Global Warming* (New York: Bloomsbury Press, 2010).
59. Naomi Oreskes and Erik M. Conway, "Science Isn't Enough to Save Us" (*New York Times*, October 17, 2018).
60. Vanessa Ogle, "Archipelago Capitalism: Tax Havens, Offshore Money, and the State, 1950s–1970s," *American Historical Review* 122 (2017): 1431–58.
61. Van Bavel, *Invisible Hand?*; Gran, *Rise of the Rich*.
62. For instance, in the late twentieth century at the national level in Iran, the Philippines, Haiti, and South Africa.
63. Patrick Manning, "Songs of Democracy," http://manning.pitt.edu.
64. Aiqun Hu, *China's Social Insurance in the Twentieth Century: A Global Perspective* (Leiden: Brill, 2015).

65. Adam J. Tooze, *Crashed: How a Decade of Financial Crises Changed the World* (New York: Viking, 2018).
66. J. R. McNeill and Peter Engelke, *The Great Acceleration: An Environmental History of the Anthropocene since 1945* (Cambridge, MA: The Belknap Press of Harvard University Press, 2014).
67. Gregory Cushman has emphasized the general implications of the current focus on extraction of the lithosphere. Gregory T. Cushman, *Guano and the Opening of the Pacific World: A Global Ecological History* (New York: Cambridge University Press, 2013).
68. He noted concerns over the "time constant" or the time that it took Gaia to reach equilibrium after changes in its conditions, and the "loop gain" or the specific magnitude of feedback, and whether it had positive or negative effects on Gaia's adjustments. Lovelock, *The Revenge of Gaia*; see also Toby Tyrrell, *On Gaia: A Critical Investigation of the Relationship between Life and Earth* (Princeton, NJ: Princeton University Press, 2013).
69. Hansen, James, *Storms of My Grandchildren: The Truth about the Coming Climate Catastrophe and Our Last Chance to Save Humanity* (London: Bloomsbury, 2009).
70. Thomas Piketty, trans. Arthur Goldhammer, *Capital in the Twenty-first Century* (Cambridge, MA: The Belknap Press at Harvard University Press, 2014); Branko Milanovic, *Global Inequality: A New Approach for the Age of Globalization* (Cambridge, MA: The Belknap Press at Harvard University Press, 2016).
71. Patrick Manning, "Inequality: Historical and Disciplinary Approaches," *American Historical Review* 122 (2017): 1–22; Patrick Manning, *Big Data in History* (Houndmills, UK: Palgrave Macmillan, 2013). The CHIA project, described in this report, has been able to identify problems in building a global dataset but has not achieved substantial funding. In this regard it is a successor to the work of Jay Forrester. Forrester, *World Dynamics.*
72. James E. Hansen, "Scientific Reticence and Sea Level Rise," *Environmental Research Letters* 2 (May 2007).
73. The administration of South African President Thabo Mbeki, 1999–2008, denied that HIV virus causes the symptoms of AIDS, with costly results; elsewhere, the widespread recurrence of measles because of refusal to vaccinate is another case of denial. On the more encouraging case of Ebola in Nigeria, see Katherine Harmon Courage, "How Did Nigeria Quash Its Ebola Outbreak So Quickly?" *Scientific American*, October 18, 2014.
74. David Wallace-Wells, *The Uninhabitable Earth: Life After Warming* (New York: Tim Duggan Books, 2019).
75. In a further irony, another source of petrochemicals awaits exploitation: the methane hydrates that have accumulated under permafrost and on underwater continental slopes. Efforts to exploit them will serve further to raise temperatures. See Gerald R. Dickens, "Methane Hydrates and Abrupt Climate Change," *Geotimes* (November 2004), www.geotimes.org/nov04/feature_climate.html.
76. Health conditions are converging – 2018 life expectation averaged 79 years in North America and 76 years in Latin America; literacy rates for the two

regions are also very close. But numbers in poverty remain unchanged while concentration of wealth increases.

77. See the indications of transfers of wealth, since 1980, to the wealthiest as given in "World Inequality Report, 2018," https://wir2018.wid.world/files/download/wir2018-summary-english.pdf
78. "Social sciences and the humanities, especially in their most prestigious bastions, showed themselves scarcely more attuned to the advent of the Anthropocene than governments floundering with energy policy and climate politics. The intellectual flight from reality made it slightly easier for those in positions of power to avoid facing up to it." McNeill and Engelke. *The Great Acceleration,* 208–09.

HOPE FOR ADAPTATIONS

1. Specialists in biological and cultural evolution have developed elaborate and explicit models for the reproduction and behavior of humans according to individual-level logic. In recent years, they have sought to show that individual-level logic can lead to the development of collaborative behavior. Social scientists (economists and sociologists) have explored individual-level logic, relying on simple models. Historians and some anthropologists have assumed a mix of group and individual behavior, but without modeling the specifics in any detail. The proponents of these three individual-level approaches to human behavior have had little discussion with each other.
2. As I have noted, Raimo Tuomela has arguably demonstrated through game theory that we-groups function at a level that is not reducible to individual-level behavior. Tuomela, *Social Ontology*, 179–213.
3. This argument, that popular culture is raising the prominence and relative strength of commoners in society, brings up the question of the relative strength of commoners over time. Is there a way to track this issue for the last few millennia?
4. The medallion was created in 1787 and the engraving of the slave ship was completed in 1789; both were approved and circulated by the Society for Effecting the Abolition of the Slave Trade, based in London.
5. Seymour Drescher, *Abolition: A History of Slavery and Antislavery* (Cambridge: Cambridge University Press, 2009); Sylviane Diouf, ed. *Fighting the Slave Trade: West African Strategies* (Athens, OH: Ohio University Press, 2003).
6. Nile Green, "Persian Print and the Stanhope Revolution: Industrialization, Evangelicism, and the Birth of Printing in Early Qajar Iran," *Comparative Studies of South Asia, Africa and the Middle East* 30 (2010): 473–90.
7. Manning, *African Diaspora*, 209–82.
8. "In the many art-producing centers in the rest of the world outside Euroamerica, a variety of local negotiations between indigeneity, tradition, modernity, and globalization led, first, to the forging of distinct kinds of modern art, and then, in artistic exchanges within nearby regions and with distant centers, the emergence of specific kinds of contemporary art." Terry Smith, "Contemporary Art: World Currents in Transition Beyond

Globalization," in Hans Beiting, Andrea Buddensteig, and Peter Weibel (eds.), *The Global Contemporary and the Rise of New Art Worlds* (Cambridge, MA: MIT Press, 2013), 186–92, at 192.

9. John Dower, *War without Mercy: Race and Power in the Pacific War* (New York: Pantheon, 1986).
10. One could use the term "information" as well as "knowledge," though "information" tends to refer to data that can come under proprietary control in digital form. In either case, one could measure the volume of knowledge in book publications for the nineteenth century or in terabytes of stored information today.
11. One easy and popular linkage was to create literary works that drew on the novelties in science and technology: Jules Verne was a successful practitioner of this approach.
12. An important but unsuccessful effort to identify and limit the privileging of corporate interests over societal interests was the Club of Rome, formed in 1968, and its argument for "limits to growth." It gained wide attention but was ideologically defeated. Its research project made some advances in systems dynamics but did not gain deep involvement of social scientists and historians. One of many negative responses was *Models of Doom.* H. S. D. Cole et al. (eds.), *Models of Doom: A Critique of the Limits to Growth* (New York: Universe Books, 1973). The volume includes a detailed and valuable response of the authors of *Limits to Growth* to this critique, pp. 217–40. See also Meadows, *Limits to Growth.*
13. In its September 2018 issue, *Scientific American* focused on "A Singular Species: The Science of Being Human."
14. Mahmood Mamdani, *Citizen and Subject: Contemporary Africa and the Legacy of Late Colonialism* (Princeton, NJ: Princeton University Press, 1996); Frederick Cooper, *Citizenship, Inequality, and Difference: Historical Perspectives* (Princeton, NJ: Princeton University Press, 2018).
15. The most populous seven nations are China, India, United States, Indonesia, Brazil, Pakistan, and Nigeria, all with more than a projected 200 million population in 2020. The next most populous seven nations are Bangladesh, Russia, Japan, Mexico, Japan, Philippines, and Ethiopia, all with more than a projected 100 million population in 2020. Nigeria is expected to become the world's third most populous nation by 2050, with 400 million inhabitants.
16. Benedict Anderson's classic study of the spread of nationalism might reasonably be revised again, to account for a world in which nations, large and small, alternate between inclusion and exclusion. Benedict R. Anderson, *Imagined Communities: Reflections on the Origin and Spread of Nationalism*, rev. ed. (London: Verso, 2006).
17. Of the top 20 philanthropic foundations listed in 2019 by Wikipedia, 11 were based in the United States, 5 in Europe, 3 in Asia, and 1 in Canada.
18. Social movements are not limited to commoners. As a clear example, supporters of the ideology of neoliberalism formed alliances throughout business communities and beyond in the late twentieth century, achieving great advances in their campaign to reduce regulation of corporations. This group too participates in global democratic discourse.

19. Rumors of the "extinction" of the Taino and other populations of the Caribbean are greatly exaggerated, especially in that many Taino women had children with immigrant men and raised their children with attention to Taino culture.
20. The Arctic Council (https://arctic-council.org), established in 1996 as an intergovernmental organization including eight nations, meets with the Arctic Council Indigenous Peoples Secretariat (www.arcticpeoples.com), established in 1994 by six indigenous organizations.
21. Kloppenberg's insights on democracy, discussed briefly for the eighteenth century in Chapter 8, are worthy of extension for the Anthropocene epoch. As I see it, the question is whether it will be possible to match, at the global level, those principles that have been helpful at the national level: popular sovereignty, autonomy, equality, deliberation, pluralism, and reciprocity. Kloppenberg, *Toward Democracy.*
22. Examples include the Gates Foundation, the UN Development Program, and Amnesty International.
23. Prasenjit Duara, *The Crisis of Global Modernity: Asian Traditions and a Sustainable Future* (Cambridge: Cambridge University Press, 2015).
24. Gran, *Rise of the Rich.*
25. Patrick Manning, "1789–1792 and 1989–1992. Global Interactions of Social Movements," *World History Connected* 3, 1 (2005) http://worldhistoryconnected.org/3.1/manning.html .
26. Discussion of such issues could conceivably take place at two institutions that gather leaders of contending outlooks: the World Economic Forum at Davos and the World Social Forum, based at Porto Alegre. Are there conditions under which both might challenge the dogma of growth, seeking ways to get benefits of growth but on a smaller scale? Or conditions under which both might challenge the dogma of self-regulation and explore new means of regulating institutions in order to advance social welfare?

Glossary

Terms are listed alphabetically, as given in the text. Where possible, terms are followed by a bracketed letter indicating the field or discipline within which each term is used. Some glossary terms are more general and apply to multiple fields. In some cases a term is used in two fields: for instance, [C, S]

Key:
B = biological evolution
C = cultural evolution
G = Gaia
H = Human System
S = social evolution

Archive, cultural. [C] The archive in cultural evolution is the brains of individuals in which learning is stored and from which it is transferred to others.

Archive, genetic. [B] The archive in genetic evolution consists of DNA molecules, distributed throughout the cells and organisms of humanity.

Archive, social. [S] The archive in social evolution is distributed throughout the organisms, institutions, and material resources of humanity. It dependably preserves information that can generate social behavior. Because the human social order exists at many levels, the archive must also exist in many places and at multiple levels. The archive includes both materiality and behavior: it includes brain cells; parental practices; the wisdom of senior persons, institutions, rituals; and doubtless more.

Beneficiary population. [S] The population whose social welfare is the basis for assessing the fitness of a social institution.

Capitalism. [S] A set of institutions with economic objectives, working at multiple scales, including enterprises to achieve profit in production of goods and services in local and global trade, in combination with associations of proprietors and managers to achieve influence over states, workers, and communities at national and global levels.

Characteristic. [B, C, S] The term "trait" is commonly used in biology, cultural evolution, and anthropology to refer to constituents of genetic or social makeup. Because "traits" have too often been assumed to be modular and fixed, I have chosen the more general term "characteristic" to refer to specific elements of genetic, cultural, and social makeup.

Coevolution. [B, C, S] In biological evolution, the evolution of at least two species [or categories of evolution], which occurs in a mutually dependent manner. In cultural evolution, the mutually dependent evolution of genes and culture. In social evolution, concepts of coevolution may apply to as many as five orders of social change – biological, cultural, social, environmental, and systemic, so that coevolution is the interaction of these different processes with each other and with the environment.

Collective intentionality. [S] The expression of a we-perspective in which individuals expect to act in group ethos, belong and commit to one or more groups, and seek to achieve group objectives.

Community. [S] A speaking community satisfies the conditions for an unstructured "we-group": its members commit themselves to the group through the effort of learning their shared language.

Confederation. [S] The social institution of confederation arose as closely related communities, perhaps three to five in number, combined to form a larger yet decentralized institution that enabled larger-scale sharing of resources.

Cooperation. [C] Cooperation occurs when an individual incurs a cost to provide a benefit to one or more individuals.

Culture, expressive. [S] In societies, representations that offer commentaries on the world through various media including rhetoric, poetry, music, dance, or theatre.

Culture, group-level. [S] The exchange and creation of ideas, information, representation, etc. among members of a community or larger institution.

Culture, individual-level. [C] Transmission of behavior from one individual to another through imitation or instruction, with retention of the learning in the brain.

Culture, material. [S] In societies, material resources constructed by human effort.

Culture, reflective. [S] In societies, conceptualization of the world and of social relations within it.

Culture, societal. [S] In societies, the organization of society as constructed and sustained by its members, including families, organizations, ethnicities, working groups, and governmental structures.

Dual heritage. [C] In cultural evolution, the coevolution of genetic evolution with the results of social learning – where the latter is the passage of knowledge across generations of brains through learning and instruction.

Earth System. [G] The constituents of the natural world, at all levels: Earth's interacting physical, chemical, and biological processes.

Evolution, biological. [B] The change of animal species through modification of individual genomes by the process of natural selection, yielding descent with modification.

Evolution, cultural. [C] The change in practices of individuals among *Homo sapiens* through social learning and its coevolution with biological change.

Evolution, human. [H] The interaction of biological, cultural, and social evolution, in the context of influences of the natural world, in *Homo sapiens.*

Evolution, social. [S] The change in practices among *Homo sapiens* through creation and regulation of social institutions by social groups.

Fitness, biological. [B] Success of an individual organism and its characteristics as measured by the number of surviving offspring.

Fitness, inclusive. [B, C] Success of an individual organism, its close relatives, and their characteristics as measured by the number of their surviving offspring.

Fitness, social. [S] Success of a social institution and its characteristics as measured by the improvements in the social welfare of a designated beneficiary population, where social welfare could be measured by number of offspring, wealth of offspring, or another measure.

Gaia. [G] The robust global system in which the biosphere interacts with inorganic substances to bring stability to the temperature of both atmosphere and oceans, thus enabling life to continue and thrive; the term "Gaia" first referred to the Greek goddess of the Earth.

Global democratic discourse. [S] This informal discourse encompasses nations, societies, communities, and organizations to sustain a discourse on human priorities. It draws on the long history of local debate and governance, in some cases invoking the well-known Athenian model of democracy.

Group. [B, C, S] Several individual humans. (See also I-group and we-group.)

Hierarchy. [B, S] See I-hierarchy, we-hierarchy.

Holocene. [G] As defined in this book, the geological era from 11,500 years ago to 1800 CE.

Human nature. [B, S] In biological evolution, a multiscale pattern of behavior, ranging from the genetic level to the level of individual and group expression, but with no consensus on the details of the scales and the mechanisms. In social evolution, behavioral predispositions arising at the phenotypical level; expression of emotions.

Human System. [H] An open system, as defined by the fact that it ingests materials from its environment and extrudes materials back into the environment. It is a historical and adaptive system, in that it functions not only according to an initial plan but also undergoes change in response to external and internal influences.

Ideology. [S] Articulations of social priorities, arising from discourse and commonly associated with the perspective, available information, and interests of members of a social group.

I-group. [C, S] A group composed of participants who each focus on their own interests.

I-hierarchy. [C, S] An I-group organized by rank, in which participants are classified into subgroups defined as superior or subordinate to each other.

I-network. [C, S] In an I-network, relationships among participating nodes are those of propinquity, communication, and reciprocity, but not of explicit agreements.

Inheritance. [H] Inheritance of an archive (biological, cultural, social) from one generation to the next.

Institution. [S] An organizational form with particular objectives, involving human activities and behavior, consciously constructed and supported by members of a community. Each institution has inherent dynamics arising from its purpose, materials, and activities.

Knowledge, general. [S] Knowledge and skills on a wide range of subjects, shared by families, communities, through elementary and secondary schools, and via popular culture.

Knowledge, specialized. [S] Theoretical and practical knowledge organized by discipline in the natural sciences, social sciences, humanities and arts, and associated technical disciplines.

Language. [B, C, S] In its Basic Property, according to Berwick and Chomsky, language is a finite computational system yielding an infinity of expressions, each of which has a definite interpretation in thought and sound.

Language phylum (plural phyla). [S] An ancestral language family that gives strong evidence of having existed for at least 15,000 years, with families (or subfamilies) within it that formed later.

Level. [H] A level of elevation within a set of scales.

Multilevel selection. [C] Natural selection as described by the Price Equation includes a series of nested levels, with equivalent results at levels from genes to individuals to groups. Multilevel selection thus provides a way to discuss selection at the level of groups, while remaining consistent with individual-level selection.

Network. [B, C, S] A set of entities and the connections existing among them. In graphic terms, entities are referred to as nodes or vertices and ties are referred to as edges or arcs.

Phylum (plural phyla). See Language phylum.

Pleistocene. [G] The geologic era from 2.6 million years ago to 11,500 years ago.

Pliocene. [G] The geologic era from 5.3 million years ago to 2.6 million years ago.

Popular culture. [S] Cultural practices and ideas shared across boundaries of family and ethnicity, especially through electronic media.

Price Equation. [C] A statement of the coevolution of the frequency of a characteristic in evolution with the fitness of the individual.

Prosocial behavior. [C] Cooperation, supported by development of altruistic genes, plus altruistic punishment. The degree of cooperative and altruistic-punishment behavior in individuals is assumed to be influenced by genetic inheritance of those proclivities.

Reductionism. An approach to scientific theory in which a theory developed at one level of aggregation is applied to a higher level of aggregation or complexity.

Regulation. [S] Social regulation and reform, in which institutional members or social groups modify existing institutions.

Representation. [S] By representation is meant the portrayal, modeling, conceptualization, translation, or recapitulation of a phenomenon, projecting the portrayal from one arena to another.

Scale. [H] In the study of complex systems and hierarchies, the concept of scale refers to the combination of (1) the level of analysis (for example, analyzing the whole or a specific component of the system); and (2) the level of observation (for example, observing a system as an external viewer or as an internal participant).

Selection, natural. [B] Darwin: "Owing to this struggle for life, any variation, however slight and from whatever cause proceeding, if it be in any degree profitable to an individual of any species, in its infinitely complex relations to other organic beings and to external nature, will tend to the preservation of that individual, and will generally be inherited by its offspring. . . . I have called this principle, by which each slight variation, if useful, is preserved, by the term of Natural Selection."

Selection, personal. [C] The level of the individual organism in the framework of Multilevel selection.

Selection, social. [S] An analog to natural selection, in which new institutions survive or fail through various processes according to their social fitness. In addition, social selection is more complex than the dichotomous choice of success or failure: institutions may also undergo *regulation* and reform, in which institutional members or social groups modify existing institutions.

Social welfare. [S] In determination of institutional fitness, social welfare is a specified measure of the institution's success.

Society. [S] The term encompasses social orders from localized agricultural, pastoral, and fishing groups to large-scale states. Many evolving societies emphasized decentralization, convening leadership groups only when the occasion required major decisions. The forms of society grew to quite different sizes according to their institutions: formation of states was a possibility but not the destiny for all.

Speech. [S, H] Vocalized, syntactic language.

Subsystem. [B, C, S] A self-contained system within a larger system.

we-group. [S] A group consisting of members who meet three criteria: a *group reason* for participating together; a *collectivity condition* in which members agree they share the same situation; and *collective commitment* in their willingness to act on the other two criteria.

we-hierarchy. [S] A hierarchical network for which members express collective intentionality and make an explicit and common agreement as to the objective of the network and the roles of participants.

we-network. [S] A network for which members express collective intentionality and make an explicit and common agreement as to the objective of the network.

Zeitgeist. [S] A common expression of outlook and ideology linked to certain institutions and to behavior and emotions that are central to those institutions.

References

Online Documents

"Language Resources." www.cambridge.org/Humanity

Patrick Manning, "Songs of Democracy," http://manning.pitt.edu/pdf/Democratization%201989-1992.pdf

Books and Articles

Abu-Lughod, Janet. *Before European Hegemony: The World System AD 1250–1350*. New York: Oxford University Press, 1989.

Acemoglu, Daron, and James A. Robinson. *Why Nations Fail: The Origins of Power, Prosperity, and Poverty*. London: Profile Books, 2012.

Adolphs, Ralph, and David J. Anderson. *The Neuroscience of Emotion: A New Synthesis* Princeton, NJ: Princeton University Press, 2018.

Aiello, Leslie C., and R. I. M. Dunbar. "Neocortex Size, Group Size, and the Evolution of Language." *Current Anthropology* 34 (1993): 184–93.

Alfani, Guido, trans. Christine Calvert. *Calamities and the Economy in Renaissance Italy: The Grand Tour of the Four Horsemen of the Apocalypse*. New York: Palgrave Macmillan, 2013.

Allen, Jim, and James F. O'Connell. "Getting from Sunda to Sahul." In Geoffrey Clark, Foss Leach, and Sue O'Connor (eds.), *Islands of Inquiry: Colonisation, Seafaring, and the Archaeology of Maritime Landscapes*, 31–46. Canberra: Australian National University Press, 2009.

Allsen, Thomas T. *Culture and Conquest in Mongol Eurasia*. Cambridge: Cambridge University Press, 2011.

al-Musawi, Muhsin J. *The Medieval Islamic Republic of Letters: Arabic Knowledge Construction*. Notre Dame, IN: University of Notre Dame Press, 2015.

Alston, Eric, Lee J. Alston, Bernardo Mueller, and Tomas Nonnenmacher. *Institutional and Organizational Analysis: Concepts and Applications*. Cambridge: Cambridge University Press, 2018.

Ambrose, S. H. "Late Pleistocene Human Population Bottlenecks, Volcanic Winter, and Differentiation of Modern Humans." *Journal of Human Evolution* 34, 6 (1998): 623–51.

Amin, Samir. *L'accumulation à l'échelle mondiale: critique de la théorie du sous-développement*. Dakar: IFAN, 1970.

Anderson, Benedict. *Imagined Communities: Reflections on the Origin and Spread of Nationalism*, rev. ed. London: Verso, 2006.

Anon. "Ancient Krasnyi Yar." American Museum of Natural History, www.amnh.org/exhibitions/horse/domesticating-horses/ancient-krasnyi-yar.

Anthony, David W. *The Horse, the Wheel, and Language: How Bronze-age Riders from the Eurasian Steppes Changed the World*. Princeton, NJ: Princeton University Press, 2007.

Arrighi, Giovanni. *The Long Twentieth Century: Money, Power, and the Origins of Our Times*. London: Verso, 1994.

Aslanian, Sebouh David. *From the Indian Ocean to the Mediterranean: The Global Trade Networks of Armenian Merchants from New Julfa*. Berkeley: University of California Press, 2011.

Atkinson, Quentin D. "Phonemic Diversity supports a Serial Founder Effect Model of Language Expansion from Africa." *Science* 332, 6027 (2011): 346–49.

Aubert, M., A. Brumm, M. Ramli et al. "Pleistocene Cave Art from Sulawesi, Indonesia." *Nature* 514 (2014): 223–27.

Aubert, M., P. Setiawan, A. Oktaviana et al. "Palaeolithic Cave art in Borneo." *Nature* 564 (2018): 254–57.

Austin, Gareth, and Kaoru Sugihara, eds. *Labour-Intensive Industrialization in Global History*. London: Routledge, 2013.

Axelrod, Robert. *The Evolution of Cooperation*. New York: Basic Books, 1980.

Ayala, Francisco Jose, and Theodosius Dobzhansky, eds. *Studies in the Philosophy of Biology: Reduction and Related Problems*. London: Macmillan, 1974.

Bacharach, Michael. *Beyond Individual Choice: Teams and Frames in Game Theory*, edited by Natalie Gold and Robert Sugen. Princeton, NJ: Princeton University Press, 2006.

Bairoch, Paul. *Le Tiers-Monde dans l'impasse: de démarrage économique du XVIIIe au XXe siècle*. Paris: Gallimard, 1971.

Ballan, Mohamad. "The Scholar and the Sultan: A Translation of the Historic Encounter between Ibn Khaldun and Timur." https://ballandalus.wordpress.com/2014/08/30/the-scholar-and-the-sultan-a-translation-of-the-historic-encounter-between-ibn-khaldun-and-timur/

Bandura, Albert. *Social Learning Theory*. Morristown, NJ: General Learning Press, 1971.

Bandura, Albert. "Self-efficacy: Toward a Unifying Theory of Behavioral Change." *Psychological Review* 84 (1977): 191–215.

Bandura, Albert, and Richard H. Walters. *Social Learning and Personality Development*. New York: Holt, Rinehart and Winston, 1963.

Barendse, R. J. "The Feudal Mutation: Military and Economic Transformations of the Ethnosphere in the Tenth to Thirteenth Centuries." *Journal of World History* 14 (2003): 503–29.

Barrett, Lisa Feldman. *How Emotions Are Made: The Secret Life of the Brain*. Boston: Houghton Mifflin Harcourt, 2018.

Bartlett, Kenneth R. "Burckhardt's Humanist Myopia: Machiavelli, Guicciardini and the Wider World." *Scripta Mediterranea* 16–17 (1995–96): 17–30.

Batchelor, Robert. "The Global and the Maritime: Divergent Paradigms for Understanding the Role of Translation in the Emergence of Early Modern Science." In Patrick Manning and Abigail Owen (eds.), *Knowledge in Translation: Global Patterns of Scientific Exchange, 1000–1800 CE*, 75–90. Pittsburgh: University of Pittsburgh Press, 2018.

Beckert, Sven. *Empire of Cotton: A Global History*. New York: Alfred A. Knopf, 2014.

Beecroft, Alexander. *An Ecology of World Literature: From Antiquity to the Present Day*. London: Verso, 2015.

Bellwood, Peter, James J. Fox, and Darrell Tryon. *The Austronesians: Historical and Comparative Perspectives*. Canberra: Australian National University, 1995.

Benton, Lauren A. *Law and Colonial Cultures: Legal Regimes in World History, 1400–1900*. Cambridge: Cambridge University Press, 2002.

Berg, Maxine. "Skill, Craft, and Histories of Industrialisation in Europe and Asia." *Transactions of the Royal Historical Society* 24 (2014): 127–48.

Berwick, Robert, and Noam Chomsky. *Why Only Us: Language and Evolution*. Cambridge, MA: MIT Press, 2016.

Biasutti, Renato, ed. *Le Razze e I popoli della terra*, 4 vols. Turin: Unione Tipografico – Editrice Torinese, 1967 [1941].

Bickerton, Derek. *Language and Species*. Chicago: University of Chicago Press, 1981.

Black, Jeremy. *War and Its Causes*. London: Rowman and Littlefield, 2019.

Blau, Peter M. *Exchange & Power in Social Life*. New Brunswick, NJ: Transaction Publishers, 1986.

Borowy, Iris. "Science and Technology for Development in a Postcolonial World: Negotiations at the United Nations, 1960–1980." *NTM Journal of the History of Science, Technology and Medicine* 26 (2018): 31–62.

Borowy, Iris, and Matthias Schmelzer, eds. *History of the Future of Economic Growth: Historical Roots of Current Debates on Sustainable Degrowth*. Abingdon, Oxon, UK: Routledge, 2017.

Boyd, Robert, and Peter J. Richerson. *Culture and the Evolutionary Process*. Chicago: University of Chicago Press, 1985.

Boyd, Robert, and Peter J. Richerson. *The Origin and Evolution of Cultures*. New York: Oxford University Press, 2005.

Boyd, Robert, and Peter J. Richerson. "Culture and the Evolution of Human Cooperation." *Philosophical Transactions of the Royal Society* 364 (2009): 3281–88.

Brandon, Pepijn. *War, Capital, and the Dutch State (1588–1795)*. Leiden: Brill, 2015.

Broadberry, Stephen, Hanhui Guan, and David Daokui Li. (2018), "China, Europe and the Great Divergence: A Study in Historical National Accounting, 980–1850." *Journal of Economic History*, 78 (2018): 955–1000.

Brooke, John L. *Climate Change and the Course of Global History: A Rough Journey*. New York: Cambridge University Press, 2014.

Brooks, Alison S., John E. Yellen, Richard Potts et al. "Long-Distance Stone Transport and Pigment Use in the Earliest Middle Stone Age." *Science* 360 (6 April, 2018): 90–94.

Bulliet, Richard. *Cotton, Climate, and Camels in Early Islamic Iran: A Moment in World History*. New York: Columbia University Press, 2011.

Bulliet, Richard. *The Wheel: Inventions and Reinventions*. New York: Columbia University Press, 2016.

Burbank, Jane, and Frederick Cooper. *Empires in World History: Power and the Politics of Difference*. Princeton, NJ: Princeton University Press, 2010.

Burke, Edmund III. "Islam at the Center: Technological Complexes and the Roots of Modernity." *Journal of World History* 20 (2009): 165–86.

Burke, James, and Robert Ornstein. *The Axemaker's Gift: A Double-Edged History of Human Culture*. New York: G. P. Putnam's Sons, 1995.

Burkhardt, Richard W., Jr. *Patterns of Behavior: Konrad Lorenz, Niko Tinbergen, and the Founding of Ethology*. Chicago: University of Chicago Press, 2005.

Campbell, Bruce M. S. *The Great Transition: Climate, Disease and Society in the Late-Medieval World*. Cambridge: Cambridge University Press, 2016.

Campbell, Donald T. "On the Conflicts between Biological and Social Evolution and between Psychology and Moral Tradition" (presidential address, American Psychological Association). *American Psychologist* (December 1975): 1103–1126; reprinted as Donald T. Campbell, "On the Conflicts between Biological and Social Evolution and between Psychology and Moral Tradition." *Zygon* 11, 3 (1976): 167–208.

Campbell, Michael C., and Sarah A. Tishkoff, "The Evolution of Human Genetic and Phenotypic Variation in Africa." *Current Biology* 20 (2010): R166–73.

Cann, Rebecca L., Mark Stoneking, and Allan C. Wilson. "Mitochondrial DNA and Human Evolution." *Nature* 325 (1987): 31–36.

Carneiro, R. L. "Scale Analysis, Evolutionary Sequences, and the Rating of Cultures." In R. Naroll and R. Cohen (eds.), *A Handbook of Method in Cultural Anthropology*. Garden City, NY: Doubleday, 1970, 846.

Carson, Rachel. *Silent Spring*. New York: Fawcett Crest, 1962.

Cavalli-Sforza, L. L. *Genes, Peoples, and Language*. New York: North Point Press, 2000.

Cavalli-Sforza, L. Luca, Paolo Menozzi, and Alberto Piazza. *The History and Geography of Human Genes*, abridged. Princeton, NJ: Princeton University Press, 1994.

Chakraborty, Titas. *Work and Society in the East India Company Settlements in Bengal, 1650–1757*. PhD dissertation, University of Pittsburgh, 2016.

Chaloupka, George. *Journey in Time: The World's Longest Continuing Art Tradition*. Chatswood, AU: Reed, 1993.

Chase-Dunn, Christopher, and Bruce Lerro, *Social Change: Globalization from the Stone Age to the Present*. Boulder, CO: Paradigm Publishers, 2014.

Chaudhury, Sushil. *From Prosperity to Decline, Eighteenth Century Bengal*. New Delhi: Manohar, 1995.

Chaudhury, Sushil. "The Asian Merchants and Companies in Bengal's Export Trade ca. Mid-eighteenth Century." In Sushil Chaudhury and Michel Morineau, *Merchants, Companies and Trade. Europe and Asia in the Early Modern Era*. Cambridge: Cambridge University Press, 1999.

Childe, V. Gordon. *What Happened in History*. Harmondsworth, UK: Pelican Books, 1942.

Ching, Francis D. K., Mark Jazombek, and Vikramaditya Prakash. *A Global History of Architecture*, 3rd ed. Hoboken, NJ: John Wiley & Sons, 2017.

Chomsky, Noam. *Syntactic Structures*. 's-Gravenhage: Mouton, 1957.

Chomsky, Noam. *The Minimalist Program*. Cambridge, MA: MIT Press, 1995.

Chouin, Gérard. "Reflections on plague in African history (14th–19th c.)." *Afriques: débats, méthodes, et terrains d'histoire* 9 (2018), https://journals.openedition.org/afriques/2084.

Christian, David. *Maps of Time: An Introduction to Big History*. Berkeley: University of California Press, 2003.

Clark, Gregory, and David Jacks. "Coal and the Industrial Revolution, 1700–1869." *European Review of Economic History* 11 (2007): 39–72.

Cline, Eric. *1177 BC: The Year Civilization Collapsed*. Princeton, NJ: Princeton University Press, 2014.

Clottes, Jean, trans. Olivier Y. Martin and Robert T. Martin. *What Is Paleolithic Art? Cave Paintings and the Dawn of Human Creativity*. Chicago: University of Chicago Press, 2016.

Cohen, Ronald, and Elman R. Service. *Origins of the State: The Anthropology of Political Evolution*. Philadelphia: Institute for the Study of Human Issues, 1978.

Cole, H. S. D., Christopher Freeman, Marie Jahoda, and K. L. R. Pavitt, eds. *Models of Doom: A Critique of the Limits to Growth*. New York: Universe Books, 1973.

Conley, Robert J. *The Cherokee Nation: A History*. Albuquerque: University of New Mexico Press, 2008.

Conrad, Sebastian. "Enlightenment in Global History: A Historiographical Critique." *American Historical Review* 117 (2012): 999–1027.

Cook, Noble David. *Born to Die: Disease and New World Conquest, 1492–1650*. Cambridge: Cambridge University Press, 1998.

Cooper, Frederick. *Citizenship, Inequality, and Difference: Historical Perspectives*. Princeton, NJ: Princeton University Press, 2018.

Croft, William. *Joseph Harold Greenberg, 1915–2001: A Biographical Memoir*. Washington, DC: National Academy of Sciences, 2007.

Crosby, Alfred W. *The Columbian Exchange: Biological Consequences of 1492*. Westport, CT: Greenwood Publishers, 1972.

Curtin, Philip D. *Cross-Cultural Trade in World History*. Cambridge: Cambridge University Press, 1984.

Cushman, Gregory T. *Guano and the Opening of the Pacific World: A Global Ecological History*. New York: Cambridge University Press, 2013.

Darwin, Charles. *On the Origin of Species by Means of Natural Selection*. London: John Murray, 1859.

Darwin, Charles. *The Descent of Man and Selection in Relation to Sex*, 2nd ed. New York: A. L. Burt, 1874.

Darwin, John. *After Tamerlane: The Global History of Empire after 1405*. London: Bloomsbury Press, 2008.

Davis, Mike. *Late Victorian Holocausts: El Niño Famines and the Making of the Third World*. London: Verso, 2001.

Dawkins, Richard. *The Selfish Gene*. Oxford: Oxford University Press, 1979.

"Deglaciation." After Arthur S. Dyke, "An Outline of North American Deglaciation," 2004. www.youtube.com/watch?v=wbsURVgoRD0.

Degler, Carl N. *In Search of Human Nature: The Decline and Revival of Darwinism in American Social Thought*. New York: Oxford University Press, 1991.

DeLancy, Scott, and Victor Golla. "The Penutian Hypothesis: Retrospect and Prospect." *International Journal of American Linguistics* 63 (1997), 171–202.

De Vries, Jan. "Connecting Europe and Asia: A Quantitative Analysis of the Cape Route Trade, 1497–1795." In Dennis Flynn, Arturo Giráldez, and Richard von Glahn (eds.), *Global Connections and Monetary History, 1470–1800*, 35–106. London: Ashgate, 2003.

de Zwart, Pim. "Globalization in the Early Modern Era: New Evidence from the Dutch-Asiatic Trade, c. 1600–1800." *Journal of Economic History* 76 (2016): 520–58.

de Zwart, Pim, and Jan Luiten van Zanden. *The Origins of Globalization: World Trade in the Making of the Global Economy, 1500–1800*. Cambridge: Cambridge University Press, 2018.

Diamond, Jared. *Guns, Germs, and Steel: The Fates of Human Societies*. New York: Norton, 1997.

Di Cosmo, Nicola. "Climate Change and the Rise of an Empire." *The Institute Letter* (Institute for Advanced Study), Spring 2014, 1, 15.

Dillehay, Tom D., Carlos Ocampo, José Saavedra et al. "New Archaeological Evidence for an Early Human Presence at Monte Verde, Chile." *PLoS ONE* (November 18, 2015): 1–27, DOI:10.137/journal.pone.0141923.

Dingle, Hugh. *Migration: The Biology of Life on the Move*. New York: Oxford University Press, 1996.

Diouf, Sylviane, ed. *Fighting the Slave Trade: West African Strategies*. Athens, OH: Ohio University Press, 2003.

Dobzhansky, Theodosius. *Genetics and the Origin of Species*, introduced by Stephen Jay Gould. New York: Columbia University Press, 1982 [1937].

Dolgopolsky, Aharon, with an introduction by Colin Renfrew. *The Nostratic Macrofamily and Linguistic Palaeontology*. Cambridge: McDonald Institute for Archaeological Research, 1998.

Douka, Katerina, Viviane Slon, Zenobia Jacobs et al. "Age Estimates for Hominin Fossils and the Onset of the Upper Paleolithic at Denisova Cave." *Nature* 565 (2019): 640–44.

Dower, John. *War without Mercy: Race and Power in the Pacific War*. New York: Pantheon, 1986.

Drescher, Seymour. *Abolition: A History of Slavery and Antislavery*. Cambridge: Cambridge University Press, 2009.

Duara, Prasenjit. *The Crisis of Global Modernity: Asian Traditions and a Sustainable Future*. Cambridge: Cambridge University Press, 2015.

Dunbar, R. I. M. "Mind the Bonding Gap: Constraints on the Evolution of Hominin Societies." In Stephen Shennan (ed.), *Pattern and Process in Cultural Evolution*, 223–34. Berkeley: University of California Press, 2009.

Dunbar, Robin. *Grooming, Gossip and the Evolution of Language*. London: Faber and Faber, 1996.

Dunbar, Robin I. M., and Richard Sosis, "Optimising Human Community Sizes." *Evolution and Human Behavior* 39 (2018): 106–111.

Durham, William H. *Coevolution: Genes, Culture, and Human Diversity*. Stanford, CA: Stanford University Press, 1991.

Ehret, Christopher. *Reconstructing Proto-Afroasiatic (Proto-Afrasian): Vowels, Tone, Consonants, and Vocabulary*. Berkeley: University of California Press, 1995.

Ehret, Christopher. "Nostratic - or proto-Human?" In Dolgopolsky, *The Nostratic Macrofamily*.

Ehret, Christopher. *A Historical-Comparative Reconstruction of Nilo-Saharan*. Köln: Rüdiger Köppe Verlag, 2001.

Ehret, Christopher. *History and the Testimony of Language*. Berkeley: University of California Press, 2011.

Ehret, Christopher. "Early Humans: Tools, Language, and Culture." In David Christian(ed.), *The Cambridge World History*. Vol. 1: *Introducing World History, to 10,000 BCE*, 339–61. Cambridge: Cambridge University Press, 2015.

Ehret, Christopher. "Africa from 48,000 to 9500 BCE." In David Christian (ed.), *Cambridge World History*, 1: 362–93.

Ehret, Christopher. "Khoesan Languages and Late Stone Age Archaeology." Forthcoming.

Ehrlich, Paul R., and Peter H. Raven. "Butterflies and Plants: A Study in Coevolution," *Evolution* 18 (1964): 586–608.

Eibl-Eibesfeldt, Irenäus. *Human Ethology*. New York: Aldine de Gruyter, 1989.

El Hamel, Chouki. *Black Morocco: A History of Slavery, Race, and Islam*. Cambridge: Cambridge University Press, 2013.

Elias, Norbert, trans. Stephen Mennell and Grace Morrissey. *What Is Sociology?* New York: Columbia University Press, 1978.

Erlandson, Jon. "Ancient Immigrants: Archaeology and Maritime Migrations." In Jan Lucassen, Leo Lucassen, and Patrick Manning (eds.), *Migration History in World History: Multidisciplinary Approaches*, 191–214. Leiden: Brill, 2010.

Erlandson, Jon M., Michael H. Graham, Bruce J. Bourque, Debra Corbett, James A. Estes, and Robert S. Steneck, "The Kelp Highway Hypothesis: Marine Ecology, the Coastal Migration Theory, and the Peopling of the Americas." *The Journal of Island and Coastal Archaeology* 2 (2007): 161–74.

Eshelman, Jason A., R. S. Malhi, J. R. Johnson et al. "Mitochondrial DNA and Prehistoric Settlements: Native Migrations on the Western Edge of North America." *Human Biology* 76 (2004), 55–75.

Fahmy, Khaled. *All the Pasha's Men: Mehmed Ali, His Army and the Making of Modern Egypt*. Cambridge: Cambridge University Press, 1997.

Felsenfeld, Gary. "A Brief History of Epigenetics." *Cold Spring Harbor Perspectives in Biology* (2014). DOI: 10.1101/cshperspect.a018200.

Fernández-Armesto, Felipe. *A Foot in the River: Why Our Lives Change–and the Limits of Evolution*. Oxford: Oxford University Press, 2015.

Fisher, Ronald A. *The Genetical Theory of Natural Selection*. Oxford: Clarendon Press, 1930.

Fitch, W. Tecumseh. *The Evolution of Language*. Cambridge: Cambridge University Press, 2010.

Fitch, W. Tecumseh. "Noam Chomsky and the Biology of Language." In Oren Harman and Michael R. Dietrich (eds.), *Outsider Scientists: Routes to Innovation in Biology*, 201–22. Chicago: University of Chicago Press, 2013.

Flannery, Kent, and Joyce Marcus. *The Creation of Inequality: How Our Prehistoric Ancestors Set the Stage for Monarchy, Slavery, and Empire*. Cambridge, MA: Harvard University Press, 2012.

Flynn, Dennis O., and Arturo Giráldez. "Born with a 'Silver Spoon': World Trade's Origin in 1571." *Journal of World History* 6 (1995): 201–21.

Forrester, Jay W. *World Dynamics*. Cambridge, MA: Wright-Allen Press, 1971.

Foster, William Z. *Outline History of the World Trade Union Movement*. New York: International Publishers, 1956.

Fourshey, Catherine Cymone, Rhonda M. Gonzales, and Christine Saidi. *Bantu Africa: 3500 BCE to Present*. New York: Oxford University Press, 2017.

Frank, Andre Gunder. *World Accumulation 1492–1789*. New York: Monthly Review Press, 1978.

Frank, Andre Gunder. *ReOrient: Global Economy in the Asian Age*. Berkeley: University of California Press, 1998.

Frank, Andre Gunder, edited and with an introduction by Robert A. Denemark. *ReOrienting the Nineteenth Century: Global Economy in the Continuing Asian Age*. Boulder, CO: Paradigm Publishers, 2014.

Fressoz, Jean-Baptiste, and Christophe Bonneuil. "Growth Unlimited: The Idea of Infinite Growth from Fossil Capitalism to Green Capitalism." In Iris Borowy and Matthias Schmelzer (eds.), *History of the Future of Economic Growth: Historical Roots of Current Debates on Sustainable Degrowth*, 52–68. London: Routledge, 2017.

Fu, Qiaomei, M. Meyer, X. Gao et al. "DNA Analysis of an Early Modern Human from Tianyuan Cave, China." *PNAS* 110 (2013): 2223–27.

Garba, Abubakar. "The Architecture and Chemistry of a Dug-out: The Dufuna Canoe in Ethno-archaeological Perspective." *Berichte des Sonderforschungsbereichs* 268 (1996): 193–200.

Gerritsen, Anne. "Ceramics for Local and Global Markets: Jingdezhen's Agora of Technologies." In Dagmar Schäfer (ed.), *Cultures of Knowledge: Technology in Chinese History*, 161–84. Leiden: Brill, 2011.

Gillot, Jacques, and Jean-François Niort. *Le Code Noir*. Paris: Editions le Cavalier Bleu, 2015.

Gills, Barry K., and Andre Gunder Frank. "World System Cycles, Crises, and Hegemonic Shifts, 1700 BC to 1700 AD." In Andre Gunder Frank and Barry K. Gills (eds.), *The World System: Five Hundred Years or Five Thousand?* 143–99. London: Routledge, 1992.

Gimbutas, Marija. *The Goddesses and Gods of Old Europe, 6500–3500 BC, Myths and Cult Images*, new and updated ed. Berkeley: University of California Press, 1982.

Gobineau, Arthur, comte de, trans. Adrian Collins. *The Inequality of Human Races*. Los Angeles: Noontide Press, 1966.

Goodman, Dena. *The Republic of Letters: A Cultural History of the French Enlightenment*. Ithaca, NY: Cornell University Press, 1994.

Goudsblom, Joop. *Fire and Civilization*. London: Penguin, 1992.

Gould, Stephen Jay. *Ontogeny and Phylogeny*. Cambridge, MA: Belknap Press of Harvard University Press, 1977.

Gould, Stephen Jay *Bully for Brontosaurus*. New York: W. W. Norton, 1991.

Gould, Stephen Jay. *The Structure of Evolutionary Theory*. Cambridge, MA: Belknap Press of Harvard University Press, 2002.

Goveia, Elsa. "The West Indian Slave Laws of the Eighteenth Century." *Revista de Ciencias Sociales* 4 (1960): 75–105.

Grafe, Regina, and Oscar Gelderblom. "The Rise and Fall of Merchant Guilds: Re-thinking the Comparative Study of Commercial Institutions in Premodern Europe." *Journal of Interdisciplinary History* 40 (2009): 477–511.

Gran, Peter. *The Rise of the Rich: A New View of Modern World History*. Syracuse, NY: Syracuse University Press, 2009.

Green, Monica. "Taking 'Pandemic' Seriously: Making the Black Death Global." *The Medieval Globe* 1, 1 (2014), https://scholarworks.wmich.edu/tmg/vol1/iss1/4

Green, Monica. "Putting Africa on the Black Death Map: Narratives from Genetics and History." *Afriques: débats, méthodes, et terrains d'histoire* 9 (2018), https://journals.openedition.org/afriques/2084

Green, Nile. "Persian Print and the Stanhope Revolution: Industrialization, Evangelicism, and the Birth of Printing in Early Qajar Iran." *Comparative Studies of South Asia, Africa and the Middle East* 30 (2010): 473–90.

Greenberg, J. H. *Language in the Americas*. Stanford, CA: Stanford University Press, 1987.

Greenberg, Joseph H. *Studies in African Linguistic Classification*. New Haven, CT: Compass Publishing Company, 1955.

Greenberg, Joseph H. *The Languages of Africa*. Bloomington: Indiana University Press, 1963.

Greenberg, Joseph H., ed. *Universals of Language*, 2nd ed. Cambridge, MA: MIT Press, 1965.

Greenberg, Joseph H. *Indo-European and Its Nearest Neighbors: The Eurasiatic Language Family*, 2 vols. Stanford, CA: Stanford University Press, 2000.

Greenberg, Joseph H., edited and introduced by William Croft. *Genetic Linguistics: Essays on Theory and Method.* Oxford: Oxford University Press, 2018.

Greenberg, Joseph H. "The Indo-Pacific Hypothesis (1971)." In Greenberg (ed. William Croft), *Genetic Linguistics: Essays on Theory and Method*, 193–276. Oxford: Oxford University Press, 2005.

Greenberg, Joseph H., Christy G. Turner II, Stephen L. Zegura et al. "The Settlement of the Americas: A Comparison of the Linguistic, Dental, and Genetic Evidence [and Comments and Reply]." *Current Anthropology* 27 (1986): 477–97.

Groves, Anna. "There's a Second Impact Crater under Greenland." *Astronomy Magazine*, February 13, 2019, www.astronomy.com.

Guerrero, Saul. *The Environmental History of Silver mining in New Spain and Mexico, 16 c to 19 c: A Shift of Paradigm*. PhD dissertation, McGill University, 2015.

Hall, Kenneth R. "Ports-of-Trade, Maritime Diasporas, and Networks of Trade and Cultural Integration in the Bay of Bengal Region of the Indian Ocean: c. 1300–1500." *Journal of the Economic and Social History of the Orient* 53: 109–45.

Hamilton, W. D. "The Genetical Evolution of Social Behaviour, I." *Journal of Theoretical Biology* 7 (1964): 1–16.

Hamilton, W. D. "The Genetical Evolution of Social Behaviour, II." *Journal of Theoretical Biology* 7 (1964): 17–52.

Hansen, James. *Storms of My Grandchildren: The Truth about the Coming Climate Catastrophe and Our Last Chance to Save Humanity*. London: Bloomsbury, 2009.

Hansen, James E. "Scientific Reticence and Sea Level Rise." *Environmental Research Letters* 2 (May 2007), DOI:10.1088/1748-9326/2/2/024002.

Harari, Yuval Noah. *Sapiens: A Brief History of Humankind*. New York: Harper, 2015.

Harman, Oren. "On the Importance of the Parvenu: The Amazing Case of George Price in Evolutionary Biology." In Oren Harman and Michael R. Dietrich (eds.), *Outsider Scientists: Routes to Innovation in Biology*, 312–30. Chicago: University of Chicago Press, 2013.

Harpending, Henry, and Alan Rogers. "Genetic Perspectives on Human origins and Differentiation." *Annual Review of Genomics and Human Genetics* 1 (2000): 361–85.

Hawas, May, ed. *The Routledge Companion to World Literature and World History*. London: Routledge, 2018.

Hayek, Friedrich A. *Studies in Philosophy, Politics and Economics*. Chicago: University of Chicago Press, 1967.

Heine, Bernd, and Derek Nurse. *African Languages: An Introduction*. Cambridge: Cambridge University Press, 2000.

Hellie, Richard. *Slavery in Russia, 1450–1725*. Chicago: University of Chicago Press, 1982.

Henrich, Joseph. *The Secret of Our Success: How Culture Is Driving Human Evolution, Domesticating Our Species, and Making Us Smarter*. Princeton, NJ: Princeton University Press, 2016.

Henrich, Joseph. "Cooperation, Punishment, and the Evolution of Human Institutions." *Science* 312 (2006); 60–61.

Henrich, Joseph, and R. McElreath. "The Evolution of Cultural Evolution." *Evolutionary Anthropology*, 12 (2003): 123–35.

Henrich, Natalie, and Joseph Henrich. *Why Humans Cooperate: A Cultural and Evolutionary Explanation*. Oxford: Oxford University Press, 2007.

Hindriks, Frank. "Constitutive Rules, Language, and Ontology." *Erkenntnis* 71 (2009): 253–75.

Hirbo, Jibril, A. Ranciaro, and S. A. Tishkoff. "Population Structure and Migration in Africa: Correlations between Archaeological, Linguistic and Genetic Data." In *Causes and Consequences of Human Migration: An Evolutionary Perspective*, C.B.C. (2009): 135–71. 10.1017/CBO9781139003308.011.

Hochschild, Adam. *King Leopold's Ghost: A Story of Greed, Terror, and Heroism in Colonial Africa*. Boston: Houghton Mifflin, 1998.

Hoerder, Dirk. *Cultures in Contact: World Migrations in the Second Millennium*. Durham, NC: Duke University Press, 2002.
Hofmeester, Karin, Jan Lucassen, Leo Lucassen, Rombert Stapel, and Richard Zijdeman. "The Global Collaboratory on the History of Labour Relations, 1500–2000: Background, Set-up, Taxonomy, and Applications." *Working Paper*, October 2013. http://hdl.handle.net/10622/40GRAD
Hofmeester, Karin, and Marcel van der Linden, eds. *Handbook Global History of Work*. Berlin: de Gruyter, 2018.
Hopkins, A. G. "Globalization with and without Empires: From Bali to Labrador. In A. G. Hopkins (ed.), *Globalization in World History*, 221–43. New York: W. W. Norton, 2002.
Hopkins, A. G. *American Empire: A Global History*. Princeton, NJ: Princeton University Press, 2018.
Hu, Aiqun. *China's Social Insurance in the Twentieth Century: A Global Perspective*. Leiden: Brill, 2015.
Husemann, M., F. E. Zachos, R. J. Paxton, and J. C. Habel. "Effective Population Size in Ecology and Evolution." *Heredity* 117 (2016): 191–92.
Ibn Fadlan, translated with commentary by Richard N. Frye. *Ibn Fadlan's Journey to Russia: A Tenth-Century Traveler from Baghdad to the Volga River*. Princeton, NJ: Markus Wiener, 2005.
Ibn Khaldun. *The Muqaddimah: An Introduction to History*, 3 vols., translated from the Arabic by Franz Rosenthal. Princeton, NJ: Princeton University Press, 1958.
Ibn Mājid al-Sa'dī, Ahmad. *Kitāb al-fawā'id fī uṣūl 'ilm al-baḥr wa-al-qawā'id*. Damascus: al-Maṭba'ah al-Ta'āwunīyah, 1971.
Irwin, Robert. *Ibn Khaldun: An Intellectual Biography*. Princeton, NJ: Princeton University Press, 2018.
Israel, Jonathan I. *Dutch Primacy in World Trade, 1585–1740*. Oxford: Clarendon Press, 1989.
Jablonski, N. G., and G. Chaplin. "The Evolution of Skin Coloration." *Journal of Human Evolution* 39 (2000): 57–106.
Jablonski, Nina G., and Leslie C. Aiello, eds. *The Origins and Diversification of Language*. San Francisco: California Academy of Sciences, 1998.
Jacob, François. *"Evolution and Tinkering." Science* 196 (1977): 1161–66.
Janis, Irving L. *Groupthink: Psychological Studies of Policy Decisions and Fiascoes*, 2nd ed. Boston: Houghton Mifflin Co., 1982.
Johnson, Cara Roure, and Sally McBrearty. "500,000 Year Old Blades from the Kapthurin Formation, Kenya." *Journal of Human Evolution* 58 (2010): 193–200.
Johnson, Douglas, and David Anderson, eds. *The Ecology of Survival: Case Studies from Northeast African History*. London: Lester Crook Academic Publishing, 1988.
Johnstone, Paul. *The Sea-Craft of Prehistory*. Cambridge, MA: Harvard University Press, 1980.
Jones, E. L. *The European Miracle: Environments, Economies and Geopolitics in the History of Europe and Asia*, 2nd ed. Cambridge: Cambridge University Press, 1987.
Jones, P. D., and M. E. Mann. "Climate over Past Millennia." *Reviews of Geophysics* 42, 2 (2004), DOI:10.1029/2003RG000143.

Kamrani, Kambiz. "Tianyuan Man Genome Reveals the Nuances of Asian Prehistory." *Blog, Physical Anthropology*, October 14, 2017.

Kawai, Kaori, ed. *Institutions: The Evolution of Human Sociality*, trans. Minako Sato. Kyoto: Kyoto University Press, 2017 [2013].

Kea, Ray A. *Settlements, Trade, and Polities in the Seventeenth-Century Gold Coast*. Baltimore: Johns Hopkins University Press, 1982.

Keita, Shomarka. "A Brief Introduction to a Geochemical Method Used in Assessing Migration in Biological Anthropology." In Lucassen et al., *Migration History in World History*, 59–74.

Keita S. O. Y., and A. J. Boyce. "Genetics, Egypt, and History: Interpreting Geographical Patterns of Y Chromosome Variation." *History in Africa* 32 (2005): 221–46.

Kelly, Robert L. *The Fifth Beginning: What Six Million Years of Human History Can Tell Us about Our Future*. Berkeley: University of California Press, 2016.

Kennedy, Paul. *The Rise and Fall of the Great Powers*. New York: Random House, 1988.

Kenrick, Douglas. *Jomon of Japan: The World's Oldest Pottery*. London: Kegan Paul International, 1995.

Keulman, Kenneth, ed. *Critical Moments in Religious History*. Macon, GA: Mercer University Press, 1993.

Kjaer, Kurt H. "A Large Impact Crater beneath Hiawatha Glacier in Northwest Greenland." *Science Advances* 4, 11 (2018), DOI:10.1126/sciadv.aar8173.

Kloppenberg, James. *Toward Democracy: The Struggle for Self-Rule in European and American Thought*. New York: Oxford University Press, 2016.

Knaap, Gerrit. "Shipping and Trade in Java, c. 1775: A Quantitative Analysis." *Modern Asian Studies* 33 (1999): 405–20.

Knaap, Gerrit. "All About Money: Maritime Trade in Makassar and West Java around 1775." *Journal of the Economic and Social History of the Orient* 49 (2006): 482–502.

Knaap, Gerrit, and Heather Sutherland. *Monsoon Traders. Ships, Skippers and Commodities in Eighteenth-Century Makassar*. Leiden: Brill, 2004.

Kocka, Jürgen, trans. Jeremiah Riemer. *A Short History of Capitalism*. Princeton, NJ: Princeton University Press, 2016.

Kohler, Timothy A., Sarah Cole, and Stanca Ciupe. "Population and Warfare: A Test of the Turchin Model in Pueblo Societies." In Shennan, ed., *Pattern and Process in Cultural Evolution*, 277–96.

Kuper, Adam. *The Invention of Primitive Society: Transformations of an Illusion*. London: Routledge, 1988.

Kuper, Adam. *Culture: The Anthropologists' Account*. Cambridge, MA: Harvard University Press, 1999.

Kuroda, Akinobu. "The Eurasian Silver Century, 1276–1359: Commensurability and Multiplicity." *Journal of Global History* 4 (2009), 245–69.

Lachance, Joseph, Benjamin Vernot, Clara C. Elbers et al. "Evolutionary History and Adaptation from High-Coverage Whole-Genome Sequences of Diverse African Hunter-Gatherers." *Cell* 150 (2012), 457–69.

"Language Vitality and Endangerment." UNESCO Programme Safeguarding of Endangered Languages," Paris, March 10–12, 2003. www.unesco.org/new/fileadmin/MULTIMEDIA/HQ/CLT/pdf/ Language_vitality_and_endangerment_EN.pdf (accessed July 19, 2018).

Levtzion, N., and J. F. P. Hopkins, eds. *Corpus of Early Arabic Sources for West African History*. Cambridge: Cambridge University Press, 1981.

Li, Heng, and Richard Durbin. "Inference of Human Population History from Individual Whole-Genome Sequences." *Nature* 475 (2011): 493–96.

Lieberman, Philip. "The Evolution of Human Speech: Its Anatomical and Neural Bases." *Current Anthropology* 48, 1 (2007), 39–66.

Linebaugh, Peter, and Marcus Rediker. *The Many-Headed Hydra: Sailors, Slaves, Commoners, and the Hidden History of the Revolutionary Atlantic*. Boston: Beacon, 2000.

Liu Xinru. *The Silk Road in World History*. New York: Oxford University Press, 2010.

Lomax, A., and N. Berkowitz, "The Evolutionary Taxonomy of Culture." *Science* 177 (21 July, 1972): 230, 233.

Lombard, Marlize. and Laurel Phillipson. "Indications of Bow and Arrow Use 64,000 Years Ago in KwaZulu-Natal, South Africa." *Antiquity* 84 (2010): 635–48.

López, Saioa, Lucy van Dorp, and Garrett Hellenthal. "Human Dispersal Out of Africa: A Lasting Debate." *Evolutionary Bioinformation Online* 11, Suppl 2 (2015): 57–68.

Lopreato, Joseph. *Human Nature and Biocultural Evolution*. Boston: Allen & Unwin, 1984.

Lotka, Alfred J. *Elements of Physical Biology*. Baltimore: Williams & Wilkins, 1925.

Lovelock, James, *Gaia, A New Look at Life on Earth*. Oxford: Oxford University Press, 1979.

Lovelock, James. *The Revenge of Gaia: Earth's Climate Crisis and the Fate of Humanity*. New York: Basic Books, 2007.

Lucassen, Jan, Leo Lucassen, and Patrick Manning, eds. *Migration History in World History: Multidisciplinary Approaches*. Leiden: Brill, 2010.

Lucassen, Leo. "Working Together: New Directions in Global Labour History." *Journal of Global History* 11 (2016): 66–87.

Lumley, Henry de. *La grande histoire des premiers hommes européens*. Paris: O. Jacob, 2007.

Luschan, Felix von. *Völker, Rassen, Sprachen*. Berlin: Welt-Verlag, 1922.

Luukkanen, Harri, and William F. Fitzhugh. *The Bark Canoes and Skin Boats of Northern Eurasia*. Washington, DC: Smithsonian Publications, 2018.

Ly, Abdoulaye. *La Compagnie du Sénégal*. Paris: Présence africaine, 1958.

Ma, Ning. The *Age of Silver: The Rise of the Novel East and West*. New York: Oxford University Press, 2018.

MacPhee, Ross D. E. *End of the Megafauna: The Fate of the World's Hugest, Fiercest, and Strangest Animals*. New York: W. W. Norton, 2018.

Magness, Philip W. "The Pejorative Origins of the Term 'Neoliberalism'." *AIER*, www.aier.org/article,pejorative-origins-term-"neoliberalism."

Mallory, J. P. *In Search of the Indo-Europeans: Language, Archaeology, and Myth.* London: Thames and Hudson, 1989.

Mamdani, Mahmood. *Citizen and Subject: Contemporary Africa and the Legacy of Late Colonialism.* Princeton, NJ: Princeton University Press, 1996.

Manning, Patrick. *Navigating World History: Historians Create a Global Past.* New York: Palgrave Macmillan, 2003.

Manning, Patrick. "1789–1792 and 1989–1992. Global Interactions of Social Movements." *World History Connected* 3, 1 (2005). http://worldhistoryconnected.org/3.1/manning.html

Manning, Patrick. "*Homo sapiens* Populates the Earth: A Provisional Synthesis, Privileging Linguistic Data." *Journal of World History* 17 (2006): 115–58.

Manning, Patrick. "Cross-Community Migration: A Distinctive Human Pattern." *Social Evolution and History* 5, 2 (2006): 24–54.

Manning, Patrick. *The African Diaspora: A History through Culture.* New York: Columbia University Press, 2009.

Manning, Patrick. "Global History and Maritime History." *International Journal of Maritime History* 15 (2013): 1–22.

Manning, Patrick. *Big Data in History.* Houndmills, UK: Palgrave Macmillan, 2013.

Manning, Patrick. "Locating Africans on the World Stage: A Problem in World History." *Journal of World History* 26 (2015): 605–37.

Manning, Patrick. "Inequality: Historical and Disciplinary Approaches." *American Historical Review* 122 (2017): 1–22.

Manning, Patrick. "Introduction." In Manning and Owen, eds., *Knowledge in Translation*, 1–16.

Manning, Patrick. "The Life-Sciences, 1900–2000: Analysis and Social Welfare from Mendel and Koch to Biotech and Conservation." *Asian Review of World Histories* 6 (2018): 185–208.

Manning, Patrick. *Methods for Human History: Studying Social, Cultural, and Biological Evolution.* New York: Pagrave Macmillan, 2020.

Manning, Patrick, Dennis O. Flynn, and Qiyao Wang, "Silver Circulation Worldwide: Initial Steps in Comprehensive Research." *Journal of World-Historical Information* 3–4 (2016–17): 1–18.

Manning, Patrick, and Aubrey Hillman. "Climate as a Factor in Migration and Social Change, 200,000 to 5000 years ago." American Historical Association annual meeting, New Orleans, January 5, 2013.

Manning, Patrick, and Abigail Owen, eds. *Knowledge in Translation: Global Patterns of Scientific Exchange, 1000–1800 CE.* Pittsburgh: University of Pittsburgh Press, 2018.

Manning, Patrick, with Tiffany Trimmer. *Migration in World History*, 2nd ed. London: Routledge, 2012.

Marcott, Shaun A., Jeremy D. Shakun, Peter U. Clark, and Alan C. Mix. "A Reconstruction of Regional and Global Temperature for the Past 11,300 Years." *Science* 339, 6124 (2013): 1198–1201.

Markovits, Claude. *Global World of Indian Merchants, 1750–1947: Traders of Sind from Bukhara to Panama.* Cambridge: Cambridge University Press, 2000.

Maynard Smith, John, and George R. Price. "The Logic of Animal Conflict." *Nature* 246, 5427 (1973): 15–18.

Maynard Smith, John, and Eörs Szathmáry. *The Major Transitions in Evolution*. Oxford: Oxford University Press, 1995.

Mayr, Ernst. *Evolution and the Diversity of Life*. Cambridge, MA: Harvard University Press, 1975.

McBrearty, Sally. "Down with the Revolution." In Paul Mellars, Katie Boyle, Ofer Bar-Yosef, and Chris Stringer (eds.), *Rethinking the Human Revolution*, 133–51. Oxford: Oxbow Books, 2007.

McBrearty, Sally, and Alison Brooks. "The Revolution that Wasn't: A New Interpretation of the Origin of Modern Human Behavior." *Journal of Human Evolution* 39 (2000): 453–563.

McElreath, Richard, and Robert Boyd. *Mathematical Models of Social Evolution: A Guide for the Perplexed*. Chicago: University of Chicago Press, 2007.

McEvedy, Colin, and Richard Jones. *Atlas of World Population History*. Harmondsworth: Penguin, 1978.

McLean, Paul. *Culture in Networks*. Cambridge: Polity Press, 2017.

McNeill, J. R. *Something New under the Sun: An Environmental History of the Twentieth-Century World*. New York: W. W. Norton, 2000.

McNeill, J. R., and Peter Engelke. *The Great Acceleration: An Environmental History of the Anthropocene since 1945*. Cambridge, MA: Belknap Press of Harvard University Press, 2014.

McNeill, J. R., and William H. McNeill. *The Human Web: A Bird's-Eye View of Human History*. New York: W. W. Norton, 2003.

Meadows, Donnella H., Dennis L. Meadows, Jørgen Randers, and William W. Behrens II. *The Limits to Growth: A Report for the Club of Rome's Project on the Predicament of Mankind*. New York: Universe Books, 1972.

Mesoudi, Alex. *Cultural Evolution: How Darwinian Theory Can Explain Human Culture and Synthesize the Social Sciences*. Chicago: University of Chicago Press, 2011.

Milanovic, Branko. *Global Inequality: A New Approach for the Age of Globalization*. Cambridge, MA: The Belknap Press at Harvard University Press, 2016.

Miller, James G. *Living Systems*. New York: McGraw-Hill, 1978.

Mithen, Steven. *After the Ice: A Global Human History, 20,000–5000 BC*. Cambridge, MA: Harvard University Press, 2003.

Mondal, Mayukh, Jaume Bertanpetit, and Oscar Lao. "Approximate Bayesian Computation with Deep Learning Supports a Third Archaic Introgression in Asia and Oceania." *Nature Communications* 10, 246 (January 16, 2019).

Moore, Jason W. *Capitalism in the Web of Life: Ecology and the Accumulation of Capital*. London: Verso, 2014.

Moore, Jason W., ed. *Anthropocene or Capitalocene? Nature, History, and the Crisis of Capitalism*. Oakland, CA: PM Press, 2016.

Morange, Michel, trans. Matthew Cobb. *A History of Molecular Biology*. Cambridge, MA: Harvard University Press, 1998.

Moreno-Mayar, J. Victor, Ben A. Potter, Lasse Vinner et al. "Terminal Pleistocene Alaskan Genome reveals first founding population of Native Americans." *Nature* 553 (2018): 203–08.

Morillo, Stephen. "A 'Feudal Mutation'? Conceptual Tools and Historical Patterns in World History." *Journal of World History* 14, 4 (2003): 531–50.

Morris, Ian. *The Measure of Civilization: How Social Development Decides the Fate of Nations*. Princeton, NJ: Princeton University Press, 2013.

"Neolithic Site of Çatalhöyük," UNESCO World Heritage Centre, https://whc.unesco.org/en/list/1527.

Nichols, Johanna. *Linguistic Diversity in Space and Time*. Chicago: University of Chicago Press, 1992.

Nichols, Johanna. "The Origin and Dispersal of Languages: Linguistic Evidence." In Jablonski and Aiello, eds., *The Origin and Diversification of Language*, 127–70.

North, Douglass C. "Economic Performance through Time," Nobel Prize lecture, 1993.

North, Douglass C., and Robert Paul Thomas. *The Rise of the Western World*. Cambridge: Cambridge University Press, 1973.

North, Douglass C., John Joseph Wallis, and Barry R. Weingast. *Violence and Social Orders: A Conceptual Framework for Interpreting Recorded History*. New York: Cambridge University Press, 2009.

Novartis Foundation Symposium. *The Limits of Reductionism in Biology*. New York: John Wiley & Sons, 1998.

O'Fallon, Brendan, and Lars Fehren-Schmitz. "Native Americans Experienced a Strong Population Bottleneck Coincident with European Contact." *PNAS* 108, 51 (2011): 20444–48.

Ogle, Vanessa. "Archipelago Capitalism: Tax Havens, Offshore Money, and the State, 1950s–1970s." *American Historical Review* 122 (2017): 1431–58.

Okasha, Samir. *Evolution and the Levels of Selection*. Oxford: Clarendon Press, 2006.

Olson, Steve. *Mapping Human History: Genes, Race, and Our Common Origins*. Boston: Houghton Mifflin, 2003.

Oreskes, Naomi, and Erik M. Conway. *Merchants of Doubt: How a Handful of Scientists Obscured the Truth on Issues from Tobacco Smoke to Global Warming*. New York: Bloomsbury Press, 2010.

Oreskes, Naomi, and Erik M. Conway, "Science Isn't Enough to Save Us." *New York Times*, October 17, 2018.

O'Rourke, Kevin H., and Jeffrey G. Williamson. *Globalization and History: The Evolution of a Nineteenth-Century Atlantic Economy*. Cambridge, MA: MIT Press, 1999.

Osterhammel, Jürgen. *The Transformation of the World: A Global History of the Nineteenth Century*. Princeton, NJ: Princeton University Press, 2014.

Owen-Smith, Thomas, and Nathan W. Hill, eds. *Trans-Himalayan Linguistics: Historical and Descriptive Linguistics of the Himalayan Area*. Berlin: de Gruyter, 2014.

Paine, Lincoln. *Sea and Civilization: A Maritime History of the World*. New York: Alfred A. Knopf, 2013.

Palmer, Colin. *Human Cargoes: The British Slave Trade to Spanish America, 1700–1739*. Urbana: University of Illinois Press, 1981.

Park, Katharine, and Lorraine Daston, "Introduction: The Age of the New." In Katharine Park and Lorraine Daston (eds.), *The Cambridge History of Science, vol. 3: Early Modern Science*, 1–18. Cambridge: Cambridge University Press, 2008.

Parker, Geoffrey. *Global Crisis: War, Climate Change & Catastrophe in the Seventeenth Century*. New Haven, CT: Yale University Press, 2013.

Parsons, Talcott. "Social Structure and the Symbolic Media of Interchange." In Peter M. Blau, *Approaches to the Study of Social Structure*, 94–120. New York: The Free Press, 1975.

Parthasarathi, Prasannan. *Why Europe Grew Rich and Asia Did Not: Global Economic Divergence, 1600–1850*. Cambridge: Cambridge University Press, 2011.

Piketty, Thomas, trans. Arthur Goldhammer. *Capital in the Twenty-first Century*. Cambridge, MA: The Belknap Press at Harvard University Press, 2014.

Pinker, Steven. *The Blank Slate: The Modern Denial of Human Nature*. New York: Viking, 2002.

Pinker, Steven. *The Better Angels of Our Nature: Why Violence Has Declined*. New York: Viking, 2011.

Pollock, Sheldon. *The Language of the Gods in the World of Men: Sanskrit, Culture, and Power in Premodern India*. Berkeley: University of California Press, 2006.

Pomeranz, Kenneth. *The Great Divergence: China, Europe, and the Making of the Modern World Economy*. Princeton, NJ: Princeton University Press, 2000.

Potts, Richard, A. K. Behrensmeyer, J. T. Faith et al. "Environmental Dynamics during the Onset of the Middle Stone Age in Eastern Africa." *Science* 360 (2018): 86–90.

Prak, Maarten. *Citizens without Nations: Urban Citizenship in Europe and the World, c. 1000–1789*. Cambridge: Cambridge University Press, 2018.

Prakash, Om. *The Dutch East India Company in Bengal*. Princeton, NJ: Princeton University Press, 1985.

Prazniak, Roxann. "Marāgha Observatory: A Star in the Constellation of Eurasian Scientific Translation." In Patrick Manning and Abigail Owen (eds.), *Knowledge in Translation: Global Patterns of Scientific Exchange, 1000–1800 CE*, 227–43. Pittsburgh: University of Pittsburgh Press, 2018.

Preyer, Gerhard, and Georg Peter, eds. *Social Ontology and Collective Intentionality: Critical Essays on the Philosophy of Raimo Tuomela with His Responses*. Cham, Switzerland: Springer, 2017.

Provine, William B. "Ernst Mayr: Genetics and Speciation." Genetics 167, 3 (2004): 1041–46.

Raghavan, Maanasa, P. Skoglund, K. E. Graf et al. "Upper Palaeolithic Siberian Genome Reveals Dual Ancestry of Native Americans." *Nature* 505 (2014): 87–91.

Randolph-Quinney, Patrick, and Anthony Sinclair. "African Tools Push Back the Origins of Human Technological Innovation." *The Conversation*, March 16, 2018.

Reba, Meredith, Femke Reitsma, and Karen C. Seto, "Spatializing 6,000 Years of Global Urbanization from 3700 BC to AD 2000." *Scientific Data* 3 (2016), article no. 160034.

Reed D. L., J. E. Light, H. L. Dibble, J. M. Allen, J. J. Kirchman. "Pair of Lice Lost or Paradise Regained: The Evolutionary History of Anthropoid Primate Lice." *BMC Biology* 5 (2007) DOI:10.1186/1741-7007-5-7.

Reich, David. *Who We Are and How We Got Here: Ancient DNA and the New Science of the Human Past.* New York: Pantheon, 2018.

Reischauer, E. O. *Ennin's Travels in T'ang China.* New York: The Ronald Press, 1955.

Richerson, Peter J., and Robert Boyd. "Complex Societies: The Evolutionary Origins of a Crude Superorganism." *Human Nature* 10 (1999): 253–89.

Richerson, Peter J., and Robert Boyd. "The Evolution of Subjective Commitment to Groups: A Tribal Instincts Hypothesis." In Randolph M. Neese (ed.), *Evolution and the Capacity for Commitment,* 186–220. New York: Russell Sage Foundation, 2001.

Riello, Giorgio. *Cotton: The Fabric that Made the Modern World.* Cambridge: Cambridge University Press, 2013.

Rosenberg, Nathan, and L. E. Birdzell, Jr. *How the West Grew Rich: The Economic Transformation of the Industrial World.* New York: Basic Books, 1986.

Ruddiman, William F. *Earth Transformed.* New York: W. H. Freeman, 2014.

Ruhlen, Merritt. *The Origin of Language: Tracing the Evolution of the Mother Tongue.* New York: John Wiley & Sons, 1994.

Russell, Edmund. *Evolutionary History: Uniting History and Biology to Understand Life on Earth.* New York: Cambridge University Press, 2011.

Ryder, A.F.C. *Benin and the Europeans, 1485–1897.* New York: Humanities Press, 1969.

Sandgathe D. M., H. L. Dibble, P. Goldberg et al. "Timing of the Appearance of Habitual Fire Use." *Proceedings of the National Academy of Sciences of the USA* (2011) DOI:10.173/pnas.1106759108.

Saussure, Ferdinand de, trans. Wade Baskin. *Course in General Linguistics.* New York: McGraw Hill, 1966 [1916].

Scheib, Christiana Lyn, Hongjie Li, Tariq Desai et al., "Ancient human parallel lineages within North America contributed to a coastal expansion." *Science* 360 (2018): 1024–27.

Scheidel, Walter. *The Great Leveler: Violence and the History of Inequality from the Stone Age to the Twenty-first Century.* Princeton, NJ: Princeton University Press, 2017.

Schmidtz, D. "Friedrich Hayek." *Stanford Encyclopedia of Philosophy* (updated December 14, 2016), https://plato.stanford.edu/entries/friedrich-hayek/ .

Schwartz, Jeffrey H. "What's Real about Human Evolution: Received Wisdom, Assumptions, and Scenarios." In J. H. Schwartz (ed.), *Rethinking Human Evolution,* 61–91. Cambridge, MA: MIT Press, 2018.

Schweikard, David P., and Hans Bernhard Schmid. "Collective Intentionality." *The Stanford Encyclopedia of Philosophy* (Summer 2013 Edition), Edward N. Zalta (ed.). https://plato.stanford.edu/archives/sum2013/entries/collective-intentionality/

Scott, James C. *Against the Grain: A Deep History of the Earliest States.* New Haven, CT: Yale University Press, 2017.

Searle, John R. *The Construction of Social Reality.* New York: The Free Press, 1995.

Seed, Pat. "Celestial Navigation." In Manning and Owen, eds., *Knowledge in Translation*, 275–91.

Senut, Brigitte, M. Pickford, D. Gommery, and L. Ségan, "Palaeoenvironments and the Origin of Hominid Bipedalism." *Historical Biology* 30 (2018): 284–96.

Sereno, Paul C., Elena A. A. Garcea, Hélène Jousse et al. "Lakeside Cemeteries in the Sahara: 5000 Years of Holocene Population and Environmental Change." *PLoS ONE* 3(8): e2995. https://doi.org/10.1371/journal.pone.0002995

Sharp, Paul, and Jacob Weisdorf. "Globalization Revisited: Market Integration and the Wheat Trade between Noth America and Britain from the Eighteenth Century." *Explorations in Economic History* 50 (2013): 88–98.

Shea, John J., and Matthew L. Sisk. "Complex Projectile Technology and *Homo sapiens* Dispersal into Western Eurasia." *PaleoAnthropology* 10 (2010): 100–22.

Shennan, Stephen. *Genes, Memes and Human History: Darwinian Archaeology and Cultural Evolution*. New York: Thames and Hudson, 2003.

Shennan, Stephen, ed. *Pattern and Process in Cultural Evolution*. Berkeley: University of California Press, 2009.

Shryock, Andrew, and Daniel Lord Smail, *Deep History: The Architecture of Past and Present*. Berkeley: University of California Press, 2011.

Simon, Herbert. "Near Decomposability and Complexity: How a Mind Resides in a Brain." In H. J. Morowitz and J. L. Singer (eds.), *The Mind, the Brain, and Complex Adaptive Systems*, 26. Santa Fe Institute Studies in the Sciences of Compexity, Proceedings, Vol. XXII. Reading, MA: Addison Wesley, 1995.

Skoglund, Pontus, Swapan Mallick, Maria Cátira Bortolini et al. "Genetic Evidence for Two Founding Populations of the Americas." *Nature* 525 (2015): 104–8.

Smaldino, Paul E. "The Cultural Evolution of Emergent Group-Level Traits." *Behavioral and Brain Sciences* 37 (2014): 243–54.

Smil, Vaclav C. *Energy Transitions: Global and National Perspectives*, 2nd ed. Santa Barbara, CA: Praeger, 2017.

Smith, Richard J. "Buddhism and the 'Great Persecution' in China." In *Critical Moments in Religious History*, ed. Kenneth Keulman.

Smith, Terry. "Contemporary Art: World Currents in Transition Beyond Globalization." In Hans Beiting, Andrea Buddensteig, and Peter Weibel (eds.), *The Global Contemporary and the Rise of New Art Worlds*, 186–92. Cambridge, MA: MIT Press, 2013.

Spencer, Herbert. "Progress: Its Law and Cause." *Humboldt Library of Popular Science Literature*, No. 17, 233–85. New York: J. Fitzgerald, 1881.

Spier, Fred. *Big History and the Future of Humanity*, 2nd ed. Malden, MA: John Wiley & Sons, 2015.

Staubach, Suzanne. *Clay: The History and Evolution of Humankind's Relationship with Earth's Most Primal Element*. New York: Berkley Books, 2005.

Stausberg, Michael, Yuhan Sohrab-Dinshaw Vevaina, and Anna Tessmann, *The Wiley-Blackwell Companion to Zoroastrianism*. Chichester, UK: John Wiley & Sons, 2015.

Stringer, Chris. *In Search of the Neanderthals: Solving the Puzzle of Human Origins*. London: Thames & Hudson, 1993.

Stringer, Chris. "The Origin and Evolution of *Homo sapiens*." *Philosophical Transactions of the Royal Society B* 371: 2015.0237 (2016). https://doi.org/10.1098/rstb.2015.0237.

Stringer, Chris, and Peter Andrews. *The Complete World of Human Evolution*. London: Thames and Hudson, 2005.

Subrahmanyam, Sanjay. "On World Historians in the Sixteenth Century." *Representations* 91, 1 (2005): 26–57.

Sutton, J. E. G. "The Aquatic Civilization of Middle Africa." *Journal of African History* 15 (1974): 527–46.

Tattersall, Ian. *Masters of the Planet: The Search for our Human Origins*. New York: Palgrave Macmillan, 2012.

Telegin, Dmitry Yakolevich, trans. V. A. Pyatkovskiy, ed. J. P. Mallory. *Dereivka. A Settlement and Cemetery of Copper Age Horse Keepers on the Middle Dnieper*. Oxford: BAR International Series, 1986.

Thompson, E. P. *The Making of the English Working Class*. New York: Pantheon, 1964.

Thorp, Robert L. *China in the Early Bronze Age: Shang Civilization*. Philadelphia: University of Pennsylvania Press, 2006.

Tomasello, Michael. "Human Culture in Evolutionary Perspective." In M. Gelfand (ed.), *Advances in Culture and Psychology*, 5–51. Oxford: Oxford University Press, 2011.

Tomasello, Michael. *A Natural History of Human Thinking*. Cambridge, MA: Harvard University Press, 2014.

Tomasello, Michael. *Becoming Human: A Theory of Ontogeny*. Cambridge, MA: The Belknap Press of Harvard University Press, 2019.

Tomasello, M., Kruger, A., and Ratner, H. "Cultural Learning." *Behavioral and Brain Sciences*, 16, 3 (1993), 495–511. DOI:10.1017/S0140525X0003123X.

Tooze, Adam J. *Crashed: How a Decade of Financial Crises Changed the World*. New York: Viking, 2018.

Toups, Melissa A., Andrew Kitchen, Jessica E. Light, and David L. Reed. "Origin of Clothing Lice Indicates Early Clothing Use by Anatomically Modern Humans in Africa." *Molecular Biology and Evolution* 28 (2011): 29–32.

Tunscherer, Michel. "Coffee in the Red Sea Area from the Sixteenth to the Nineteenth Century." In William Gervase Clarence-Smith and Steven Topik (eds.),*The Global Coffee Economy in Africa, Asia, and Latin America*, 50–66. Cambridge: Cambridge University Press, 2003.

Tuomela, Raimo. *Philosophy of Social Practices: A Collective Acceptance View*. Cambridge: Cambridge University Press, 2002.

Tuomela, Raimo. *Social Ontology: Collective Intentionality and Group Agents*. Oxford: Oxford University Press, 2013.

Turchin, Peter. *Historical Dynamics: Why States Rise and Fall*. Princeton, NJ: Princeton University Press, 2003.

Turchin, Peter. *Ultra Society: How 10,000 Years of War Made Humans the Greatest Cooperators on Earth*. Chaplin, CT: Beresta Books, 2015.

Turchin, Peter. *Ages of Discord: A Structural-Demographic Analysis of American History*. Chaplin, CT: Beresta Books, 2017.

Turchin, Peter, Thomas E. Currie, Harvey Whitehouse et al. "Quantitative Historical Analysis Uncovers a Single Dimension of Complexity that Structures Global Variation in Human Social Organization." *PNAS* (December 21, 2017), E144–51. www.pnas.org/cgi/doi/10.1073/pnas.1708800115

Turner, Jonathan H. *Human Institutions: A Theory of Societal Evolution*. Lanham, MD: Rowman & Littlefield, 2003.

Turner, Jonathan H. *Human Emotions: A Sociological Theory*. London: Routledge, 2007.

Tylor, Edward B. *Primitive Culture*. London: J. Murray, 1871.

Tyrrell, Toby. *On Gaia: A Critical Investigation of the Relationship between Life and Earth*. Princeton, NJ: Princeton University Press, 2013.

Udehn, Lars. *Methodological Individualism: Background, History, and Meaning*. London: Routledge, 2001.

Unger, Richard W. "The Tonnage of Europe's Merchant Fleets 1300–1800." *The American Neptune* 52 (1992): 247–61.

"Unlabeled Renatto Luschan Skin color map.svg." Wikipedia Commons.

van Bavel, Bas. *The Invisible Hand? How Market Economies Have Emerged and Declined since AD 500*. Oxford: Oxford University Press, 2016.

van der Linden, Marcel. *Workers of the World: Essays toward a Global Labor History*. Leiden: Brill, 2008.

van Driem, George. "Trans-Himalayan." In Thomas Owen-Smith and Nathan W. Hill (eds.), *Trans-Himalayan Linguistics: Historical and Descriptive Linguistics of the Himalayan Area*, 11–40. Berlin: de Gruyter, 2014.

van Rossum, Matthias. *Werkers van de wereld: Globalisering, arbeid en interculturele ontmoetingen tussen Aziatische en Europese zeelieden in dienst van de VOC, 1600–1800*. Hilversum: Verloren, 2014.

van Zanden, Jan, Luiten, Joerg Baten et al. *How Was Life? Global Well-Being since 1820*. Paris: OECD Publishing, 2014.

Von Bertalanffy, Ludwig. *General System Theory: Foundations, Development, Applications*. New York: G. Braziller, 1969.

von Scheve, Christian, and Mikko Salmela. *Collective Emotions: Perspectives from Psychology, Philosophy, and Sociology*. Oxford: Oxford University Press, 2014.

Wallace-Wells, David. *The Uninhabitable Earth: Life After Warming*. New York: Tim Duggan Books, 2019.

Wallerstein, Immanuel. *The Modern World-System*, 3 vols. New York: Academic Press, 1974, 1980, 1989.

Wallis, Helen M., and E. D. Grinstead. "A Chinese Terrestrial Globe." *The British Museum Quarterly* 25 (1962): 83–91.

Warsh, Molly A. *American Baroque: Pearls and the Nature of Empire, 1492–1700*. Chapel Hill: University of North Carolina Press, 2018.

Werner, J. Jeffrey, and Pamela R. Willoughby. "Middle Stone Age Technology and Cultural Evolution at Magubike Rockshelter, Southern Tanzania." *African Archaeological Review* 34 (2017): 249–73.

White, Sam. *A Cold Welcome: The Little Ice Age and Europe's Encounter with North America*. Cambridge, MA: Harvard University Press, 2017.

Wiesner-Hanks, Merry E. *Gender in History: Global Perspectives*, 2nd ed. Malden, MA: Wiley-Blackwell, 2011.

Wiesner-Hanks, Merry E., ed. *The Cambridge World History*, 10 vols. Cambridge: Cambridge University Press, 2015.

Williams, George C. *Adaptation and Natural Selection: A Critique of Some Current Evolutionary Thought*. Princeton, NJ: Princeton University Press, 1966.

Wilson, Edward O. *On Human Nature*. Cambridge, MA: Harvard University Press, 1978.

Witzel, E. J. Michael. *The Origins of the World's Mythologies*. Oxford: Oxford University Press, 2012.

Wolbach, Wendy S., Joanne P. Ballard, Paul A. Mayewski et al. "Extraordinary Biomass-Burning Episode and Impact Winter Triggered by the Younger Dryas Cosmic Impact ~12,800 Years Ago. 1. Ice Cores and Glaciers." *Journal of Geology* 126, 2 (2018): 165–84.

Wong, R. Bin. *China Transformed: Historical Change and the Limits of European Experience*. Ithaca, NY: Cornell University Press, 1997.

Wu, Xiaohong, Chi Zhang, Paul Goldberg et al. "Early Pottery at 20,000 Years Ago in Xianrendong Cave, China." Science 336 (2012): 1696–1700.

Yang, Bin. *Cowrie Shells and Cowrie Money: A Global History*. London: Routledge, 2018.

Yücesoy, Hayrettin. "Translation as Self-Consciousness: Ancient Sciences, Antediluvian Wisdom, and the 'Abbāsid Translation Movement." *Journal of World History* 20 (2009): 523–57.

Zaccarella, Emiliano, and Angela D. Friederici, "Merge in the Human Brain: A Sub-Region Based on Functional Investigation in the Left Pars Opercularis." *Frontiers in Psychology*, November 27, 2015. https://doi.org/10.3389/fpsyg

Zahedieh, Nuala. *The Capital and the Colonies: London and the Atlantic Economy 1660–1700*. Cambridge: Cambridge University Press, 2010.

Index

Printed in the United Kingdom by TJ Clays Ltd.